SPORT, PHYSICAL CULTURE, AND THE MOVING BODY

Critical Issues in Sport and Society

Michael A. Messner, Douglas Hartmann, and Jeffrey Montez de Oca,
Series Editors

Critical Issues in Sport and Society features scholarly books that help expand our understanding of the new and myriad ways in which sport is intertwined with social life in the contemporary world. Using the tools of various scholarly disciplines, including sociology, anthropology, history, media studies and others, books in this series investigate the growing impact of sport and sports-related activities on various aspects of social life as well as key developments and changes in the sporting world and emerging sporting practices. Series authors produce groundbreaking research that brings empirical and applied work together with cultural critique and historical perspectives written in an engaging, accessible format.

Rachel Allison, *Kicking Center: Gender and the Selling of Women's Professional Soccer*

Jules Boykoff, *Activism and the Olympics: Dissent at the Games in Vancouver and London*

Diana Tracy Cohen, *Iron Dads: Managing Family, Work, and Endurance Sport Identities*

Cheryl Cooky and Michael A. Messner, *No Slam Dunk: Gender, Sport, and the Unevenness of Social Change*

Jennifer Guiliano, *Indian Spectacle: College Mascots and the Anxiety of Modern America*

Kathryn E. Henne, *Testing for Athlete Citizenship: Regulating Doping and Sex in Sport*

Jeffrey L. Kidder, *Parkour and the City: Risk, Masculinity, and Meaning in a Postmodern Sport*

Michael A. Messner and Michela Musto, eds., *Child's Play: Sport in Kids' Worlds*

Jeffrey Montez de Oca, *Discipline and Indulgence: College Football, Media, and the American Way of Life during the Cold War*

Joshua I. Newman, Holly Thorpe, and David L. Andrews, eds., *Sport, Physical Culture, and the Moving Body: Materialisms, Technologies, Ecologies*

Stephen C. Poulson, *Why Would Anyone Do That? Lifestyle Sport in the Twenty-First Century*

Nicole Willms, *When Women Rule the Court: Gender, Race, and Japanese American Basketball*

SPORT, PHYSICAL CULTURE, AND THE MOVING BODY

Materialisms, Technologies, Ecologies

EDITED BY

JOSHUA I. NEWMAN, HOLLY THORPE, AND DAVID L. ANDREWS

RUTGERS UNIVERSITY PRESS

New Brunswick, Camden, and Newark, New Jersey, and London

Library of Congress Cataloging-in-Publication Data

Names: Newman, Joshua I., 1976– editor. | Thorpe, Holly, editor. | Andrews, David L., 1962– editor.
Title: Sport, physical culture, and the moving body : materialisms, technologies, ecologies / Edited by Joshua I. Newman, Holly Thorpe, and David L. Andrews.
Description: New Brunswick, New Jersey : Rutgers University Press, [2020] | Includes bibliographical references and index.
Identifiers: LCCN 2019011263 | ISBN 9780813591810 (pbk.) | ISBN 9780813591827 (cloth) | ISBN 9780813591834
Subjects: LCSH: Sports sciences. | Human mechanics—Physiological aspects. | Physical education and training.
Classification: LCC GV558 .S683 2020 | DDC 796.01/5—dc23
LC record available at https://lccn.loc.gov/2019011263

A British Cataloging-in-Publication record for this book is available from the British Library.

♾ The paper used in this publication meets the requirements of the American National Standard for Information Sciences—Permanence of Paper for Printed Library Materials, ANSI Z39.48-1992.

www.rutgersuniversitypress.org

Manufactured in the United States of America

CONTENTS

FOREWORD

Since Diana Coole and I first labored over how to characterize an emerging mode of analysis, research in the new materialisms has undergone explosive growth. We hoped for but certainly did not anticipate the breadth and vigor of scholarship bringing substances, biological processes, artifacts, and material environments into view as not just effects but also agents of stabilization and change in the worlds in which we live.

What is distinctive in new materialist approaches is the inclusion in research of materiality in its vibrancy, effectivity, affectivity, and obduracy. Scholars who thus work to bring matters and materialities into the foreground do not ignore the patterns of signification and representation that galvanize language, culture, and politics. Rather, they explore how the material fleshiness of bodies and the materiality of objects and environments constitute the ground, the matrix, and the co-constituent agents of the discourses and symbolic systems that organize and give meaning to our lives.

New materialist scholars explore how, for instance, the responsiveness and plasticity of biological processes contributes to the making of acculturated subjectivity. They consider how things, big and small, invite or compel living beings and social and political institutions to shift or adapt how things go. And they investigate the ecological interdependencies that condition humans, nonhuman creatures, and environments in their varied persistence and decay. To my mind, new materialism as an approach to research is best conceived not a displacement of humans so much as their re-situation in the multiscalar material and temporal conditions of their existence. The very best work undertakes analysis of material, cultural, and linguistic processes together, exploring the imbrication and mutual constitution of signifying systems, embodied subjectivities, and the societies and landscapes of cultural and political economies.

Joshua Newman, Holly Thorpe, and David Andrews have pulled together an exemplary volume of new materialist scholarship on sport and physical culture. *Sport, Physical Culture, and the Moving Body* focuses on the flesh, data, and landscapes that are sculpted, honed, and enhanced for health, beauty, and competition. It shows how integral the physicality and substance of bodies, objects, and venues are to understanding the aspirations, the geopolitical jockeying, and the representations of suffering and grace that together mobilize sporting culture. In bringing into analytic view the interaction and uneven accommodation between matter and repetition, perception and norm, the acquired and the reconstructed—whether talking about bodies, technologies, or environments—the collection paints a fuller picture of the role of sport and physical culture in our lives.

Collectively, the essays illuminate the biological processes, materials, and chemicals that structure and power physical movement, whether virtuoso athleticism, grudging exercise, or tournament travel. They trace how such movements and the physical and technological preparations we make for them shape the waters, air, and ecologies that are the context and condition for them. In doing so, the essays show with detail and perspicacity how sport, exercise, and health, as practice and industry, articulate our daily lives with broader and often global cultural, political, and ecological dynamics. The pathways of analysis in these papers span kinds, species, and scales. And while such multidimensionality could seem to invite a mashed analysis of everything that matters—with fuzzy objects and innumerable relations only superficially rendered—it is with brilliant filigrees of insight that the sharply drawn papers in this volume trace the interconnectivities between the mud and the medals, the muscular and the monetary, the molecular and the macroeconomic, the mundane and the magnificent.

This impressive collection of essays covers an extensive topical terrain and draws on a diversity of theoretical frameworks that fall under the rubric of the new materialisms. Because of the collection's breadth and conceptual diversity, each essay in concert with the others serves as a superb model of new materialist research for sport and physical culture specialists as well as for scholars who are generally interested in how the world works. Together, they also provide a convincing argument for the rich insights to be gained when scholars include new materialist perspectives in their work.

Samantha Frost
Champaign, Illinois

SPORT, PHYSICAL CULTURE, AND THE MOVING BODY

INTRODUCTION

Sport, Physical Culture, and New Materialisms

JOSHUA I. NEWMAN, HOLLY THORPE,
AND DAVID L. ANDREWS

Gold. Signified as Au on the periodic table and sometimes represented as atomic number 79. This valuable and highly sought-after precious metal has for millennia been taken for vital purposes of currency, adornment, and aesthetics. Simple in its material composition, gold is complicated when made purposeful and purposed in the world. Malleable and ductile, gold is made to conform to the human world and its necessities—shaped by the goldsmith's hand, spun by the loom into fine wire, or cut into coinage by the minter's blanking press. It is one of the least reactive chemical elements and thus tends to tolerate exposure to acids and other transformative chemical agents and as such maintains its natural qualities long after extraction. For this reason, gold both symbolizes and materializes permanency. From a wedding ring to the Golden Temple of Amritsar, forged gold, it seems, is substantial matter.

However, gold is also fluid. Gold is the matter-form of both currents and currency. The seafaring pirates of the popular imagination are thought to have charted nautical courses through confluences of thermohaline circulation, prevailing winds, and prize law territorialities in pursuit of gold's glisten and the station afforded thereby. Gold, as an expression of fortune, is both fixed and in flux. In this way, it is at once material and symbolic, a double functionality perhaps best exemplified by the post–Bretton Woods severance from the gold standard.[1] Gold is connector and conduit. Once it is pursued by emperor, king, or state, gold connects otherwise disaffiliated peoples. This was the case in the past (e.g., the wars of Cyrus the Great in founding the Achaemenid Empire, ca. 559 BC) as it is now (e.g., last century's War of the Golden Stool between the British imperial government of the Ashanti Region and the Empire of Ashanti), as millions have found themselves and their lives embroiled in its acquisition. In this way, human life is subjugated to gold (and its pursuits).

It is also the case, however, that gold is subordinated to human interests in the Anthropocene. A highly efficient, noncorrosive conductor of electricity, it is commonly used in mobile telephones, laptop computers, and fiber optic cable systems. Today, our digital world is streamed in real time through an expansive network of gold-laced fiber optic cables. We might thus say that gold is the fabric of cybernetic life. Indeed, gold is *made* for human life—sometimes as its accoutrement and sometimes as its metallic substitute or supernumerary. Gold is made for and makes the human body significant on wedding days, on coronation days, or during a papal homily. In this way, gold valorizes flesh. In other contexts, flesh and bones are surrogated by gold matter—the material body replaced or augmented by golden teeth or gold-hinged orthotic prostheses. For example, scientists have recently developed a polymer designed to mimic the elastic and high-resolution sensory capabilities of real skin (Kim et al., 2014). We now have prosthetic gloves made up of layers of stretchable gold and silicon sensors that allow the user to detect heat and feel textures (Gerratt, Michaud, & Lacour, 2015).[2]

Conversely, it is also certainly the case that *the living body is made for gold.* When gold is heated and cast as a circular medallion imprinted with five overlain circles—as it is every 2 years in the form of Olympic gold medals—it produces meaningful, textual, cultural, muscular, vascular, and physiologically attendant bodies. In the realm of discourse, the case can and has been made that the gold sought by Olympian athletes is, in a Derridian sense, a most consequential of referents; the element and its medal-pursuant bodies act as free-floating signifiers that activate the conditions of emergence from which sport-based nationalism, ethnic and cultural identities, and identity politics are forged, performed, and made meaningful. As the object of discourse, the gold medal provokes a heightened national politics that the novelist George Orwell likened to "war minus the shooting" (see Beck, 2013, p. 72). As *embodiments of the nation* (Chisholm, 1999), combative Olympians dual in 4K interlaced megapixels for gold as billions of adoring, media-consuming spectator-fans look on. When draped on the neck of an individual athletic body, the golden Olympic pendant anchors a complex system of meanings and discursive practices. The mass-mediated, gold-pursuant athletic performance provokes a *multitudinal grammar* (see Virno, 2003) that articulates members of vast imagined communities (the so-called masses; see Anderson, 1983) to the individual body (the athlete or head of state), to the signifying systems of agonistic nationalism, to Au, to commercializing and mediatizing processes and outcomes.

In service of the nation, these sweat-seeping corpuses produce, through craft and adornment, a particular order of things: a sporting geopolitics conferred through lifelong sacrifice to the pursuit of national gold. Gold here serves as a confirmatory object, fixing a unity in discourse between the symbolic, the corporeal, the body politic, and the spectacular. Once ceremonially draped on the neck of the flag-bearing victor (or before or after, for that matter), the gold object provokes for the masses a complex political ecology of nonequivalent actors: the athlete body, podium-posted national pride, the International Olympic Committee, imagined communities, corporatized brands, global sport-media complexes, competition as (political economic) hegemony, stadiums, images flowing through fiber optic data grids, ideas flowing through grids of intelligibility, stadium labor, water (that inhabits the arena's race pool or that was made unpotable by toxic refuse by-produced during arena construction), the one (athlete), the many, and so on.

Thus the gold medal is fixed and affixing, but in its materiality, it is also paradoxically volatile. Just as it establishes linkages and order, the medal also incites disorder. To wit, its material and discursive functions have at times been set against the dominant (state, cultural, identity) politics it has been cast to serve. For example, the gold medal won by the late Muhammad Ali (then Cassius Clay) at the 1960 Summer Olympic Games came to be destabilized from its ordered state and ordering function when, a few months after the boxer's triumph in Rome, an enraged Ali threw the medal into the Ohio River after he and a friend were refused service at a "whites-only" restaurant in his hometown of Louisville, Kentucky (Gorn, 1998). Cultural theorists remind us that in moments of such transgressive praxis, the material form (the golden medallion) gives way to the discursive and the performative, that the object in its absence produced order and, in this case, disorder—a disunity of signifier and signified, a discursive formation that unsettles dominant systems of meaning. This is certainly the case; the gold medal as *langue* (as Saussure [1916/2011] would have it) is an effective piece of (im)material culture. The act of throwing away that most treasured of sporting currency is deeply cultural, rendered and read in nonguaranteed yet politically powerful ways as an act of dissent.

Indeed, but to limit our analysis to the field of meaning would negate the materiality of Ali's golden praxis, for this scene in 1960 reveals both textual and material interconnectivities. We have the phenotypically bounded bodies occupying (or banished from) a corporeally spatialized diner; the concrete sidewalks on which Ali walked; a gold medal forged in Rome and a pugilist's body forged in the segregation-era gyms of Louisville's all-black, working-class West End neighborhood; the increasingly contaminated waters of the Ohio River; the ripple created as the gold medallion entered into and displaced water; the then 21-year-old George Rogers Clark Memorial Bridge and the automobility[3] its steel afforded to people of Louisville, Kentucky, and Jeffersonville, Indiana; and Ali's bruised hands and the kinetic events they created with the release of an upper-cut or a golden projectile.

The gold medal in this story, once set airborne (if not before), mediates a complex *assemblage* of things (our use of "assemblage" here follows DeLanda, 2006, 2013)—the architectural etchings of famed bridge architect Paul Philippe Cret, the more than 5,000 tons of steel used in its construction, the additional capital needed to acquire this much steel in the wake of the Steel Strike of 1919, the labor power of steel miners, bridge builders, and the material and symbolic outputs of a road sign painter named Cassius Marcellus Clay Sr. and those of his son—as the assembled material basis for a significant media event that would come almost a century after the bridge was built. Of course, the medal's cultural significance and politico-ideological potency are often best canvassed on the surfaces of meaning and (en/de)coding, but the gold-as-medal form also bears the imprint of a material history and object agency. It summons by way of its existence and significance operations of power, knowledge, praxis, and material action. Its materiality is contoured around the wars and mining disasters and wealth that it forged. The medal presents as its significant/ce for miners, athletes, diner staff, and sign painters a complex assemblage of labor power and wages within webs of class and racial inequalities that directly and indirectly abetted its *form*-ation.

Moreover, the materialities of the medal-tossing event connect materialities passed to those made by, and for, the future. In the summer of 2014, as in previous years, volunteers from the Louisville area conducted their annual "Ohio River Sweep." During the sweep, it is common for volunteers to pull everything from cars and boats to household machines from the bottom of the river. However, while sifting through the debris that June, Kentuckian Robert Bradbury discovered a round medallion with the impression of an ancient Roman god. In this moment, the currents of time, water, sediment, and politics brought agency back to steel, diners, miners, boxing gloves, and bruised knuckles. The material assemblage of concrete and steel and gold and pugilist anatomy and segregation economies was excavated, recycled into vibrant matter. Its renewed existence created circumstantial activity on social media–wielding electronic computer screens around the world bearing its image, populated local print and television media content, and filled the coffers of its finder (in the form of a $200,000 reward issued by the Ali family). Just as Ali's protest was cultural and performative, it was also material—matter made (it) matter. The dislocation of the object from its subjectified owner's body set in motion a new material assemblage. The relocation of the medal to its owner did much the same.

This concern for the productive qua agentic capacities of material things leads us to the project at hand. In this book, we look to "new" materialist approaches to make better sense of sporting and active bodies and the material assemblages and ecologies from and within which they are constituted and which they constitute. The book's contributors draw on feminist, object-oriented, posthumanist theories and heuristics to explore the vibrancy and agency of various forms and objects of active embodiment. These forms of the active body—the material loci of exercise, physical fitness, health activities, kinesthetic movement, leisure, recreation, dance, and sport (see Andrews & Silk, 2015)—surface as both culture (constituted in the fields of language, meaning, and discourse) and physics (matter and its motion through time and space), co-constituting the realm of *physical culture* from which the body as physiological phenomenon is made and makes social and material conditions. A materialist approach to the analysis of physical culture thus looks to explicate articulations of the physical (material, physics) and the cultural (political, systems of meaning) within and therefrom active embodiment. This is our task here: to utilize materially oriented theories and epistemological frames to find agency in the spatiotemporally bounded body and the material surfaces and objects with which it interfaces and which it traverses, alters, and displaces, for the (sporting, cultural) body as moving mass is, to rephrase Coole and Frost (2010), an excess, force, vitality, relationality, or difference that renders related matter active, self-creative, productive, and unpredictable.

Such a materialist approach seeks to account for complex and constitutive interactions in and between anatomical (e.g., muscles, bones, lungs, capillaries), physical (e.g., gentrifying stadiums, mobilizing roadways, growing grass, flying balls, waves rolling onto a beach), technological (e.g., hyperbolic chambers, Fitbits, anabolic steroids, sex-testing technologies), environmental (e.g., gold, steel, water, turfgrass, waste from whey production), political and economic (e.g., sovereignty, free markets, political ideologies, global capitalism), and cultural (gendered identities, racial politics, habitus, etc.) *actors*. In doing this, we draw inspiration from Bruno Latour (2005) and John Law (1992) in shifting our focus away from "the social" as a natural condition or order—and in this case, sport and physical culture as inherently social phenomena—and instead look to trace the associations and assemblages forged around the productive object-form (see Meillassoux, 2010). As such, we do not see the material object as merely subjugated or symptomatic of human action or scientific rationality or the active body as the unwitting subordinate of the Cartesian mind but instead look to how the material, the cultural, and the physical serve as co-constitutive "actors" that form a series of ties within complex networks and assemblages. Ontologically, we explore how a "flat approach" to actors and ties—whereby nonequivalent elements ranging from the muscle, to the rulebook, to the digital technology, to the water, to the body signifier—might provide new insights to the potentialities and politics of the material-cultural-physical assemblages that form the contemporary sport and physical culture landscape.[4]

The book features contributions from leading scholars in the fields of sport studies, sociology, history, women's studies, science and technology studies (STS), kinesiology, and cultural studies who, with varied approaches and epistemological bases, explicate the agentive, generative capacity of (moving) things (or things with the capacity to move or be set in motion by the physical impetuses of sport and physical culture). Some of the things our contributors grab hold of are inherently "sporty" (the titanium golf club, the ball, the sand trap), others less so (water, GIS technologies, labor markets), and others still get drawn into the realm of physical culture as effect, affect, or intermediation (the sensual and sensory body, homes bulldozed for stadium development projects, city streets turned into jogging tracks, the season ticket, etc.). In some cases, our contributors call into question the existential or metaphysical qualities of the thing (the thing as it exists, as it relates); in others, contributors look at how we might know the thing as a thing—as a matter of consequence.

In so doing, the book takes two subtle yet potentially substantial turns away from prevailing scholarship on the social significance of sport and physical culture. First, we take a slightly different tack to the study of "the social" than many sociologists of sport and physical activity, and particularly [physical] cultural studies scholars, have taken in recent decades. Whereas sport and physical culture scholars inspired by the representational and cultural turns, poststructuralism, Derrida, and many post-Saussurean social theories have tended to privilege discourse, language, and meaning—and whereby cultural studies scholars of the Raymond Williams and Stuart Hall traditions have tended to look to matter(s) of the moving body in the first instance as meaningful—here we explore culture as consequence of materiality. The (physical) cultural studies turn to the body resulted in research studies that often reduced the body to textual patterns, media representations, semiotic systems, and/or grand corporeal narratives or erased the researcher's own body, politics, and praxical dimensions from empirical field studies (see, e.g., Aalten, 2004; Brace-Govan, 2002; Dworkin & Wachs, 2004; Newman, 2013). This was, perhaps, the natural outcome, for the field of cultural studies writ large was going through similar iterations in the United States and elsewhere (i.e., a cultural studies of representational politics—if not convenience—rather than of political struggle that characterized its British forbearer). At the same time, it is undeniable that many such works provided critical interrogations of the body within popular/political culture and offered insight into various topics of inquiry. Our critique is that this sensibility led to "the abstraction of politically-enfleshed bodies, the disappearance of authorial bodies, and the missing empirical dialectics of the self" (Giardina & Newman, 2011, p. 44). That is, physical culture and the competitive and constitutive bodies comprised thereof are rendered in social science as discourse/representation and are predictively extrapolated without deep consideration of the material and physiological as determinants of culture.

Against such a trend, we look here to extend a study of sport and physical culture that squarely locates the body as radically embodied, fleshed, and materially consequential (see Giardina & Newman, 2011; Newman & Giardina, 2014; Newman, Shields, & McLeod, 2015; and Thorpe, 2011, 2016; and see also Carrington, 2008). Although this might not seem a very novel turn—feminist scholars have for some time called for an extended consideration of the affectivity (Ahmed, 2004; Gregg & Seigworth, 2010), fleshiness (Birke, 1999; Bordo, 2004; Grosz, 1994), and fluidity (Braidotti, 1994; Shildrick, 1997/2015) of the body—we believe that a focus on the productivities, potentialities, and connectivities brought to life by the body-object as it makes, ties, draws into contact, and moves associated actors qua actants is of some consequence for being, knowing, and acting on moving bodies.[5] In seeking to better map out these materialities as they come to form cultural embodiments, we see new heuristic pathways for exploring how physical culture as symbolic performance and practice is both productive of and produced by changes in chromosomes, cells, muscles, water molecules, natural environments, sexed and heteronormalized bodies, geopolitics, and various economies and the physical world.

Hence this is a project aimed at exploring the depths and possibilities of *physicality*. Our use of the term here refers to the physical attributes of the human and/as the object—often with particular emphasis on the human body's capacity for physiological activity, force, or expression. Physicality in this sense is the state or quality of being physical, with an allusion to the corporeal body as an expression of and articulation to its mental, cultural, and affective capacities. But we are also referring to how those human corporealities, identities, performances, and performativities are bounded to configurations of the material, juridical, economic, hyperobjective,[6] and volatile world. Simply, this is our attempt to put the physics back into physical culture (and the study thereof)—and to do so by emphasizing the object-form, its ecologies and agencies, and the broader articulations it brings to life.

ON EMBODIMENT: THE ACTIVE BODY
OF/BETWEEN CULTURE AND BIOLOGY

How then might scholars of sport and physical culture reconcile physics, physiology, and performance? Most commonly, in the sociological and cultural study of sport and human movement, we look to examine how the moving body acts as an *embodiment* of some broader political, social, or economic configuration. In its most common usage, this term *embodiment* refers to a physical entity typifying an abstraction or giving concrete form to an abstract concept. The term is often used in anthropomorphic ways to instill human or agentive qualities to a thing (i.e., "the Confederate flag embodies racist ideologies"). In other instances, embodiment refers to a human who enfleshes an image or conception (i.e., "Tiger Woods embodies American multiculturalism"). And in still other cases, the term refers to the material thing's ability to evoke reference to other things (or, in some instances, ideas). In its general use, *embodiment* is utilized in heuristics to link the abstract or immaterial to the object or material (body).

What, then, does it mean to embody, to be (an) embodiment? And for our purposes, how does one go about studying the material bases of embodiment? As a baseline, it would seem that as a construct, embodiment deals in some ways with *surfacing*—that is, a shift in shape or a change in form from material to immaterial, from static to fluid, from concrete to abstract, from backstage to frontstage. For *something* to be embodied, it needs to become, or to surface as, *some thing*. The distinction of things as objects and things as representations here is critical. Representational thought would have us believe that the discursive thing overdetermines that state of the material thing—that, to very slightly rephrase Derrida, there is *no-thing outside of the text*. While such an ontology has served scholars of the active, moving body well in explaining identity politics, circuits of culture, and politics of representation brought to life by systems and moments of embodiment, in certain camps social scientists and scholars of physical culture have lost sight of the physics and kinetics that inexorably ground spheres of embodiment and representation. As we can see, however, this is not the whole story of embodiment.

First, let us discuss the notion of embodiment as it tends to be taken up in the so-called hard sciences. Consider embodiment and neurons. In his widely cited book *Embodiment and Cognitive Science*, famed psychologist Raymond Gibbs (2005) explains that embodiment in the cognitive sciences "refers to understanding the role of an agent's own body in its everyday, situated cognition. For example, how do our bodies influence the ways we think and speak?" (p. 1). Although Gibbs's second statement might cause some Cartesians (and certainly most Lacanians) among us to take pause—if for no other reason than a series of seemingly unreasonable assumptions about the interrelations of mind, body, subjectivity, and agency—this opening salvo to Gibbs's book provides insight into how some scholars of the modern positivist tradition frame the body and embodiment: that is, *the body makes the neurological and material and social*. The body is the soma of a neurological hub—and in the complex aggregate of multiple articulations, we find embodiment.

Embodiment in this way is conceived as a coupling mechanism.[7] In other key works in the field of "brain science," embodiment refers to "experience in the brain's modality-specific systems for perception, action, and introspection" (Niedenthal, Barsalou, Winkielman, Krauth-Gruber, & Ric, 2005, p. 184). Embodiment in this line of thinking is the surfacing of cognition or, put simply, "the embodiment of cognition" (Ballard, Hayhoe, Pook, & Rao, 1997, p. 723). This version of embodiment presents the body as both enabling and constricting—the active endpoint of the mind and yet bounding cognition in/as action to the body's physical limitations, spatial locations, and corporeal dispositions. At the "embodiment level," the neuron is interceded by synapse; pain, pleasure, gait, and *physios* are polarities of computation and cognitive operation (see also Goodwin, 2000; Clark, 2008; Gibbs, 2005). In other words, from this perspective the body associates with a neuron to carry out a cognitive impulse and thus an embodiment process.

Consider further embodiment and cells. Based on what we have discussed thus far, such an embodiment would seem tautological: the embodiment of bodily matter. However, for much of the literature in human anatomy and physiology, such is not the case. Embodiment in these broad fields of inquiry tends to take on a temporal, iterative, categorical, or confirmatory quality. Scientists often use phrases such as "embodiment of the intervention" to illustrate how the scientifically manipulated human body reveals a theory or confirms a hypothesis, can be classified in a particular way, or morphs in expected and unexpected ways (embodiment in this way is morphology materialized). Scientists use the term *embodiment* to explain how the body reacts and changes at different stages of an intervention (e.g., "embodies the predicted relationship/effect/behavior"). Scientists sometimes use the term to illustrate how cells in isolated and composite forms more broadly fit within a classificatory model or clinical framework.

Public health scholar Nancy Krieger (2005) takes a slightly different and perhaps more illuminating approach to embodiment. She explains that we, as humans, "are simultaneously social beings and biological organisms" and that the notion of embodiment advances three heuristic realities:

> (1) bodies tell stories about—and cannot be studied divorced from—the conditions of our existence; (2) bodies tell stories that often—but not always—match people's stated accounts; and (3) bodies tell stories that people cannot or will not tell, either because they are unable, forbidden, or choose not to tell. (p. 350)

The body, in this way, is a site of scientific revelation in the midst of the illusory social; it is object, objectifiable to contrivances of measurement and verification.

These hard science approaches stand in contradistinction to work emanating from the fields of sociology and anthropology. Consider work on embodiment and the *socius*. Social psychologists have extensively labored to explicate "social embodiment." Barsalou, Niedenthal, Barbey, and Ruppert (2003) in particular describe the notion of social embodiment as "states of the body, such as postures, arm movements, and facial expressions, [that] arise during social interaction and play central roles in social information processing" (p. 43). Or consider the stubborn fascination with embodiment in the field of anthropology (not to mention theology), where scholars from Clifford Geertz to Thomas Csordas have dedicated much of their respective careers to contemplating the micropolitics of representation and the empirical complexities of studying self-presentation in the realm(s) of everyday culture production. In this field, the body serves as inscription, expression, and locus of ideology, affect, habitus, identity, and emotion—a living and breathing representation of broader systems of thought, power, identity politics, and history.

This inverted Hegelianism is, of course, not exclusive to anthropology. It is widely taken up in the fields of political economy, sociology, and history. Let us reflect on relations of the body and capital. When we turn to political economy, such as that proffered by classical economist David Ricardo (1891/2015), capital is "embodied" labor (although he rarely made any allusion to the bodies from which that labor came). Indeed, the body has since been explicitly recognized by two dyadic economic traditions: Marxism, in which the laboring body is seen to be the foundation and source of capital accumulation, and the works of Chicago School economists such as Theodore Schultz (1960), Gary Becker (1962), and Jacob Mincer (1958), each of whom saw the human corpus as repository for human capital investment and accumulation.[8] Clarifying the embodied state (and in antithesis to the Chicago and Austrian schools), Bourdieu (1986) notes that "most of the properties of cultural capital can be deduced from the fact that, in its fundamental state, it is linked to the body and presupposes embodiment" (p. 244). Bourdieu (1978) also refers to this embodied form of cultural capital as physical or corporeal capital, which he treats as a form of capital in its own right.

Since the mid-1980s, embodiment has become a major domain for sociological inquiry, and it remains "one of the most contested concepts in the social sciences" (Shilling, 2005, p. 6). As Bryan Turner (2008), one of the preeminent body theorists, explains,

> Sociological theory can be said to be organized around a number of perennial contrasts—agency and structure, the individual and society, nature and culture, mind and body. . . . The importance of the sociology of the body is that it lies at the axis of these theoretical tensions and it is thus a necessary component of any genuine sociology. The difficulty of providing a coherent account of what we mean by "the body" is an effect of these theoretical problems. (p. 209)

The analysis of the body and embodiment has produced "an intellectual battlefield" over which "the respective claims of post-structuralism and post-modernism, phenomenology, feminism, socio-biology, sociology and cultural studies have fought" (Shilling, 2005, p. 6; also see Csordas, 1994; Davis, 1997; Featherstone, Hepworth, & Turner, 1991; Shilling, 2007; Turner, 2008; Williams & Bendelow, 1998). Increasingly, social science scholars from various fields have drawn on an array of theoretical and methodological perspectives to shed new light on various aspects of the body. Yet, as Williams and Bendelow (1998) suggest, this has led to some conceptual ambiguity and disciplinary fragmentation:

> Recent years have witnessed a veritable explosion of interest in the body within social theory. . . . [But] the more the body is studied and written about the more elusive it becomes: a fleshy organic entity and a natural symbol of society; the primordial basis of our being-in-the-world and the discursive product of disciplinary technologies of power/knowledge; an ongoing structure of lived experience and the foundational basis of rational consciousness; the well-spring of human emotionality and the site of numerous "cyborg" couplings; a physical vehicle for personhood and identity and the basis from which social institutions, organizations and structures are forged. The body, in short, is all these things and much more. At best this has served to capture the multifaceted nature of the body in society. At worst it has led to a fragmentation of perspectives and a dispersal of approaches which, for the most part, talk *past* rather than *to* each other. (pp. 1–2)

Despite numerous theoretically sophisticated and empirically nuanced analyses of particular dimensions of the body and embodiment, contemporary sociological studies have yet to provide a theoretical approach that satisfactorily explains the material dimensions of the body. Rather, in each of these studies, particular aspects of embodiment are foregrounded while others "fade into the background" (Shilling, 2005, p. 6).

What we have, then, is the construct of embodiment being taken up across the sciences in epistemologically and ontologically nonessentialist ways. From the very brief survey on offer above, we see that the body tends to be considered as a *surface* upon which the neurological, biological, economic, ideological, and social materializes. In this way, the body *is* object—subordinated to the mind, to science, to biological capacity, to history, to the social. But what, then, is the nature of this body-*as*-object? The simple answer would be that the body is material, cellular, fleshy, and fluid (literally). *And* that it is simultaneously political, social, representational, and cultural (see Burkitt, 1999); it is a conduit *for* identity and affect just as it is a vessel *of* pain, pleasure, and experience. Each of the dualistic paradigms is fairly sound in reason: The material body exists—we can touch it, feel it, cut it, and so on—*and* it has the capacity to produce (representations, material outcomes, social relations, affect). The representational body also exists—we can perceive it, locate its locutions, point to its discursive politics—*and* it can point back at us.

A more critical reading of this review might suggest that "embodiment" is a literary or epistemological device; that is, it is metaphysical shorthand referring to a number of different relations as constructed in science: to give form, to abstract, to express, to make material, to couple along paradigmatic doxa. The body is produced in language, certainly, but as Michael Serres (2011) declares with resolve in *Variations of the Body*, it is also the case that language exists in the service of the body. He coyly writes,

> The genuine writer's craft demands a solitary engagement from the entire body and its sole singularity. Gymnastic exercise, a rather austere diet, life in the open air, a thousand practices of strength and flexibility, on the whole, alpine climbs, for writing, are as good as ten libraries. (p. 12)

In this sense, "embodiment" relies on surrounding disciplinary discourse to clarify the intended meaning. For instance, in literature and economics, the "em" rather than the "bodied" seems most relevant—at least as we tend to understand the latter—whereas in cellular biology, it appears that a treatment is embodied most essentially in the body. But the body is more than language, more than discourse, and more than neurons. It is more and it is moored—anchored to spatiotemporal boundaries, material conditions, and environments, moving about land and air and viral microbes as it transmits, transforms, and translates anew.

As we have argued elsewhere (Thorpe, 2014), our own bodies can "rudely remind" us of "the importance of the biological" that is "beyond our sociological gaze" (p. 667). In and between language and biology, we find an uncertainty best captured in Csordas's (1990; 1994) distinction between "embodiment as representation" and "embodiment as being-in-the-world." Csordas cautions social scientists venturing just past the cultural turn to take seriously the idea that "culture is grounded in the human body" (p. 6). He suggests that we avoid the "representationalist trap" and—as Roland Barthes did—make the distinction (and mindfully attend to the articulation) between the body as a material, biological entity and embodiment as the field defined by "perceptual experience and the mode of presence and engagement in the world" (p. 12).

Of course, as a surface, the body is discursive—in fact, those of the Derridian tradition might say that is *all* it ever can be (i.e., an assemblage of texts, the transcendental signifier). From this tradition—a tradition (physical) cultural studies seems so deeply rooted to (perhaps intentionally so, we would concede)—the body is in the first instance as in the last *cultural*. Embodiment is the surfacing of the corpus within systems of meaning (again with deference to Raymond

Williams as well as Richard Johnson and Stuart Hall). There is certainly a metaphysical coherence to the representationalist perspective: How can we ever know the body, and certainly embodiments of another, in any way other than through the communicative exchange? The cultural body is given to encoding and decoding, a prisoner to the circuit(s) of culture. This perspective—a perspective that came to hold sway at the cultural turn—seems to privilege a particular hermeneutics and ontology. With regard to hermeneutics, it is given that once we concede that to know the body (and its capacity for embodiment) is and can only be to know it as discourse (scientific, cultural, social text, etc.), then we can set out to demonstrate the body as ontologically subjected. The body and its capacity for movement, affect, expression, death, and living are made by and within systems of language. It seems, following Roberto Esposito (2015), that "the thing has been somehow 'decorporealized' by being dissolved into the idea or into the word" (p. 5). But how, then, does the scholar of the cultural body account for human experience? As discursive formation? How do we concede to human action or choice? To pain and pleasure? Affect and impulse? Capillaries and biomolecules?

We believe one possible way forward from this dilemma is found in the work of new materialists and those arguing for (a) new empiricism (e.g., Sara Ahmed, Karen Barad, Rosi Braidotti, Jane Bennett, Manuel DeLanda, Samantha Frost, Elizabeth Grosz).[9] Ahmed (2006), for example, writing in *Queer Phenomenology: Orientations, Objects, and Others*, expressly links the phenomenology of Husserl and Merleau-Ponty (among others) with feminist and critical theory to offer a queer model of *orientations*. This concept of orientation, she maintains, "allows us then to rethink the phenomenality of space—that is, how space is dependent on bodily inhabitance" (p. 6). In one especially poignant example (see pp. 109–111), she draws attention to Frantz Fanon's (1952/2008) account of wanting to smoke a cigarette and the bodily actions he (knows he) must make in order to do so while sitting at a table with a (white) man—reaching out with his hand, taking the pack of cigarettes, retrieving the matches in a particular drawer, and leaning back—all moments, Fanon writes, made "not out of habit, but of implicit knowledge. A slow construction of my self as a body in a spatial and temporal world—such seems to be the schema" (p. 91). But, Ahmed points out, the "corporeal schema" of action is in conversation with what Fanon refers to as the "historic-racial" schema interceding on his action *from below*: "The racial and historical dimensions are beneath the surface of the body described by phenomenology, which becomes, by virtue of its own orientation, a way of thinking the body that has surface appeal" (p. 110). What Ahmed is driving at, then, is the need to *dis*orient ourselves so that we may find our orientation. In so doing, we may find new planes (of affectivities) on which to act, new ways of seeing (if not experiencing) our reflected reflexion.

Of course, it should be noted that these materialist orientations do not start here or now. In ancient Chinese society, for example, the relations of matter, senses, and social organization were of principle concern to thinkers. Indeed, much of the line of thinking about the monisms and materialisms of ancient civilizations—from the Aztec concept of *Ometeotl* (the unity of the universe) to Maori principles regarding soil, forests, mountains, and water as *taonga* (treasures) to be safeguarded by and held up as identifier for the people (*tangata whenua*)—is owed (if not always acknowledged as such) to precolonial modes of knowing the world. As Kyla Wazana Tomkins (2016) suggests, "new materialism" articulates with onto-epistemologies that stand in opposition to mainstream Euro-Western philosophy and scholarship (specifically humanism and nature/culture dualism). Indeed, Tomkins (2016) finds familiarity between "new materialism"—in the vitalism and a privileging of ecocentrism over anthropocentrism—with some First Nation and Indigenous knowledge traditions (see also Sundberg, 2014; Todd, 2016). However, Zoe Todd (2016) reminds us that we must be mindful of how this "new" in new materialisms requires a deep consideration of the Indigenous knowledges and anticolonial world views that came before (more on this near the end of this chapter and in chapter 15 by Richard Pringle). The point is that much of the "new" in new materialism is in fact well established in non-Western modes of knowledge formation.

As such, it might be more accurate to refer to this shift not as something "new," per se, but as a *turn* or *re-turn* (perhaps a turn away from the so-called cultural turn). But what are new materialists turning away from? Ahmed and her feminist contemporaries (Haraway, Braidotti, Barad, Frost, Kirby, Grosz, Bordo, and Bennett among them) seek to extend the framework for understanding the *organized body* (structured in language, social order, economies of embodiment, *communitas*, and perhaps even dimensions of time and space) by looking to the *organ-ized body* (the body as composition and composing [and decomposing]). Building on and at times departing from Derrida,[10] they pluralistically posit a break from structural and representational determinism that renders embodiment as discourse in/as/about context—a shift away from representationalism and toward a radical rethinking of the agentic object—framed as "vital materiality" for Bennett (2010), "agential realism" for Barad (2007), "situated epistemology" for Haraway (1988), or "neo-vitalism" for Braidotti (2000). As John Protevi (2001) explains in his compelling *Political Physics*, this line of thinking signals a shift from the Derridian (or perhaps Hallian) representationalism that inspires many scholars of the social and the cultural:

Although the work of Jacques Derrida is a magnificent achievement and a lasting contribution to the tradition of post-phenomenological European philosophy, it is, while still necessary to any progressive philosophical and political practice, primarily of propaedeutic value in the reflection on and intervention into the convergent fields assuming the highest importance in the material structuring of the current global system of bodies politic. (p. 2)

Owing a considerable debt to Derrida's "play of significants" and Foucault's hermeneutical genealogy, a new materialist approach looks past a body that is made real by language (as opposed to the opposition) and constituted by the history of ideas and ideology. This is an impossible, implicit body. Protevi (2009) and a generation of scholars in anthropology, feminist studies, and posthuman studies and those working in contemporary continental philosophy camps haunted by Deleuze, Guattari, Braidotti, Agamben, Haraway, Barad, and Bordo are now looking beyond the body as a representation of ideas—the cultural body as articulated to the body politic—and toward the "organ-ized and organ-izing body politic."

One result of such a turn has seen scholars take up Deleuze (1988; 1994; Deleuze & Guattari, 1988) to realize Derrida's implicit body. Whereas a Derridian deconstructive approach moves us beyond "the self-grounding of a rational, meaningful sign system—the book of nature—to the inscription of marks in a world of (physiological/physical) force and signification—the 'general text,'" the Deleuzean injunction allows us to "conduct a material analysis of forceful bodies politic" (Protevi, 2001, p. 2). It is our position here and in what follows in this book that scholars of the active body might benefit from following Deleuze's shift away from "the transcendental from the conditions of possibility investigated by phenomenology" to the "patterns and triggers of actual bodies, thus implicitly articulat[ing] a research programme for the investigation of forceful bodies politic" (p. 2; see also Markula, 2019).

In *Nietzsche and Philosophy*, Deleuze (1983/2006) describes the production of forceful bodies in this fashion: "Every relationship of forces constitutes a body—whether it is chemical, biological, social, or political" (p. 457). Hence, Protevi (2001) argues, the study of *forceful* bodies requires both "a politicized physics (paying attention to the political ground of such basic physics terms as 'law')" and a "physicalized politics (paying attention to the physical ground of such basic political terms as 'force')," as both are needed "in order to understand politics as the forceful organ-ization of bodies" (p. 3). This means that

forceful bodies (i.e., persons, families, groups, parties, gangs, corporations, races, sects, nations, worlds) are thus particular force-arrangements of chemical, biological and social bodies, themselves force-arrangements: they are forceful bodies politic. In its most radical Deleuzean moment, political physics, in thinking forceful bodies politic, thus moves both "above" and "below" the level of the individual as classically conceived in liberal humanism, opening ways to investigate both "social machines" (inter alia, tribalism, monarchism, liberalism, fascism, and the experimental immanent self-orderings Deleuze and Guattari call "war machines") and the "molecular flows" of matter (somatic fluids, of course—milk, sweat, sperm, urine, blood—but also steel, electricity, concrete) they order into forceful bodies politic. The classical modern notion of life, restricted to the organic individual or perhaps the species as collection of individuals, is thus too restricted for the scope of political physics. (p. 3)

Such an approach to force and forceful bodies leads to a more complex reading of the body/embodiment, body politics, and the body politic: the body as surface and embodiment as (a) surfacing. This is not a surfacing of meaning but the becoming of a Body without Organs (as Deleuze and Guattari explain across both volumes of Schizophrenia and Capitalism).[11] Coole and Frost (2010) make this claim carefully and clearly in their introduction to their edited collection titled New Materialisms:

> One could conclude, accordingly, that "matter becomes" rather than "matter is." It is in these choreographies of becoming that we find cosmic forces assembling and disintegrating to forge more or less enduring patterns that may provisionally exhibit internally coherent, efficacious organization: objects forming and emerging within relational fields, bodies composing their natural environment in ways that are corporeally meaningful to them, and subjectivities being constituted as open series of capacities or potencies that emerge hazardously and ambiguously within a multitude of organic and social processes. (p. 10)

Here our concern is with what lies in and between flesh bodies and the material conditions from which they are made and which they make. Before turning to an outline of the book, let us first offer three brief discussions on the materiality of sport and physical culture. We first look to broaden the discussion on the moving, sporting body as vibrant matter—that is, how do the human body's anatomical characters determine its cultural, political, and economic conditions? We entertain these questions: What things make it, and of what is it made? We then explicate how scholars of the active and sporting body, such as those contributing to this book, might adopt "new" materialist epistemologies—nay, ontologies—in rethinking and reevaluating the constitutive interactions (that make sport and physical culture) of muscle, steel, water, and culture. We do so not under a singular umbrella of "new materialism" but rather under the guise of complex intersectional (nay, matrixial) approaches to interrelations and associations of objects to other objects and nonobject formations and systems. This involves navigating the messy ontological, methodological, and empirical-form intermixing of social theories, political philosophies, scientific paradigms, data, politics, and historical materialities. So we would like to move forward by next locating the various camps from which "new" materialist approaches to studying the moving body—here and elsewhere—have emerged and look to how certain metaphysical alignments around the study of bodies and/as things might help shape these and future explorations.

NEW MATERIALISMS AND THE MOVING BODY AS THING

Riding these tides, we look to renew the materialisms of old (Marx, Nietzsche, Spinoza, etc.) and chart new courses of study for the sporting and physically cultural body and its linkages. We follow Coole and Frost (2010) in seeking to excavate materiality in the traditions of "material culture, geopolitical space, critical realism, critical international political economy, globalization, and environmentalism, and in renewed calls for materialist feminism, or more materialist queer theory or postcolonial studies" (p. 2). We do so under the assumption that "the more textual approaches associated with the so-called cultural turn are increasingly being deemed inadequate for understanding contemporary society, particularly in light of some of its most urgent challenges regarding environmental, demographic, geopolitical, and economic change" (pp. 2–3). So against the tides of the cultural turn, we look here to draw attention to "a new materialist predilection for a more phenomenological approach to embodiment" (p. 19). Our emphasis is on how moving matter sets power in motion and on the way that power constitutes and is reproduced by bodies. This turn to matter, as we have sought to make clear, is aligned with various established and emerging traditions within the philosophical and social sciences. To understand power, materiality, and the object assemblage and to move beyond the anthropocentrism and privileging of representationalism noted above, our contributors look to various strands of new materialist thought and toward new epistemological frames through which we can understand matters of physical culture.

The study of the material world, as we know (it), corresponds with philosophy itself. Ancient philosophers from Greece (such as Epicurus and Democritus), to India (such as Ajita Kesakambali and Kanada), to China (from Xunzi to Yang Xiong), and beyond focused on the object, human bodies, and the interrelations between them. The modern materialist project is perhaps most frequently tied to Feuerbach's humanist atheism and Marx's redevelopment of his dialectical materialism. Marx, in flipping the world of ideas upside down, sought to unearth how the material conditions of labor might determine (or perhaps produce) social and ideological conditions of a capitalist society. All this work, of course, continues to haunt philosophical explorations into the materiality of things.

That being said, in recent years we have seen an increased focus on materiality. Much of this work is inspired by the thinking of 20th-century intellectual Georges Bataille (1949/1991; 1949/1993). Bataille's rereading of Spinoza's (1677/2002) anti-Cartesian *Ethics* twinned with a reinvigoration of a post-/anti-Kantian correlationism has inspired a recent wave of continental philosophy oriented toward the transempirical study of the thing as agent. By calling into question the anthropocentric bases of most social and philosophical science, scholars from Bruno Latour to Levi Bryant, Manuel DeLanda to Quentin Meillassoux, and Wendy Brown to Jane Bennett (to name but a few) are cacophonously exploring relations of thinking and being and of being as thing and being as (co)relation. This line of inquiry—in respect to the earth, its surfaces and ecosystems, the material forms it holds in place, and the bodies to which it gives life—looks to science, technology, physics, and environmental sciences for both critique and firm ground upon which we might reimagine the thing-as-being. They point out that the thing, in its very being, is illuminative; the thing tells a story of its relation to other things—and not just a story of how it is made or subjugated to human consciousness but how human thinking (and being/becoming) is in many ways contoured around the material object.

In *Tool-Being: Heidegger and the Metaphysics of Objects*, Graham Harman (2002) maps out one position from which we might imagine the relations of things-humans and the knowledge formations thereof (see also Harman, 2005; 2009; 2011):

> If the human perception of a house or a tree is forever haunted by some hidden surplus in the things that never becomes present, the same is true of the sheer causal interaction between rocks or raindrops. Even inanimate things only unlock each other's realities to a minimal extent, reducing one another to caricatures. . . . Even if rocks are not sentient creatures, they never encounter one another in their deepest being, but only *as present-at-hand*; it is only Heidegger's confusion of two distinct senses of the as-structure that prevents this strange result from being accepted. (p. 2)

How, then, might we develop the study of things-as-being in ways that are empirically verifiable—especially given the very loose articulations the inquirer must forge to connect microbes to cement to meaning to knowledge? How do we develop a thermodynamics of nonequivalent agents? How do we know what is outcome or what is intention? What is "natural" or what is purposeful? Karen

Barad, in her important book *Meeting the Universe Halfway*, seeks to attend to such uncertainties and speculative realities in her use of the poststructuralism of Foucault, the feminist theory of Butler, and Bohr's quantum physics. By working backward using theories from "hard sciences" (namely, physics), she explores the intra-activity between matter and human knowledge to illustrate how we might take up materialist metaphysics to reconceptualize agency as a relation or set of relations (not something bestowed upon the thing or nothing).

We would argue that the most significant striation of new materialist scholarship has come from feminist scholars. Feminists have been integral to the multiple assemblages of new materialism (Pitts-Taylor, 2016; Hird, 2004). Over the past three decades, a critical mass of feminist scholars has been working to remedy "feminism's anti-biologism" (Davis, 2009) and seeking to develop new ways of understanding "the complex interactions through which the social, the biological, and the physical emerge, persist, and transform" (Frost, 2014, p. 307). In so doing, they have increasingly drawn on the physical and biological sciences to "repudiate the notion that biology and matter are passive or inert and instead recognize the agency of biology or matter in worldly phenomena and social and political behavior" (Frost, 2011, p. 69). Some examples include Donna Haraway's (1991) deliberations on the immune system and the biopolitics of postmodern bodies, Lynda Birke's (1999, 2003) feminist biology, Anne Fausto-Sterling's (2000) critical reflections in *Sexing the Body*, Elizabeth Grosz's (1994) corporeal feminism, Barad's (2007) agential realism, and Wilson's (2004) feminist engagement with neurological theories. Others are engaging the scholarship of "new materialist" feminist theorists (e.g., Butler, 1993; Barad, 1998; Mol, 2002) and/or developmental systems theory (see Meynell, 2008) to produce fascinating investigations of sex hormones, and more recently pheromones, that directly challenge "the dualisms of culture/nature, sex/gender, science/humanities" (Sieben, 2011, p. 264; also see Roberts, 2003).

Authors such as Fausto-Sterling (2000), Oudshoorn (1994), and Roberts (2002, 2003, 2007), for example, offer lucid analyses that demonstrate the extent of the "entanglement of heteronormative discourses ... with hormonal 'realities'" (Sieben, 2011, p. 264) via an array of cases ranging from sex testing at the Olympic Games, to hormone-replacement therapies, to the effects of the increase of "environmental estrogens" on men and women's sex hormones. Engaging a queer feminist materialist science studies manifesto, Willey (2016) examines the biology of the erotic to "decanter assumptions about sexuality and human nature that shape the field of gene-brain-behavior research" and, in so doing, elaborates "a theory of *biopossibility*" (p. 553, emphasis in original). A few feminist sport scholars have expressed concern about biophobia in research on the moving body and are engaging new materialist feminist approaches to examine the biocultural dimensions of depression (Fullagar, 2008; also see chapter 8 of this book), amenorrhea (Thorpe, 2016), and relative energy deficiency (Thorpe & Clark, 2019) among exercising women. According to Frost (2014), what is at issue in such projects is "a refiguration of the body, of encultured biological organisms, of biocultural creatures" (p. 323). In other words, as "projects of refiguration," the work of feminist new materialists seeks to "rearrange or reorganise" our understanding of the relationship between nature and culture as dynamic and "the materiality of the body as an organism rather than its materiality as a subject" (p. 307).

Importantly, not all feminist scholars are in support of such approaches. For example, Sara Ahmed (2008) has criticized feminist new materialists (and she includes Barad, Elizabeth Grosz, Vicki Kirby and Elizabeth Wilson in this category) for what she considers false accusations of the long-standing biophobia evident in feminism. In response to such accusations, however, Davis (2009) argues that new materialist usage of biophobia reveals that they are "proposing a particular conceptualization of what an engagement with the biological means":

> They theorize an entanglement and nonseparability of the biological with/in sociality, and what they criticize in much feminism is the conventional assumption that the biological and the social are two separate and discrete systems that then somehow interact. (p. 67)

This is a challenge taken up by Frost (2011), who similarly applauds the efforts by feminist theorists engaging with new materialism to more fully comprehend biological bodies "not only as they are formed by the forces of language, culture and politics, but also as they are formative" (p. 7). In reconsidering the turn to biology in feminist theory, Frost (2014) advocates a critical, even philosophical, engagement with the life sciences "not only as a factual resource, but also as a figural resource":

> In shifting our conceptual orientation to biological science from fact to figure, feminists will be able to give theoretical life to scientific findings about the ways in which social environments and material habitats are processes integral to our development, growth, and social and political well-being. (p. 307)

Continuing, she suggests that one way forward is for feminists to think of the life sciences as producing something akin to what Nikolas Rose (2013, p. 23) describes as a "philosophy of life," a philosophy that encourages a reimagining of "what we think we are as living creatures" (Frost, 2014, p. 309). According to Frost (2014), rethinking how we engage with the life sciences will support feminists to "creatively re-imagine the fields and fora of feminist politics" (p. 307).

Another key contribution of feminist new materialists is in the "queering of biology" (Alaimo, 2008, p. 241; Willey, 2016). For too long, biology—like nature—has "been drafted to serve as the armory for racist, sexist, and hetero-sexist norms," such that it is "crucial that feminists invoke a counter-biology to aid our struggles" (Alaimo, 2008, p. 241). Within this wave of feminist (and queer, antiracist, postcolonial) approaches that are "taking matter seriously" (p. 243), important questions are being asked as to how the embrace of materialism and ontology not only challenges long-standing and problematic binaries of sex/gender, nature/culture, and science/feminism but also shifts understandings of power, politics, agency, and resistance. As Pitts-Taylor (2016) writes, "Feminist orientations demand we ask certain questions of matter/ing and its interlocution":

> In what ways is matter involved in, or shot through with, sex/gender, class, race, nation, citizenship, and other stratifications? How are these power relations involved in the understanding and management of biology or "life itself," and how do they materialize in bodies, corporeal processes, and environments? (p. 2)

Similarly, Clare (2016) asks, "What is a new materialist approach to politics? What counts as 'politics' within new materialist thought?" (p. 61). For Clare (2016), the notion of politics that emerges in new materialisms is at its best "not when it redefines politics so as to include non-human or more-than-human forces" but rather when it "reworks understandings of human subjectivity, making it clear how the human is always enmeshed in more-than-human worlds" (p. 61). In pursuing such lines of flight, feminist scholars across an array of disciplines are exploring new materialism and practices of colonialism (Roy & Subramaniam, 2016), the complex relationships between Indigenous ways of knowing and the "ontological turn" (Salmond, 2014, 2017; Todd, 2016), and the implications of Indigenous materialisms for posthuman, postcolonial, and new materialist ways of knowing humans, texts, and objects (Ravenscroft, 2018).

In their chapter "Matter in the Shadows: Feminist New Materialism and the Practices of Colonialism," Roy and Subramaniam (2016) remind us of the need to critically examine "the matter that lurks in the deep, dark shadows of the 'old' feminist critiques of science of earlier feminisms, the silhouettes of matter being brought forward in the feminist new materialisms, and the glints of attention to matter that have recently surfaced in postcolonial science studies" (p. 24). Continuing, they conclude that materiality should always be "understood as a 'process' rather than a product—matter makes, is made, and remade" (p. 39). To understand the processes (and politics) through which matter is made (and made known), they call for "greater interdisciplinarity across fields and generations to draw on the insights and wisdom that multiple fields have to offer":

Drawing on analyses from multiple locations sheds light on the shadows within our disciplines, rendering more perceptible the complexity of how matter and biologies come to be. We urgently need to develop models of a situated materiality and by doing so, turn to improperly buried bodies that perhaps still matter. (p. 39)

As well as working across disciplines, generations, and locations, understanding materiality as situated and in process also requires us to consider the multiple and diverse ways that cultures make meaning of matter—bodies, environments, objects, and artefacts—differently (Ravenscroft, 2018; Salmond, 2017). A growing body of scholarship is seeking to decolonize new materialisms by focusing on the contestations and various forms of marginalization and violence that have resulted when a particular group seeks to rewrite the meanings of matter and materiality or erase the processes in which matter has come to matter to different groups. A special issue of *darkmatter* titled "Subjects, Objects, Others: Materialisms from the Enslaved and Colonized," for example, promises a timely and necessary focus on "objects laden with stories of slavery, colonization and their aftermaths" with the aim to bring "materialisms from the dispossessed, oppressed and exploited into conversation with new materialism . . . to encourage us to think these subjects and the worlds they encounter in new terms" (Karina & Coffey, 2017).

Others have explored what new materialisms might mean for thinking about identity politics and particularly racial identity politics. Hames-García (2008) advocates the value of new materialist approaches for encouraging "racial identity projects that do not simply reaffirm what race is and has been, but rather, seek a transformation of race into something new" (p. 331). Continuing, she argues,

We need creative racial identity projects more than we need philosophical arguments about race. Race is not more or less a fiction than nation or ethnicity, and it is just as unlikely to go away. One cannot easily dismiss it as "only cultural" or "just ideological" because it emerges from the intra-action of history, culture, economics, and material human bodies. . . . Racial hierarchies and biological determinist assumptions about racial difference will not evaporate without a significant reorganization of economic, political, and cultural relations at the personal and societal levels as well as accounting for how different bodies intra-act with racial categories and ideas. (p. 331)

Asking similar questions in different ways, a special issue of *Postcolonial Studies* focused on the growing field of postcolonial science studies critically considers "what an infusion of materialist science studies might do for postcolonial theory" (Seth, 2009, p. 381), while others have explored how Chicana and other feminist scholars' own bodily and embodied experiences of illness, disability, and loss—for example, diabetes (Anzaldúa, 2002), premature childbirth (Moraga, 1997), cancer (Singer, 1993), bereavement (Butler, 2003)—have inspired new ways of thinking and knowing beyond the discursive and toward different understandings of community and activism beyond identity politics (Bost, 2008). In her article "Diffracting Diffraction: Cutting Together Apart," Barad (2014) cites an early conference paper by Trinh Minh-ha titled "Not You/like You" that "trouble[s] particular notions of identity and difference defined through a colonizing logic . . . [that] entails the setting of an absolute boundary, a clear dividing line, a geometry of exclusion that positions the self on one side, and the other—the not-self—on the other" (p. 169). For Minh-ha, what is needed is a disruption of difference premised on binary thinking, or what Barad (2014) refers to as "a way to figure difference differently" (p. 170). Working across spacetimematterings and acknowledging "a thick legacy of feminist theorizing about difference" and particularly influential Chicana feminists, Barad (2014) then offers her concept of diffraction as an approach that "queers binaries and calls out for a rethinking of the notions of identity and difference" (p. 171). As will be seen in various chapters in this book, the ontological and epistemological struggles and debates of feminist new materialists offer scholars of sporting and physical cultural bodies much in terms of "creatively reimagining" the politics of the moving body as vibrant matter always entrenched within (not beyond) power, politics, knowledge, and discourse.

While there is some dissonance between different feminist engagements with new materialism, Barad's notion is that the making of the object (and that which is made by the object) and its uptake in the feminist materialist literatures is echoed in the work of object-oriented ontologists such as Harman and continental materialists such as Meillassoux. For many new materialists—namely, those aligning with the schools of object-oriented ontology (OOO) and speculative realism—the agency of matter is of critical concern if we are to understand how changes to environmental, ecological, and economic surfaces come about and bring about interrelated transmutation. Much can be learned by drawing on emerging new materialist scholarship on matter(s) ranging from the environment (Connolly, 2002, 2011), to the broader political ecology (Morton, 2013; Robbins, 2011; Rocheleau, Thomas-Slayter, & Wangari, 2013), to the media (Bogost, 2012; Shaviro, 2003), and to the political economy (Cammack, 2003; Coole, 2013). When applied to the corporealities made by and meaningful through sport, we believe these new materialisms will provide important new insights into the ecological effects and material potentialities of that which is otherwise often treated as banal or ludic.

SPORT, PHYSICAL CULTURE, AND MATERIAL ASSEMBLAGES

Let us finish this introduction where we started, with Olympic gold. Let's consider the *matter* of the Olympic Games, but let us turn our attention away from the metallic round and toward the bodies and environments it makes. Just as Ali's gold medal described above, the athletic body—such as that featured during such Olympic fetes—presents itself as both assembled (built for competition, articulated to state and identity politics) and unpredictable. The athletic body, as Grant Farred (2014) reminds us, is not just a constitutive corpus of the event (say, the Olympic Games) but in itself an event. Pulling Badiou and Heidegger into the sport arena, in his book *In Motion, At Rest*, Farred explains how flesh and bones and muscles—when set into motion in sporting contexts—are expected to be extraordinary. This is in some ways sport's affective *potenza*, to provide context for the expected unexpected corporeal event. But in the dialectic of expectations, we find a body tied to cultural, racial, and gender politics; systems of accumulation; and scientific and technocratic regimes.

Consider the event of the women's 800-meter running final at the London 2012 Summer Olympic Games. One of the major "events" of the quadrennial athletics competition, the 2012 edition produced (and continues to produce) an important series of interrelated body-events with significant political associations and effects. That year, Mariya Savinova from Russia ran an extraordinary time of 1 minute and 56.19 seconds to win the gold medal. Savinova's time was a second faster than silver medalist Caster Semenya (representing South Africa) and surprise bronze medalist and fellow Russian Olympian Ekaterina Poistogova.[12] Culturally, this event was significant in the United Kingdom, as thousands of attendees and millions of television viewers watched Savinova pull away from the field in the final meters of one of the London Games' hallmark events. In Russia, millions more Olympic enthusiasts celebrated the excellent performance of their star athletes and countrywomen.

But what about the chemistry and physics of this athletic event? We now know that Savinova's performing body was supplemented with a complex admixture of performance-enhancing (namely, the banned steroid Oxandrolone) and test-masking chemical agents that abetted her gait. Russian athletics senior endurance coach Aleksey Melnikov told investigators that "they poured so much [drugs] into her [Savinova]" and were able to do so because her coaches had "very good contacts to the doping control laboratory" (quoted in McGarry, 2015, p. 1). Just as chemistry made the body, so too did sport make chemistry. Russian Olympic athletes, like those of most nations, have for decades been subjects of experimentation and manipulation justified by the national and personal lust for gold.

Just as the body was subjected to physics (speed, biomechanics, etc.) and chemistry (doping), this event also reveals a body subjected to biology. Consider further the race's runner-up, Caster Semenya. At the race's conclusion, the prerace favorite and second-place finisher Semenya was considered back in her native South Africa to have intentionally lost the race so as to not garner more media attention. Just 3 years earlier, Semenya had burst onto the international track scene as the surprise winner of the 800-meter race at the 2009 World Championships in Athletics in Berlin. In the intervening years, Semenya had garnered considerable media attention due to the IOC's and IAAF's (International Association of Athletics Associations) extensive gender testing, which inconclusively "determined" Semenya's eligibility to compete in women's sporting contests (for a critique of these heinous tests, see Cooky, Dycus, & Dworkin, 2013; Nyong'o, 2010; Camporesi & Maugeri, 2010; Shultz, 2011).

In this single track event in London in 2012, we see a vast and complex web of science, human action and rationality, culture, capital, and spectacle—all subordinated to materiality. This complex assemblage of performing bodies (muscles, capillaries, chemicals, genitalia), consuming bodies (cloaked in national fabric, Nike shoes, and Turbospeed suits), international governing bodies (heads of state, a well-furnished IOC electorate, IAAF members, WADA officials), technologies (the cutting-edge design of the Mondo running track, electronic starting blocks reading pressure rather than movement), a shifting geopolitical order (the hardships carried to London and elsewhere by migrants laborers, displaced refugees, globetrotting capitalists; ongoing racial tensions in post-Apartheid South Africa; the election of Vladimir Putin in Russia), and an increasingly unsettled ecosystem (the polluted waters of the nearby Thames, high levels of nitrogen dioxide in the air in London) came to be anchored around the realities presented by and through the chemical, material, and physical objects present(ed) therein.

What remains of this book looks to similar assemblages and the material anchor points from which experience, meaning, thermodynamics, and realities are constituted. Our contributors take a broad view of sport and physical culture, and an even broader view of "new materialism," to find the agency of things in sport, fitness, and movement culture. We believe such an exploration tells us quite a bit about the order of things (flesh, matter, knowledge, science, subjectivity)—namely, how and why sport's material bases should be of critical interest to social scientists and how the material effects of structured physical activity remain central to that realm's exigency in contemporary society. In part 1 of the book, contributing authors outline the epistemological and theoretical bases of new materialist, political ecology, developmental systems theory, and/or feminist new materialist approaches used herein to examine the actors and assemblages of movement-based material, political, and economic production. In so doing, these chapters will illustrate the potential of new materialist approaches for exploring the extent to which a focus on the fleshed body and its material conditions might bring forth new insights or onto-logical and epistemological innovation to the sociology of sport and physical activity.

In chapter 1, Pirkko Markula maps the recent emergence of the materialist approach(es) to the body—namely, by drawing from Maurice Merleau-Ponty's phenomenology of perception, Michel Foucault's disciplinary techniques of the body, Gilles Deleuze's rhizomatics and the Body without Organs, and Bruno Latour's critique of modernist fabrications of the body. In her metatheorization, Markula highlights the moving body as a social and political force and maps out the politico-epistemological bases on which scholars doing work in the new materialist tradition draw on these foundational continental philosophers.

In chapter 2, Mary Louise Adams considers the promotion of physical activity for health in the current political and social context, which prioritizes health as the goal rather than the backdrop of a good life. She argues that instrumentalist notions of physical activity and the body diminish human freedom. While her analysis is informed by critical public health literature, which has drawn usefully on Foucauldian notions of governmentality and understandings of risk, her goal in the chapter is to ask questions that are less Foucauldian and more phenom-enological, inquiring as to what the constant promotion of physical activity for health might mean for how people experience their moving bodies and therefore how they experience the world around them. Using the case of fitness tracking devices, she asks, What are the implications for contemporary forms of embod-ied subjectivity of the proliferation of instrumentalist understandings of move-ment and the body? What might be the implications of such understandings for the human capacity to act and to transform the self and the world, a capacity that, phenomenologists would say, is rooted in the body itself?

In chapter 3, Christopher McLeod and Matthew Hawzen offer a critical engagement with object-oriented ontology (OOO), a philosophy noted for its critiques of nonrealist and other realist approaches. First, they review the major contributions to the field. Next, given that OOO scholars have taken a firm stance against process philosophies and relational philosophies and given that those philosophies have received the most attention in studies of the mov-ing body, they focus specifically on the areas of contention between OOO and other approaches, such as Jane Bennett's vital materialism. Finally, they consider what an object-centered philosophy can tell us about recent debates surround-ing American football and its "concussion crisis" in an effort to determine the relevance of OOO to the field and encourage movement scholars to ask how we might best conceive of and tackle the question of movement.

What has materialized the "material" turn? To round out part 1 of the book, in chapter 4, Gavin Weedon provides a method for pursuing this question by reading the emergence of "new" materialisms diffractively—that is, *through* the conditions of their emergence and, in doing so, to bring their forebears into sharper analytical focus. Weedon turns his focus to telomere biology, taking seriously DNA's role in determining bodily movements and the kinesiological sciences that make sense of them. In this spirit of philosophical kinesiology, a few questions thread through Weedon's chapter: What happens when moving bodies can no longer be pursued as so many objects compelled to motion by cultural forms, biopolitical imperatives, and the desires they elicit and oblige or as biological vessels knowable only by a law unto themselves? When some of our most cherished ideas about bodies and societies, nature and culture, history and humanity are put at risk and put us at risk, do we have the analytical tools to respond? He traverses important theoretical and epistemological terrain in rounding out the first part of the book, mainly in bringing hard science and its objects back into focus for scholars of the cultural body.

Part 2 of the book serves in some ways as a series of responses to these and other questions raised in part 1. That is, contributors to part 2 provide theoretically and empirically grounded new materialism–inspired analyses on how the sporting and active body and material assemblages thereof are activated as, and subjugated to, various forms of corporeal technology. In chapter 5, Shannon Jette, Katelyn Esmonde, David L. Andrews, and Carolyn Pluim provide an analysis of the role that digital technologies now play in producing corporeal materialities in U.S. schools. Physical education classes across the United States now include digital technologies, most notably through the revamped President's Youth Fitness Test (PYFT), an initiative of the President's Council on Fitness, Sports and Nutrition. In this chapter, Jette and colleagues work at the intersection of physical cultural studies and science, technology, and society (STS) studies to examine FitnessGram, a web-based fitness assessment tool being used to monitor the physical activity of millions of students across the United States. In so doing, they reveal FitnessGram as a "black box" intimately connected to numerous other "black boxes" that work together to render invisible the operations of power that have produced and are enabling such technologies to measure, represent, and label the bodies of children and youths as ab/normal.

In chapter 6, Kiri Baxter provides a study of the sporting object—boxing gloves—through the frame of "entangled feminism." Within sport and contact with sporting objects, the skin is forced into a history, a being with objects' past users. It is not an indifferent exchange between autonomous beings; rather, it is within the moments of touch with objects that the skin enters into a relationship, an entangled engagement. She draws from an ethnographic case study conducted over a period of four months in the Brighton area in the South East of England to provide new insights into how the boxing object associates with broader politics of gender, performativity, and embodiment. Putting boxing gloves at the center of inquiry, she discusses the feminist engagement with the ontological realities of "new" materialism and considers their impact on feminist politics.

Drawing on the insights of new materialist feminism, in chapter 7, Simone Fullagar contributes to debates that connect different lines of thinking about the biopolitics of mental health and recovery and the materiality of gendered embodiment as a posthumanist concern. She explores three main areas of debate across the poststructuralist-inspired mental health literature that contribute to the turn toward ontological politics. The broader theoretical terrain of materialist feminism is opening up new ways of thinking about the matter of embodied life (vitality, permeability, lively relationality) as well as the materialization of power (difference marked through gender, class, race, sexuality, age, disability). Moving beyond entrenched mind-body, reason-emotion oppositions, this chapter explores how the embodied experience of moving through depression is produced through a recovery assemblage. Drawing on research undertaken with 80 Australian women, Fullagar fleshes out the biomedical, therapeutic, and social assemblage of recovery practices that regulate and transform the depressed self. In the closing pages of the chapter, Fullagar provides an interesting exploration of the different ways of representing her data to evoke (rather than describe) these complex assemblages.

In chapter 8, Samantha King proposes a multispecies analysis of (non) human actors in contemporary dietary and exercise-based regimes of accumulation. Her object of analysis is the protein powder that has now infiltrated the dietary regimes of many fitness- (and nonfitness-) seeking human actors. Her analysis of protein powder is located at the intersection of four fields—sport studies, animal studies, environmental studies, and food studies. In this chapter, she works toward an approach that moves beyond the human/animal binary as she seeks to provide a posthumanist introspective account to frame a preliminary discussion of this more-than-human foodstuff from a multispecies perspective. To wit, this chapter trends toward a "political ecology of protein powder" and in so doing gives new perspectives on how the body and its movements are bounded to broader systems of biological, organismic, and social life and how scholars of sport and physical culture might benefit from locating these systems and the matters constitutive thereof in their analyses.

In chapter 9, Marianne Clark explains how increasing scholarly dialogue around new materialisms and new empiricisms is creating vitalized spaces in which to rethink the generative capacity of the material body and how it may be studied. Clark outlines the ontological and epistemological dissonance encountered in her Foucauldian study of how adolescent girl ballet dancers construct a self through the dancing body. Using insights provided by new materialist feminism, and particularly the work of Karen Barad, to examine this dissonance, Clark reimagines the project through an ontological lens aligned with new materialist feminism. Clark outlines what this reimagining might look like with two evocative examples and considers the possibilities new materialism opens for theorizing and knowing the moving, dancing body.

In part 3 of the book, the contributors turn toward a more ecological approach to studying the sporting body, its associations, and the multiplicitous lines of temporal and material influence that constitute broader ecologies of active life (human and nonhuman). In chapter 10, Simon Darnell explores Sport for Development and Peace (SDP) and how recent appraisals of sport-for-development research have found it to be highly deterministic and ideological and often focused on measuring and evidencing the supposed "positive" outcomes of programs (see Coalter, 2013). Notably, such criticisms dovetail with key elements of actor-network theory (ANT), in which power is "the final result of a process, and not a reservoir" for social analysis (Latour, 2005, p. 63). Embracing this invective, in this chapter, Darnell attempts to "reassemble" our understanding of SDP by examining its constitutive elements in relation to their controversies, networks, and agencies. Investigating the assemblages of these actants facilitates less deterministic understandings of sport in the service of development and the social implications of SDP while productively complicating critical analyses offered to date.

In chapter 11, Douglas Booth starts with an analysis of the plethora of studies that analyze the beach as a socially constructed site of hedonism. He points out how these studies help us understand power relations at beaches and the long struggles over what constitutes legitimate physical activities, pleasures, and representations of the body. He argues that social constructionist approaches, however, are typically silent about the ways in which we understand and interact with the material dimensions of beaches—sand, ocean, surf, weather, climate, geomorphology, geology. Social constructionism largely conceptualizes such matter as malleable for human meaning and use. In this chapter, Booth reconceptualizes the material realities of the beach and affords them their own lives and agency. Using a case study of Bondi Beach, one of Australia's best-known playgrounds, he argues that such an approach to materialism has the potential to change the way we think about ourselves as agents and to reconsider our relationships with nature. Such thinking may also foster a new, more productive politics of physical and cultural pursuits in environmentally vulnerable spaces.

In chapter 12, Kyle Bunds and Michael Giardina utilize Barad's method of diffraction to understand the intra-action of water as a material object with the body, sport, profit, and politics. In examining the intra-action among water, sport, and the body politic as *relata*, they draw on four years of research (including interviews and multisited ethnographies in four countries) with international water charities that raise money in "developed" countries for the purpose of building water systems in "developing" countries. The purpose of this chapter is to "entail the processing of differences" (Taguchi, 2012, p. 263), particularly between water in developing countries (specifically Ethiopia, where Bunds has worked) and water in developed countries as *dirty made clean* by external and internal processes, including fundraisers in the United States, Canada, and United Kingdom; organizations practicing corporate social responsibility; the sporting body in "developed" countries; the body "in need" in "developing" countries; and the political processes working for and against each *relatum*. In so doing, this chapter encourages new ways of thinking about sport-related charities as well as the relation of researchers to such projects.

In chapter 13, Mary McDonald and Jennifer Sterling draw on feminist materialisms to reveal a complex narrative of the polluted waters of the 2016 Rio de Janeiro Olympic and Paralympic Games. They develop insights from feminist new materialisms to analyze unsustainable practices experienced and discussed in regards to mega sporting events, particularly in relationship to the water pollution at the 2016 Summer Olympic Games. These insights—as well as similar notions found within feminist ecocriticism and science and technology studies scholarship—also demand a reworking of notions of "agency" beyond dominant understandings of rational human intentionality toward complex understandings of the interactivity of the "environment," humans, and nonhumans. The authors provide a feminist ecocritism that, as read through the case of the Rio de Janeiro Olympics, additionally allows for fresh ways to think about social change as "the possibility of a politics and poetics of nature through a broader perception of reality that includes ethics, genetics, chemistry, politics, and biotechnological advances" (Oppermann, 2013, p. 81). In so doing, they illustrate how among inequalities, corruption scandals, and failed promises, dominant narratives promoted through the Olympic movement articulate particular notions of humanism and human exceptionalism.

In chapter 14, Oliver Rick and Jacob Bustad discuss the necessity to engage the active body as part of the city in its essential movements and argue for a particular theoretical and methodological approach to do so. They propose studying the city as an assemblage of human and nonhuman actors that together "enact" (Latour, 2005) the city as a form of social life and in which the active body—and its attendant cultures and lived experiences—takes on specific and intensified dimensions. They make the case that "flattening the city" reflects a particular point of reorientation to studying the active city that shares affinities with affect theory, nonrepresentational theory, posthumanism, new materialism, and most centrally, assemblage thinking. In particular, they incorporate the concepts of assemblage urbanism into a three-stage analysis of the cultural materialities of the active body in/as constitutive of urban space.

In chapter 15, the closing coda of the book, Richard Pringle reflects on the contributions made within *Sport, Physical Culture, and the Moving Body* and the implications these materialist approaches, theories, and empiricizations might hold for future studies of the moving and sporting body. He draws on established and emerging threads of continental philosophy to explain how material, ecological approaches to the study of the moving body might generate new (or renewed) heuristics and modalities of inquiry for scholars across the social sciences and humanities. In the afterword, Grant Farred provides a series of final thoughts about the contributions to the book and how this collection might contribute to present and future studies of sport and physical culture.

In sum, this collection offers an initial foray into the many and diverse possibilities of new materialisms for rethinking the matter of moving bodies. In offering multiple interpretations of various strands of new materialisms, we hope this work highlights new materialisms as "not one thing" and always "in process" (St. Pierre, Jackson, & Mazzei, 2016, p. 99). With Farred, Pringle, and others, we acknowledge the controversial nature of the "new" in new materialisms, but for us (and many of the authors in this work), "the descriptor 'new' does not necessarily announce something new but serves as an alert that we are determined to try to think differently" (St. Pierre et al., 2016, p. 100). Our embrace of new materialisms is not because it is "cool and trendy" (Smelik, 2017) but rather because "our encounters with the world can no longer be explained or justified by orthodox thinking" and/or because we have had "new problems overtake us that demand our attention, our finest curiosity, and urgent 'experimentation in contact with the real' (Deleuze & Guattari, 1980/1987, p. 12)" (St. Pierre et al., 2016, p. 100). We are grateful to our authors for embracing this challenge to rethink how we know what we know and to approach new ways of understanding and imagining the moving body.

Yet we also recognize that this is just an initial step toward new materialisms, and there is still much to be done, particularly at the intersections of new materialisms and new empiricisms (and/or what some refer to as postqualitative inquiry; St. Pierre, 2011; also see Fullagar, 2017; Giardina, 2017; Ray, 2019). As an ontological turn, engaging with new materialist approaches requires us to rethink the nature of being as well as the ways we "do" research: "The empirical and the material are so imbricated they must change together, and with those changes comes a rethinking of ontology. . . . As we rethink matter, we must rethink the empirical (about knowledge) and ontology (about being)"

(St. Pierre et al., 2016, p. 99). As a first step toward new materialist ways of knowing moving bodies, this anthology admittedly focuses largely on the onto-logical and less on the empirical. The chapters focus predominantly on how new materialist approaches might influence the types of questions we are asking about the matter of moving bodies, with less attention given to how new materi-alisms might inform the doing of empirical research on/with moving bodies that matter (for exceptions, see chapters 7 and 9 in this volume). Such absences are understandable. St. Pierre (2015) acknowledges the challenges of working with the "new" (new materialisms, new empiricisms, posthumanism), noting it is "dif-ficult to think because we have to learn a new language that is incompatible with the ontological grids of intelligibility that structure humanist methodologies. And it's also difficult to do because there are no prescribed 'methods' to follow" (p. 78). Similarly, Fox and Allred (2015) recognize the methodological and ethi-copolitical challenges facing those applying new materialist ontology to social research and note that the diversity of work under the new materialist umbrella necessarily leads to highly "eclectic" approaches to research design and meth-ods (p. 410). Closer to home, Fullagar (2017) recently considered how these new styles of thought might reorient both "our onto-epistemological assump-tions and theory-method approaches" in sport and physical cultural research (p. 247), and Markula (2019) issues a call to sociologists of sport and physical culture to explore "post-qualitative methodologies and ways of representation that include the material world" (p. 1; for recent examples of such efforts, see Clark & Thorpe, in press; Monforte, 2018; and McKnight, 2016).

In exploring the "epistemological consequences of adopting a materialist ontology," some are proposing particular sets of "principles for new materialist research designs and methods" (Fox & Alldred, 2015, p. 399) and advocating the potential of mixed methods as part of the research assemblage (Fox & All-dred, 2018). In applying new materialist ontology to social research, Fox and Alldred (2015) explore the research-assemblage as comprising "the bodies, things and abstractions that get caught up in social inquiry, including the events that are studied, the tools, models and precepts of research, and the research-ers" (p. 400), and suggest that this micropolitical approach "enables designs and methods to be engineered from the bottom up" (p. 411). Others are embrac-ing transcendental empiricisms and calling for more experimental and open approaches driven by "conceptual practices" (see special issue by St. Pierre et al., 2016; St. Pierre, 2015). For example, St. Pierre (2015) argues that we should avoid textbook approaches to new materialist methods, always taking particular care not to "force our new empirical, new material, posthuman, postqualitative studies into the structure of conventional humanist qualitative methodology" (p. 92). In moving beyond ontologies that privilege the "speaking subject," our responsibility is "no longer to the privileged human but to the assemblage, which is always more-than-human and always becoming" (p. 88). A recent special issue of *Cultural Studies ↔ Critical Methodologies* focused on "New Empiricisms and New Materialisms" features an array of papers from authors across the disciplines (i.e., art education, youth studies, early childhood education, digital media studies) who are working at this intersection to explore various highly creative and experimental approaches to "doing" research differently (see Hein, 2016; Lather, 2016; Mazzei, 2016; Schulte, 2016; Taguchi, 2016; Tesar & Arndt, 2016). Of particular interest to scholars of sport and physical culture might be Youngblood Jackson's (2016) paper, in which she draws on the ideas of Deleuze and Guattari to reexamine a research encounter that involved a cheerleading dance, ultimately leading her to trace the backflip "along a line of flight to describe its ontological force of creation and newness" (p. 183).

We are excited about the many possibilities for scholars of the moving body as they build on and extend the foundational works presented in our anthology and beyond, to imagine "how/if methodology can be thought in the 'new'" (St. Pierre et al., 2016, p. 99). After studying the philosophical underpinnings that enable this "new," our next challenge is to consider what our "conceptual practices" might look like if/when we begin thinking and living with concepts like diagram, rhizome, Bodies without Organs, entanglement, intra-action, and vital matter (St. Pierre, 2015, p. 92) and how such conceptual practices might enable new ways of knowing embodiment and moving bodies as more than human and always becoming.

A growing body of scholarship is working to decolonize new materialisms, with important work engaged in critically rethinking relationships between new materialism and feminism, queer theory, and antiracist philosophy. There is also a handful of scholars seeking to use new materialist insights to think through antiracism (e.g., Chen, 2012; Hames-García, 2008; Saldanha, 2006), with others exploring the multiple, diverse, and important ways that matter is made meaningful by different cultures and ethnicities. Yet this is an area that is severely underexplored in this anthology, and we recognize it as deserving further scholarly attention in the field of sport and physical cultural studies. What are the objects of sport and physical culture that "live their own lives . . . that come to encapsulate a position that refuses a norm . . . [or that] are overdetermined by meanings ascribed to black, indigenous, and other historically enslaved and colonized subjects" (Karina & Coffey, 2017)? How might new materialisms encourage different ways of understanding the "situated materialities" of race, sex, gender, sexuality, and class in sporting and physical cultural contexts? How might Indigenous sport and physical cultural studies scholars contribute to, expand, and challenge new materialisms by encouraging ways of knowing how sporting matter comes to matter differently to whom as well as when and why? To quote Philip (2004), "We must ask who speaks for nature [and matter], why and what political economic networks they are caught up in" (p. 194). We firmly believe that many in the broader sociology of sport and physical culture community are well positioned to take up these conversations and help move them forward in more meaningful and culturally nuanced ways.

Clearly, we are at the beginning of this work, but we do so in "a particularly rich, experimental moment" with scholars across an array of different disciplines also "taking up the different projects that are overtaking them as the plane opens up their thinking, provokes their curiosity, and enables them to experiment and invent new concepts and/or re-invent old concepts" (St. Pierre et al., 2016, p. 103; also see Salmond, 2012; Taguchi, 2012; St. Pierre, 2016; Youngblood Jackson, 2016; Schadler, 2017). We thank our bold authors for their "attempts at creation and experimentation, all grounded in the ethical imperative to rethink relations of being" and hope this collection inspires others to embrace the challenges and opportunities of the "new" (new materialisms, new empiricisms, etc.) to address the "encounters in their own lives that overtake them and pose problems that demand their best work" (St. Pierre et al., 2016, p. 108). We warmly welcome others to help us take this work forward and into more lively encounters with the complexities of moving bodies in the past, present, and future and the multiple sporting, physical, and cultural worlds that we occupy today and tomorrow.

NOTES

1. Specifically, we are referring to the bilateral accord adopted by Western political intermediaries (from the United States, Canada, Western Europe, Australia, and Japan) that effectively fixed many nations' currency exchange rates to the U.S. dollar. Following Bretton Woods, central banks could exchange dollar holdings at a set rate relative to the price of gold. However, by 1971, under France's aggressive acquisition of gold (and reduction of dollar reserves) and the escalating costs of the Vietnam War, the U.S. Federal Reserve under the presidency of Richard Nixon ended international convertibility of the U.S. dollar to gold—effectively doing away with the "Gold Standard" (Eichengreen & Flandreau, 1997).
2. Just as it makes life, so too gold takes life away—as it did with Crassus in ancient Rome in 53 BC, a Spanish governor in early colonial Ecuador in 1599, and more recently in the mythical land of Westeros (from George R. R. Martin's series A Song of Ice and Fire).
3. Here we are referring to the concept of automobility as conceived in the theories of John Urry (2006), Raymond Williams (1964), Jeremy Packer (2008), and Donna Haraway (1991) in particular. As the portmanteau implies, "automobility can be understood as a patterned system which is predicated in the most fundamental sense on a combination of notions of autonomy and mobility" (Böhm, Jones, Land, & Paterson, 2006, p. 4). Drawing from the work of Donna Haraway (1991) and Nigel Thrift (1996), John Urry (2006)—perhaps the leading scholar on matters automobile—explains the use of the prefix "auto" this way:

On the one hand, "auto" refers to reflexivity to the humanist self, such as the meaning of "auto" in autobiography or autoerotic. On the other hand, "auto" refers to objects or machines that possess a capacity for movement, as expressed by automatic, automation and especially automobile. This double resonance of "auto" is suggestive of how the cardriver is a "hybrid" assemblage, not simply of autonomous humans but simultaneously of machines, roads, buildings, signs and entire cultures of mobility. (p. 18)

4. Such an approach is outlined quite nicely in Ian Bogost's (2012) *Alien Phenomenology*, where the author extends the work of Levi Bryant (2011a, 2011b), Manuel DeLanda (2006), and Bruno Latour (2005) to explain how a broadened view of the *object* might benefit the study of corporeality: "For Bryant (and for Latour), the term object enjoys a wide berth: corporeal and incorporeal entities count, whether they be material objects, abstractions, object of intention, or anything else whatsoever" (p. 12).

5. The term *actant*, borrowed from literature by Callon and Latour (1992), is a heuristic tool that helps the scholar of actor networks give like qualities to ideas, things, people, expressions, emotions, and so on. It is in the first sense a linguistic exercise whereby the storyteller treats dissimilar characters (a person, an object, an emotion) with normative consequence but as an epistemological exercise strips the term *actor* of its singularly anthropomorphic qualities.

6. Our reference to "hyperobjects" here draws on the work and object-oriented ontologies of Timothy Morton. For Morton (2013), hyperobjects are those material configurations that are so massively distributed across time and space that we often fail to grasp their assembled and ecological significance (think global warming, the ecosystem, or radioactivity). He outlines the five characteristics of such hyperobjects, explaining that they are

1) viscous (they adhere to any other object they touch, no matter how hard an object tries to resist); 2) molten (so massive that they refute the idea that spacetime is fixed, concrete, and consistent): 3) nonlocal (massively distributed in time and space to the extent that their totality cannot be realized in any particular local manifestation); 4) phased (they occupy a higher-dimensional space than other entities can normally perceive); and 5) interobjective (formed by relations between more than one object). (p. 1)

7. In his 2003 keynote address to the Cognitive Science Society, Tom Ziemke lays out this approach: "While many nowadays would agree that humans are embodied cognizers, there is much less agreement on what kind of artifact could be considered embodied." He goes on to suggest that there are six contrasting notions of embodiment in the cognitive sciences, each gaining momentum within the scientific community:

1) structural coupling between agent and environment, 2) historical embodiment as the result of a history of structural coupling, 3) physical embodiment, 4) organismoid embodiment, i.e. organism-like bodily form (e.g., humanoid robots), 5) organismic embodiment of autopoietic, living systems, and 6) social embodiment. (p. 1)

While we do not have the space to unpack each of these in the detail that Ziemke does, it is worth noting that each conceptual rendering shares a common schematic—in the coupling, the structural (history, social relations, systems) materializes in the flesh form of the human corpus.

8. See also the special issue "Investment in Human Beings" in the *Journal of Political Economy* (1962). Topics include on-the-job and off-the-job training, migration, information, and health.

9. Here we view new materialism as "a cultural theory that does not privilege matter over meaning or culture over nature" (Dolphijn & van der Tuin, 2012, p. 86). Yet we must be cautious, for even Judith Butler (1993) acknowledged that "in trying to consider the materiality of the body," she "kept losing track of the subject" (p. ix). Thus it might be said that our task is troubled by the burden of seeking to find the impossible object of knowledge, the body.

10. The epistemological and ontological disjunctures of feminist materialism and Derridian grammatology are perhaps best explained in the opening chapter of Vicky Kirby's (2011) *Quantum Anthropologies*. She picks up from Protevi to explore the place of nature and anthropomorphism in Derrida's thinking, charting a more pointed approach to the culpability of nature and the human corpus in the production of everyday life.

11. For Deleuze and Guattari (1977; 1988), each human body has a limited catalog of traits, habits, movements, and affects. At the same time, every actual body also has a "virtual" dimension: a vast reservoir of potential traits, connections, affects, movements, and potentialities. This collection of potentials is what Deleuze (and later Deleuze and Guattari) calls the Body without Organs. To make oneself a body without organs, then, is to explore in thought and practice how one might activate these virtual potentials. Given that these potentials are activated (or "actualized") through contact with other bodies, the Body without Organs is always in a state of becoming. In this way, becoming is a type of being, whereby becoming represents a hylomorphic interchange of matter and potential.

12. Critics have pointed out that since women were first included in the Olympic program in 1952, the Soviet Union (now Russia) programmatically shifted the national training regimen to focus on winning more medals by sending talented female athletes. "There was an instrumental quality. It doesn't represent women's place in Russian society, but it was very useful," writes Russian history scholar Robert Edelman (quoted in Curry, 2015, p. 2).

REFERENCES

Aalten, A. (2004). "The moment when it all comes together": Embodied experiences in ballet. *European Journal of Women's Studies, 11*(3), 263–276.

Ahmed, S. (2000). *Strange encounters: Embodied others in post-coloniality.* New York, NY: Psychology Press.

Ahmed, S. (2004). Affective economies. *Social Text, 22*(2), 117–139.

Ahmed, S. (2006). *Queer phenomenology: Orientations, Objects, Others.* Durham, NC: Duke University Press.

Ahmed, S. (2008). Open forum imaginary prohibitions: Some preliminary remarks on the founding gestures of the "new materialism." *European Journal of Women's Studies, 15*(1), 23–39.

Alaimo, S. (2008). Trans-corporeal feminisms and the ethical space of nature. In S. Alaimo & S. Hekman (eds.), *Material feminisms* (pp. 237–265). Bloomington: Indiana University Press.

Anderson, B. (1983). *Imagined communities: Reflections on the spread and origins of nationalism.* London, England: Verso.

Andrews, D. L., & Silk, M. L. (2015). Physical cultural studies on sport. In R. Giulianotti (ed.), *Routledge handbook of the sociology of sport* (pp. 107–117). New York, NY: Routledge.

Anzaldúa, G. E. (2002). Now let us shift . . . the path of conocimiento . . . inner work, public acts. In G. Anzaldúa and A. L. Keating (eds.), *This bridge we call home: Radical visions for transformation* (pp. 540–578). New York, NY: Routledge.

Ballard, D. H., Hayhoe, M. M., Pook, P. K., & Rao, R. P. (1997). Deictic codes for the embodiment of cognition. *Behavioral and Brain Sciences, 20*(4), 723–742.

Barad, K. (1998). Getting real: Technoscientific practices and the materialization of reality. *Differences: A Journal of Feminist Cultural Studies, 10*(2), 87–126.

Barad, K. (2007). *Meeting the universe halfway: Quantum physics and the entanglement of matter and meaning.* Durham, NC: Duke University Press.

Barad, K. (2014). Diffracting diffraction: Cutting together-apart. *Parallax, 20*(3), 168–187.

Barsalou, L. W., Niedenthal, P. M., Barbey, A. K., & Ruppert, J. A. (2003). Social embodiment. *Psychology of Learning and Motivation, 43*, 43–92.

Bataille, G. (1991). *The accursed share: An essay on general economy. Volume I: Consumption.* New York, NY: Zone Books. (Original work published 1949)

Bataille, G. (1993). *The accursed share* (Vols. 2 and 3). New York, NY: Zone Books. (Original work published 1949)

Beck, P. J. (2013). "War minus the shooting": George Orwell on international sport and the Olympics. *Sport in History, 33*(1), 72–94.

Becker, G. S. (1962). Investment in human capital: A theoretical analysis. *Journal of Political Economy, 70*(5), 9–49.

Bennett, J. (2010). *Vibrant matter: A political ecology of things.* Raleigh, NC: Duke University Press.

Birke, L. (1999). *Feminism and the biological body.* Edinburgh, Scotland: Edinburgh University Press.

Birke, L. (2003). Shaping biology: Feminism and the idea of "the biological." In S. J. Williams, L. Birke, and G. Bendelow (eds.), *Debating biology: Sociological reflections on health, medicine and society* (pp. 39–52). New York, NY: Taylor and Francis.

Bogost, I. (2012). *Alien phenomenology, or, what it's like to be a thing.* Minneapolis: University of Minnesota Press.

Böhm, S., Jones, C., Land, C., & Paterson, M. (2006). Introduction: Impossibilities of automobility. In S. Böhm, C. Jones, C. Land, & M. Paterson (eds.), *Against automobility* (pp. 3–16). Malden, MA: Blackwell/Sociological Review.

Bordo, S. (2004). *Unbearable weight: Feminism, Western culture, and the body.* Berkeley: University of California Press.

Bost, S. (2008). From race/sex/ etc. to glucose, feeding tube, and mourning: The shifting matter of Chicana feminism. In S. Alaimo & S. Hekman (eds.), *Material feminisms* (pp. 340–372). Bloomington: Indiana University Press.

Bourdieu, P. (1978). Sport and social class. *Social Science Information, 17*(6), 819–840.

Bourdieu, P. (1986). The forms of capital. In J. G. Richardson (ed.), *Handbook of theory and research sociology of education* (pp. 241–258). New York, NY: Greenwood Press.

Brace-Govan, J. (2002). Looking at bodywork: Women and three physical activities. *Journal of Sport & Social Issues, 26*(4), 403–420.

Braidotti, R. (1994). *Nomadic subjects: Embodiment and sexual difference in contemporary feminist theory.* New York, NY: Columbia University Press.

Braidotti, R. (2000). Once upon a time in Europe. *Signs: Journal of Women in Culture and Society, 25*(4), 1061–1064.

Braidotti, R. (2013). *The posthuman.* New York, NY: John Wiley & Sons.

Bryant, L. R. (2011a). *The democracy of objects.* London, England: Open Humanities Press.

Bryant, L. R. (2011b). The ontic principle: Outline of an object-oriented ontology. In L. R. Bryant, N. Srnicek, & G. Harman (eds.), *The speculative turn: Continental materialism and realism* (pp. 261–278). Melbourne, Australia: Re.Press.

Burkitt, I. (1999). *Bodies of thought: Embodiment, identity and modernity.* London, England: Sage.

Butler, J. (1993). *Bodies that matter: On the discursive limits of "sex."* New York, NY: Routledge.

Butler, J. (2003). Afterword: After loss, what then? In D. L. Eng and D. Kazanjian (eds.), *Loss: The politics of mourning* (pp. 467–473). Berkley: University of California Press.

Callon, M., & Latour, B. (1992). Don't throw the baby out with the bath school! A reply to Collins and Yearley. *Science as Practice and Culture, 343,* 368.

Cammack, P. (2003). The governance of global capitalism: a new materialist perspective. *Historical Materialism, 11*(2), 37–59.

Camporesi, S., & Maugeri, P. (2010). Caster Semenya: Sport, categories and the creative role of ethics. *Journal of Medical Ethics, 36*(6), 378–379.

Carrington, B. (2008). "What's the footballer doing here?" Racialized performativity, reflexivity, and identity. *Cultural Studies ↔ Critical Methodologies, 8*(4), 423–452.

Chen, M. Y. (2012). *Animacies: Biopolitics, racial mattering, and queer affect.* Durham, NC: Duke University Press.

Chisholm, A. (1999). Defending the nation: National bodies, US borders, and the 1996 US Olympic women's gymnastics team. *Journal of Sport & Social Issues, 23*(2), 126–139.

Clare, S. (2016). On the politics of "new feminist materialisms." In V. Pitts-Taylor (ed.), *Mattering: Feminism, science and materialism* (pp. 58–72). New York: New York University Press.

Clark, A. (2008). *Supersizing the mind: Embodiment, action, and cognitive extension.* New York, NY: OUP USA.

Clark, M., & Thorpe, H. (In press). Towards diffractive ways of knowing women's moving bodies: A Baradian experiment with the Fitbit/motherhood entanglement. *Sociology of Sport Journal.*

Coalter, F. (2013). *Sport for development: What game are we playing?* London, England: Routledge.

Connolly, W. E. (2002). *Neuropolitics: Thinking, culture, speed.* Minneapolis: University of Minnesota Press.

Connolly, W. E. (2011). *A world of becoming.* Durham, NC: Duke University Press.

Cooky, C., Dycus, R., & Dworkin, S. L. (2013). "What makes a woman a woman?" versus "Our first lady of sport": A comparative analysis of the United States and the South African media coverage of Caster Semenya. *Journal of Sport & Social Issues, 37*(1), 31–56.

Coole, D. (2013). Agentic capacities and capacious historical materialism: Thinking with new materialisms in the political sciences. *Millennium, 41*(3), 451–469.

Coole, D., & Frost, S. (2010). Introducing the new materialisms. In D. Coole & S. Frost (eds.), *New materialisms: Ontology, agency, and politics* (pp. 1–43). Durham, NC: Duke University Press.

Csordas, T. J. (1990). Embodiment as a paradigm for anthropology. *Ethos, 18*(1), 5–47.

Csordas, T. J. (1994). *Embodiment and experience: The existential ground of culture and self.* Cambridge: Cambridge University Press.

Curry, C. (2015). The face of Russia's doping scandal: Mariya Savinova. *Women in the world.* Retrieved from https://womenintheworld.com/2015/11/10/the-face-of-russias-doping -scandal-mariya-savinova/

Davis, K. (1997). Embody-ing theory: Beyond modernist and postmodernist readings of the body. In K. Davis (ed.), *Embodied practices: Feminist perspectives on the body* (pp. 1–23). London, England: Sage.

Davis, N. (2009). New materialism and feminism's anti-biologism: A response to Sara Ahmed. *European Journal of Women's Studies, 16*(1), 67–80.

DeLanda, M. (2006). *A new philosophy of society: Assemblage theory and social complexity.* New York, NY: A&C Black.

DeLanda, M. (2013). *Intensive science and virtual philosophy.* New York, NY: A&C Black.

Deleuze, G. (1983/2006). *Nietzsche and philosophy.* New York, NY: Columbia University Press.

Deleuze, G. (1988). *Foucault.* Minneapolis: University of Minnesota Press.

Deleuze, G. (1994). *Difference and repetition.* New York, NY: Columbia University Press.

Deleuze, G., & Guattari, F. (1977). *Anti-Oedipus: Capitalism and schizophrenia.* New York, NY: Viking.

Deleuze, G., & Guattari, F. (1988). *A thousand plateaus: Capitalism and schizophrenia.* New York, NY: Bloomsbury.

Dolphijn, R., & Tuin, I. V. D. (2012). *New materialism: Interviews and cartographies.* Ann Arbor, MI: Open Humanities Press.

Dworkin, S. L., & Wachs, F. L. (2004). "Getting your body back": Postindustrial fit motherhood in *Shape Fit Pregnancy* magazine. *Gender & Society, 18*(5), 610–624.

Eichengreen, B. J., & Flandreau, M. (1997). *The gold standard in theory and history.* New York, NY: Psychology Press.

Esposito, R. (2015). *Persons and things: From the body's point of view.* New York, NY: John Wiley & Sons.

Fanon, F. (1952/2008). *Black skin, white masks.* New York: Grove Press.

Farred, G. (2014). *In motion, at rest: The event of the athletic body.* Minneapolis: University of Minnesota Press.

Fausto-Sterling, A. (2000). *Sexing the body: Gender politics and the construction of sexuality.* New York, NY: Basic Books.

Featherstone, M., Hepworth, M., & Turner, B. (eds.). (1991). *The body: Social processes and cultural theory.* Newbury Park, CA: Sage.

Fox, N. J., & Alldred, P. (2015). New materialist social inquiry: Designs, methods and the research-assemblage. *International Journal of Social Research Methodologies, 18*(4). 399–414.

Fox, N. J., & Alldred, P. (2018). Mixed methods, materialism and the micropolitics of the research-assemblage. *International Journal of Social Research Methodology, 21*(2), 191–204.

Frost, S. (2011). The implications of the new materialisms for feminist epistemology. In H. E. Grasswick (ed.), *Feminist epistemology and philosophy of science: Power in knowledge* (pp. 69–83). Dordrecht, Netherlands: Springer.

Frost, S. (2014). Re-considering the turn to biology in feminist theory. *Feminist Theory, 15*(3), 307–326.

Fullagar, S. (2008). Sites of somatic subjectivity: E-scaped mental health promotion and the biopolitics of depression. *Social Theory and Health, 6,* 323–341.

Fullagar, S. (2017). Post-qualitative inquiry and the new materialist turn: Implications for sport, health and physical culture research. *Qualitative Research in Sport, Exercise and Health, 9*(2), 247–257.

Gerratt, A. P., Michaud, H. O., & Lacour, S. P. (2015). Elastomeric electronic skin for prosthetic tactile sensation. *Advanced Functional Materials, 25*(15), 2287–2295.

Giardina, M. D. (2017). (Post?)qualitative inquiry in sport, exercise, and health: Notes on a methodologically contested present. *Qualitative Research in Sport, Exercise and Health, 9*(2), 258–270.

Giardina, M. D., & Newman, J. I. (2011). What is this "physical" in physical cultural studies? *Sociology of Sport Journal, 28*(1), 36–63.

Gibbs, R. W., Jr. (2005). *Embodiment and cognitive science.* London, England: Cambridge University Press.

Goodwin, C. (2000). Action and embodiment within situated human interaction. *Journal of Pragmatics, 32*(10), 1489–1522.

Gorn, E. J. (1998). *Muhammad Ali, the people's champ.* Champaign: University of Illinois Press.

Gregg, M., & Seigworth, G. J. (eds.). (2010). *The affect theory reader.* Durham, NC: Duke University Press.

Grosz, E. A. (1994). *Volatile bodies: Toward a corporeal feminism.* Bloomington: Indiana University Press.

Hames-García, M. (2008). How real is race? In S. Alaimo & S. Hekman (eds.), *Material feminisms* (pp. 291–307). Bloomington: Indiana University Press.

Haraway, D. (1988). Situated knowledges: The science question in feminism and the privilege of partial perspective. *Feminist Studies, 14*(3), 575–599.

Haraway, D. J. (1991). *Simians, cyborgs, and women: The reinvention of nature.* London, England: Free Association Books.

Harman, G. (2002). *Tool-being: Heidegger and the metaphysics of objects.* Chicago, IL: Open Court.

Harman, G. (2005). *Guerrilla metaphysics: Phenomenology and the carpentry of things.* Chicago, IL: Open Court.

Harman, G. (2009). *Prince of networks: Bruno Latour and metaphysics.* Prahran, Australia: Re.Press.

Harman, G. (2011). *The quadruple object.* Alresford, England: Zero Books.

Hein, S. F. (2016). The new materialism in qualitative inquiry: How compatible are the philosophies of Barad and Deleuze? *Cultural Studies ↔ Critical Methodologies, 16*(2), 132–140.

Hird, M. (2004). Feminist matters: New materialist considerations of sexual difference. *Feminist Theory, 5*(2), 223–232.

Jackson, A. Y. (2016). An ontology of a backflip. *Cultural Studies ↔ Critical Methodologies, 16*(2), 183–192.

Karina, A., & Coffey, W. S. (2017). CFP: Subjects, objects, others: materialisms from the enslaved and colonized. Retrieved from https://networks.h-net.org/node/73374/announcements/838260/cfp-subjects-objects-others-materialisms-enslaved-and-colonized

Kim, J., Lee, M., Shim, H. J., Ghaffari, R., Cho, H. R., Son, D., . . . Chu, K. (2014). Stretchable silicon nanoribbon electronics for skin prosthesis. *Nature Communications, 5,* 5747.

Kirby, V. (2011). *Quantum anthropologies: Life at large.* Durham, NC: Duke University Press.

Krieger, N. (2005). Embodiment: A conceptual glossary for epidemiology. *Journal of Epidemiology and Community Health, 59*(5), 350–355.

Lather, P. (2016). Top ten+ list: (Re)thinking ontology in (post)qualitative research. *Cultural Studies ↔ Critical Methodologies, 16*(2), 125–131.

Latour, B. (2005). *Reassembling the social: An introduction to actor-network theory.* London, England: Oxford University Press.

Law, J. (1992). Notes on the theory of the actor-network: Ordering, strategy, and heterogeneity. *Systems Practice, 5*(4), 379–393.

Markula, P. (2019). What is new about new materialism for sport sociology? Reflections on body, movement, and culture. *Sociology of Sport Journal, 36,* 1–11.

Mazzei, L. A. (2016). Voice without a subject. *Cultural Studies ↔ Critical Methodologies, 16*(2), 151–161.

McFarlane, C. (2009). Translocal assemblages: Space, power and social movements. *Geoforum, 40*(4), 561–567.

McGarry, A. (2015). Russian doping scandal: The elite athletes who were allowed to cheat the system. ABC News (Australia). Retrieved from https://www.abc.net.au/news/2015-11-10/russian-athletes-at-the-centre-of-doping-scandal/6928426

McKnight, L. (2016). Swimming lessons: Learning, new materialisms, posthumanism, and post qualitative research emerge through a pool poem. *Journal of Curriculum and Pedagogy, 13*(3), 195–205.

Meillassoux, Q. (2010). *After finitude: An essay on the necessity of contingency.* London, England: Bloomsbury.

Meynell, L. (2008). The power and promise of developmental systems theory. *Les Ateliers de l'Ethique: La Revue de CREUM, 3*(2), 88–103.

Mincer, J. (1958). Investment in human capital and personal income distribution. *Journal of Political Economy, 66*(4), 281–302.

Mol, A. (2002). *The body multiple: Ontology in medical practice.* Durham, NC: Duke University Press.

Montforte, J. (2018). What is new in new materialism for a newcomer? *Qualitative Research in Sport, Exercise and Health, 10*(3), 378–390.

Moraga, C. (1997). *Waiting in the Wings: Portrait of Queer Motherhood.* Ithaca, NY: Firebrand.

Morton, T. (2013). *Hyperobjects: Philosophy and ecology after the end of the world.* Minneapolis: University of Minnesota Press.

Newman, J. I. (2013). Arousing a [post-] Enlightenment active body praxis. *Sociology of Sport Journal, 30*(3), 380–407.

Newman, J. I., & Giardina, M. D. (2014). Moving biopolitics. *Cultural Studies ↔ Critical Methodologies, 14*(5), 419–425.

Newman, J. I., Shields, R., & McLeod, C. M. (2015). The MRSA epidemic and/as fluid biopolitics. *Body and Society.* doi:10.1177/1357034X14551844.

Niedenthal, P. M., Barsalou, L. W., Winkielman, P., Krauth-Gruber, S., & Ric, F. (2005). Embodiment in attitudes, social perception, and emotion. *Personality and Social Psychology Review, 9*(3), 184–211.

Nyong'o, T. (2010). The unforgivable transgression of being Caster Semenya. *Women & Performance: A Journal of Feminist Theory, 20*(1), 95–100.

Oppermann, S. (2013). Feminist ecocriticism: The new ecofeminist settlement. *Feminismo/s, 22,* 65–88.

Oudshoorn, N. (1994). *Beyond the natural body: An archaeology of sex hormones.* London, England: Routledge.

Packer, J. (2008). *Mobility without mayhem: Safety, cars, and citizenship.* Durham, NC: Duke University Press.

Philip, K. (2004). *Civilizing natures: Race, resources, and modernity in colonial South India.* New Brunswick, NJ: Rutgers University Press.

Pitts-Taylor, V. (2016). *Mattering: Feminism, science, and materialism.* New York: New York University Press.

Protevi, J. (ed.). (2001). *Political physics: Deleuze, Derrida and the body politic.* New York, NY: Bloomsbury.

Protevi, J. (2009). *Political affect: Connecting the social and the somatic.* Minneapolis: University of Minnesota Press.

Ravenscroft, A. (2018). Strange weather: Indigenous materialisms, new materialism, and colonialism. *Cambridge Journal of Postcolonial Literary Inquiry, 5*(3), 353–370.

Ray, J. (2019). The postqualitative turn in physical cultural studies. *Leisure Sciences*, 1–17. doi:10.1080/01490400.2018.1539681.

Ricardo, D. (2015). *On the principles of political economy, and taxation*. London, England: Cambridge Books. (Original work published 1891)

Roberts, C. (2002). A matter of embodied fact: Sex hormones and the history of bodies. *Feminist Theory*, 3(1), 7–26.

Roberts, C. (2003). Drowning in a sea of estrogens: Sex hormones, sexual reproduction and sex. *Sexualities*, 6(2), 195–213.

Roberts, C. (2007). *Messengers of sex: Hormones, biomedicine and feminism*. New York, NY: Cambridge University Press.

Robbins, P. (2011). *Political ecology: A critical introduction*. London, England: John Wiley & Sons.

Rocheleau, D., Thomas-Slayter, B., & Wangari, E. (2013). *Feminist political ecology: Global issues and local experience*. London, England: Routledge.

Rose, N. (2013). The human sciences in a biological age. *Theory, Culture and Society*, 30(1), 3–34.

Roy, D., & Subramaniam, B. (2016). Matter in the shadows: Feminist new materialism and the practices of colonialism. In V. Pitts-Taylor (ed.), *Mattering: Feminism, science, and materialism* (pp. 23–42). New York: New York University Press.

Saldanha, A. (2006). Reontologising race: The machinic geography of phenotype. *Environment and Planning D: Society and Space*, 24(1), 9–24.

Salmond, A. (2012). Ontological quarrels: Indigeneity, exclusion and citizenship in a relational world. *Anthropological Theory*, 12(2), 115–141.

Salmond, A. (2014). Tears of Rangi: Water, power and people in New Zealand. *Hau: Journal of Ethnographic Theory*, 4(3), 285–309.

Salmond, A. (2017). *Tears of Rangi: Experiments across worlds*. Auckland, New Zealand: Auckland University Press.

Saussure, F. de. (2011). *Course in general linguistics*. New York, NY: Columbia University Press. (Original work published 1916)

Schadler, C. (2017). Enactments of a new materialist ethnography: Methodological framework and research processes. *Qualitative Research*. Advance online publication. doi:10.1177/1468794117748877.

Schulte, C. M. (2016). Possible worlds: Deleuzian ontology and the project of listening in children's drawing. *Cultural Studies ↔ Critical Methodologies*, 16(2), 141–150.

Schultz, T. W. (1960). Capital formation by education. *Journal of Political Economy*, 68(6), 571–583.

Serres, M. (2011). *Variations on the body*. Minneapolis, MN: Univocal.

Seth, S. (2009). Putting knowledge in its place: Science, colonialism, and the postcolonial. *Postcolonial Studies*, 12(4), 373–388.

Shaviro, S. (2003). *Connected: Or what it means to live in the network society*. Minneapolis: University of Minnesota Press.

Shildrick, M. (2015). *Leaky bodies and boundaries: Feminism, postmodernism and (bio)ethics*. London, England: Routledge. (Original work published 1997)

Shilling, C. (2005). *The body in culture, technology and society*. London, England: Sage.

Shilling, C. (2007). Sociology and the body: Classical traditions and new agendas. *Sociological Review*, 55(s1), 1–18.

Shultz, J. (2011). Caster Semenya and the "question of too": Sex testing in elite women's sport and the issue of advantage. *Quest*, 63(2), 228–243.

Sieben, A. (2011). Heteronormative pheromones? A feminist approach to human chemical communication. *Feminist Theory, 12*(3), 263–280.

Singer, L. (1993). *Erotic welfare: Sexual theory and politics in the age of epidemic.* New York, NY: Routledge.

Smelik, A. (2017). Cool and trendy: New materialism. *Culture Weekly.* Retrieved from http://cultureweekly.tumblr.com/post/152290257466/cool-and-trendy-new-materialism

Spinoza, B. (2002). *Spinoza: The complete works.* London, England: Hackett.

St. Pierre, E. A. (2011). Post qualitative research: The critique and the coming after. In N. K. Denzin & Y. S. Lincoln (eds.), *Sage handbook of qualitative inquiry* (4th ed., pp. 611–635). Los Angeles, CA: Sage.

St. Pierre, E. A. (2015). Practices for the "new" in the new empiricisms, the new materialisms, and post qualitative inquiry. In N. K. Denzin & M. D. Giardina (eds.), *Qualitative inquiry and the politics of research* (pp. 75–95). Walnut Creek, CA: Left Coast Press.

St. Pierre, E. A. (2016). The empirical and the new empiricisms. *Cultural Studies ↔ Critical Methodologies, 16*(2), 111–124.

St. Pierre, E. A., Jackson, A., & Mazzei, L. (2016). New empiricisms and new materialisms: Conditions for new inquiry. *Cultural Studies ↔ Critical Methodologies, 16*(2), 99–110.

Sundberg, J. (2014). Decolonizing posthumanist geographies. *Cultural Geographies, 21*(1), 33–47.

Taguchi, H. L. (2012). A diffractive and Deleuzian approach to analysing interview data. *Feminist Theory, 13*(3), 265–281.

Taguchi, H. L. (2016). "The concept as method": Tracing-and-mapping the problem of the neuro(n) in the field of education. *Cultural Studies ↔ Critical Methodologies, 16*(2), 213–223.

Tesar, M., & Arndt, S. (2016). Vibrancy of childhood things: Power, philosophy, and political ecology of matter. *Cultural Studies ↔ Critical Methodologies, 16*(2), 193–200.

Thorpe, H. (2011). *Snowboarding bodies in theory and practice.* New York, NY: Palgrave Macmillan.

Thorpe, H. (2014). Moving bodies beyond the social/biological divide: Toward theoretical and transdisciplinary adventures. *Sport, Education and Society, 19*(5), 666–686.

Thorpe, H. (2016). Athletic women's experiences of amenorrhea: Biomedical technologies, somatic ethics and embodied subjectivities. *Sociology of Sport Journal, 33*(1), 1–13.

Thorpe, H., & Clark, M. (2019). Gut feminism, new materialisms and sportswomen's embodied health: The case of RED-S in endurance athletes. *Qualitative Research in Sport, Exercise and Health,* doi:10.1080/2159676X.2019.1631879.

Thrift, N. (1996). *Spatial formations.* London, England: Sage.

Todd, Z. (2016). An Indigenous feminist's take on the ontological turn: "Ontology" is just another word for colonialism. *Journal of Historical Sociology, 29*(1), 4–22.

Tompkins, K. W. (2016, Spring). On the limits and promise of new materialist philosophy. *Lateral, 5*(1).

Turner, B. (2008). *The body and society: Explorations in social theory* (3rd ed.). Los Angeles, CA: Sage.

Urry, J. (2006). Inhabiting the car. In S. Böhm, C. Jones, C. Land, & M. Paterson (eds.), *Against automobility* (pp. 17–31). Malden, MA: Blackwell/Sociological Review.

Virno, P. (2003). *A grammar of the multitude.* Los Angeles, CA: Semiotext (e).

Willey, A. (2016). Biopossibility: A queer feminist materialist science studies manifesto, with special reference to the question of monogamous behavior. *Signs: Journal of Women in Culture and Society, 41*(3), 553–577.

Williams, R. (1964). *Second generation*. London, England: Chatto and Windus.

Williams, S., & Bendelow, G. (1998). *The lived body: Sociological themes, embodied issues*. London, England: Routledge.

Wilson, E. (2004). *Psychosomatic: Feminism and the neurological Body*. Durham, NC: Duke University Press.

Ziemke, T. (2003, July). What's that thing called embodiment? In *Proceedings of the 25th Annual meeting of the Cognitive Science Society* (pp. 1305–1310). Mahwah, NJ: Lawrence Erlbaum.

PART I BODY ONTOLOGIES

1 · CONTEXTUALIZING THE MATERIAL, MOVING BODY

PIRKKO MARKULA

What does it mean to study bodies in motion in (physical) cultural contexts? Obviously, it is something different from studying the body purely as a matter of metabolism and mechanics. Accounting for culture involves an additional engagement with theoretical conceptualizations of how the material body matters within various cultural, political, environmental, and economic systems and how the body's movement acts as a social force. This is by no means a novel observation.

Sociocultural researchers of sport have been captivated by the body since the 1980s, and it is now commonly recognized that material conditions of sociality also produce sporting bodies. Certain sociopolitical circumstances solidify into recognizable dominant or marginalized identities expressed by the active body. Many sport studies scholars have critiqued how gendered, transgendered, racialized, classed, nationalized, aged, able-bodied, and/or sexualized sporting and exercising bodies reproduce or oppose social inequality through such theoretical lenses as critical disability theory, critical feminist theory, critical masculinity theory, intersectionality theory, or critical race theory.[1]

This research highlights the contradictory manner in which sporting bodies both align with dominant identities and also open up possibilities for resistance. The corporeal body emerges as a surface on which the researcher's critical gaze can read the social conditions of normalization or marginalization. As a signified for social meanings, the body is a means to analyze how larger economic, social, and political issues impact sport and physical activity. These analyses of the sporting body as an ideologically constructed surface for the forces of dominance and with corresponding agency conceptualize the body as a "text" to be analyzed. This representationalist research can be located within the "textual" or "linguistic turn" in social sciences and humanities and, as the editors of this book term it, the "cultural turn" in sport and physical cultural studies. This standpoint, while effectively highlighting how power, knowledge, language, and subjectivity are intertwined to produce the relations of dominance as they structure and construct the (physically active) body, tends to ignore the material dimensions of culture, life, and the body. As a consequence, approaches such as new materialism and posthumanism have emerged to help researchers consider new ways of thinking about the significance of matter and the process of materialization in the current cultural and political landscape. While interdisciplinary, they share several common characteristics: the need to account for humans as embodied subjects in the material world of active objects, the need to attend to how the material environment is produced and consumed, and the need to account for the achievements of natural sciences in social and cultural analyses of materiality. To then examine the human and nonhuman bodies in local and global environments of political and commercial realities, new materialists and posthumanists frequently rely on "rereading" of well-established works by such philosophers as Bergson, Deleuze, Foucault, Latour, Merleau-Ponty, Nietzsche, and Spinoza.

Against these developments, there have been several calls to include the "biological body" in sociocultural sport research to complement the abstract analysis of power structures (e.g., Silk, Andrews & Thorpe, 2017; Booth, 2012; Fullagar, 2017; Giardina & Newman, 2011a, 2011b; Larsson, 2014; Larsson & Quennerstedt, 2012; Thorpe, 2014, 2016). As the editors of this book indicate, physical cultural scholars now look for new ways to incorporate "materiality" within the social analyses of the physically active body. In this book, furthermore, several authors engage with embodiment in the material politics of global and local sport and exercise spaces using theoretical insights from Foucault, Merleau-Ponty, Latour, and feminist scholars strongly inspired by Deleuze.

So what is new about new materialist approaches for those of us who have already established traditions of phenomenological and poststructuralist research of the sporting and exercising body in diverse environments? Perhaps nothing entirely radical. However, to transcend the somewhat narrow focus on the representational body, we can look into the existing theoretical heritage through a different lens to rediscover its materialist impulse and then open up our analyses to include the moving body in the contemporary contexts of physical activity. In what follows, I contextualize possibilities for a new contemporary material/social interface by rereading well-known French theorists Maurice Merleau-Ponty, Bruno Latour, Michel Foucault, and Gilles Deleuze, whose works, when examined more closely, directly explores material bodies in motion.

MERLEAU-PONTY: MOTILITY AND MEANING

In his *Phenomenology of Perception*, Merleau-Ponty (1978) locates the moving body within the phenomenological field. While this framework includes a plethora of concepts, for the purpose of this chapter, I focus on "essence," "perceived body," and "habit," which, to my reading, are most directly related to a phenomenological understanding of the moving body.

Phenomenology and Essence

Merleau-Ponty (1978) emphasizes that fundamentally, phenomenology is a study of essences—the substance of what is studied. For example, he would ask, What is the (exercising, dancing, sporting) body by nature? What is the substance of perception or consciousness? He further clarifies that an essence, in a phenomenological sense, is a means for understanding our involvement in the world: Through the living relationship of experience, one can rediscover one's actual presence in the world. As researchers of physical culture, we can engage with movement experiences to rediscover the essence of, for example, the (athletic) self, health, competition, or enjoyment to understand involvement in the (physical cultural) world.

Essences can be found in basic human experiences grounded in the perception of the world around us. This world exists before perceptions: We can only perceive a world that is there to be perceived. In other words, the world transcends human or social construction. Eventually, however, the description of the perception of the world will tap into knowledge that is "primordial": Through perception that takes place before conscious reflection, phenomenologists can achieve direct contact with and then describe the prereflective, transcendental world (Merleau-Ponty, 1978, p. xvi).[2] Because the body provides the mechanism for perception, it is an essential aspect of phenomenological analysis.

The Perceiving Body in Space

Perception, Merleau-Ponty (1978) argues, is the way to make sense of the world, and phenomenology brings "truth into being" by looking for meanings in perceptions. Importantly, we perceive the world through the body. While our senses, however, can be studied objectively through different natural sciences, the body escapes such observations when it relates to other bodies in the world. "My body is the pivot of the world," Merleau-Ponty explains and adds, "I am conscious of the world through the medium of my body" (p. 82). It is crucial, however, not to treat the body as an object or a representation of the world. From a phenomenological perspective, the body does not represent the world but enables one to know the meanings of the world.

All bodies exist in space. Merleau-Ponty (1978) distinguishes between two types of space: bodily space and objective space. Bodily space consists of the material body itself. This body allows access to objective space that provides the background for the body to act. Bodily space and objective space thus form a practical system where both are necessary for existence and experience. However, it is bodily action, the body's movement, that brings spatiality into being: "Movement is not limited to submitting passively to space and time, it actively assumes them, it takes them up in their basic significance which is obscured in the commonplaceness of established situations" (p. 102).[3]

Abstract Movement, Concrete Movement

The body's movement thus is an important opening to primordial, prereflective knowledge. "It is never our objective body that we move," continues Merleau-Ponty (1978); it is "our phenomenal body" that has the potential to act (p. 106). This is "abstract movement": movement patterns of dance and sport or gestures and postures (e.g., smiling when happy) that are unnecessary for everyday existence. The abstract movement, unlike the concrete movement that has "the world as given" as its background (p. 111), creates its own background that, nevertheless, "is not a representation . . . but is immanent in the movement inspiring and sustaining it at every moment" (p. 110). The concepts of objective space and abstract movement bring together the physiological body of the objective space and the (psychological) consciousness of the body space.

Merleau-Ponty (1978) uses the term *content* to refer to the material body and the term *form* to refer to the consciousness that gives linguistic expression to the body and its perceptions to advocate for "a dialectic of form and content" (p. 139) where both are considered instruments for meaning making. As a form, the body *inhabits* space and time: It combines and includes them. The body uses its senses to give expression to its movements, and thus the consciousness's orientation to the world takes expression in movement. Merleau-Ponty summarizes, "Movement is not thought about movement" (p. 139) but rather acts as a basic intentionality toward the world.[4]

Meaning and Habit

Although motility gives "motor meaning" to verbal orders, movement experiences, Merleau-Ponty (1978) contends, are not knowledge. They provide us "with a way of access to the world and the objects" (pp. 140–141). The moving body, its experience in the world, makes accessible the primordial knowledge usually hidden under taken-for-granted assumptions.

Movement and consciousness (that gives movement its symbolic function or meaning) unite in *habit*: the body adopting "new" movements and then incorporating them into a new but recognizable "formula." The habit illustrates the mediating function of the body: The body expresses our being in the world, and through the body, we come to understand the world. Merleau-Ponty (1978) uses dance as an example: The habit of dancing is discovering the "formula" of movement that is recognized as dance. The dancing body "'catches' ... and 'comprehends' movement" (pp. 142–143) but also "grasps" its significance, its figurative meaning as dance. When forming a habit such as dance, the body, through its absorbed movement expression, gives new meanings to life but, according to Merleau-Ponty, also gives new meaning to "meaning": "Bodily experience forces us to acknowledge an imposition of meaning which is not the work of a universal constituting consciousness, a meaning which clings to certain contents. . . . In it we learn to know that union of essence and existence" (p. 147). The body thus has to be seen as a "unity": We are the body, not beside it or outside of it.[5]

In summary, Merleau-Ponty's (1978) analysis of the meaning of the moving body, its motility, requires a conceptual unity of the content and the form: The physiological body cannot exist without its form, and consciousness does not exist without the body, the content. These join in bodily experience, in the habit: the body's behavior and function. Movement experiences in the world are prereflective and can give us access to the primordial knowledge beyond conscious meaning making. The moving body thus acts as the window to the world, and united with consciousness, the mind gives significance to what takes place around us. In other words, movement experiences mediate between the objective world and the conscious meaning making of it. The objective world thus is a central part of understanding the essences, what we think of as truth about human existence. In the phenomenological framework, the individual is, however, the starting point of understanding the world. The primary question for such an analysis is, How can an individual be open to phenomena from the world? The phenomena transcend the individual, or are not controllable by the individual, but as Merleau-Ponty (1978) reasons, they exist because the individual can take them up and live them. Considering these points, how can Merleau-Ponty's phenomenology contextualize an analysis of the material, moving body in physical culture?

In her review of phenomenology in sport studies, Allen-Collinson (2009) notes that "relatively little use has been made *explicitly* of phenomenological approaches" (p. 284, italics original), although Merleau-Ponty's approach is particularly "well-suited in the in-depth portrayal of the corporeally grounded experience" of physical activity (p. 284). It is clear that Merleau-Ponty's (1978) conceptual philosophy does not directly explore the body or experiences of physical activity. Instead, Merleau-Ponty is interested in how the world becomes meaningful through bodily action. Movement experiences (or the physically active body), from Merleau-Ponty's point of view, are not objects of sociological, historical, philosophical, or physical cultural studies or phenomenological analysis. What is the point, then, to engage with Merleau-Ponty? One approach to his phenomenology is to illustrate the unique ability of the moving and consequently the different modalities of physical activity (exercise, dance, sport) to create knowledge of the "being-in-the-world" and thus justify them as essential phenomena for human existence in the world.[6] As the moving body exists in the world and gives it meaning, a physical cultural studies researcher engaging with Merleau-Ponty would begin by identifying a phenomenon whose essence or "nature" they plan to identify. An analysis of the actual movement experiences should then lead the researcher to this essence. In this process, the moving body acts as a mediator for an understanding of the world and also as an access to the imperceptible, primordial knowledge that ultimately determines how we understand the essence of a phenomenon. To detect how the body mediates meaning making, a detailed description of the experience is needed. For researchers to capture the world beyond perception, their own experiences should be bracketed out as carefully as possible to ensure that the description pertains to the primordial world, not only the researchers' reflection. By inhabiting the world, the physically active individual is then open to experiencing and learning about the world.

Although Merleau-Ponty (1978) acknowledges the existence of "the Human World" and the intersubjectivity of meaning making, he also investigated "true" knowledge within the primordial world. One theorist who, like Merleau-Ponty, advocates for an interdisciplinary approach to understanding the body but located his inquiry within the larger politics of knowledge production is Bruno Latour.

LATOUR: AFFECTED AND AFFECTING BODIES

Trained in philosophy, Latour, like Merleau-Ponty (1978), critiques the separation of knowledge production into physiological (or scientific) and social scientific divisions. Similarly, he also takes issue with how "the body" is constructed within this binary. Latour's conceptualization of the body becomes comprehensible when located within his general critique of the "bifurcation" of knowledge production in what he titled the "modernist constitution."

The Critique of the Modernist Constitution

The modernist constitution, according to Latour (1993), is based on "guarantees" that cement the division of knowledge production into two branches: nature and society. To illustrate Latour's critique of the modernist constitution, I detail each with a particular focus on their effects on studying the material body.

First, the modernist constitution guarantees that "nature" is real and is outside of what humans can construct. Outside of human influence, it is transcendental and should be studied objectively, uninfluenced by the researcher. These scientists study directly measurable primary qualities—such as genes, viruses, mitochondria, hemoglobin, and insulin levels—that are accessible only to the scientist's trained eye. At the same time, nature is "constructible" by science, which is able to suggest changes, for example, to the constitution of the body as one biological entity.

Second, different from nature, "the social" is entirely constructed by humans and thus should be studied only as a representation of the real. As a human creation, society is "immanent," and thus social scientists study secondary qualities that exist only in our minds, imaginations, and cultural accounts and therefore are not directly measurable by observation. For example, we study the images that represent how the body is socially constructed as a feminine or masculine sporting body. At the same time, we conceptualize society as something that has solidified into an entity, with life relatively unaffected by human efforts to steer its course.

The first two guarantee the third: Nature and society are considered entirely distinct entities and, as such, should be studied separately. In this bifurcated thought field, representation, appearance, subjectivity, and history belong to the mind and reality, objectivity, and immutability to the world (Latour, 2010a). Latour critiques this logic because it divides the body into a physiological body of receptors, muscles, bones, and nerves and a phenomenological body, "the subjective embodiment" that "will thrive on the lived-in impression provided by something 'more'" than the physiological body (Latour, 2004, p. 208). The bifurcation is limiting because, in Latour's view, the phenomenological body "will always refer only to the depth of our subjection to ourselves" (p. 208), not to any (primordial), universal reality as envisioned by Merleau-Ponty (1978). In this sense, phenomenological accounts are "representational" in the same way as other social science accounts without an ability to reveal more about "carnal" realities than other social science accounts despite their "endless series of holistic arguments to 'reconcile' the physiological and the phenomenological bodies into a single whole" (Latour, 2004, p. 208).

In addition to phenomenology, Latour (2010a) demonstrates how critical theory's conceptualization of the "real" reflects the limitations of the modernist constitution. Unlike phenomenology, critical theory does not claim to unite natural science in its discussion of materiality (of social conditions) but, as Latour (2010a) asserts, does assume a real world that is nevertheless hidden "behind the veil of appearances" (pp. 474–475). Such a thought system is based on a certainty of a real world that is *beyond* this world" (p. 475, italics original). Unlike the transcendent world of natural science, the world of the critical theorist is "constructed" as "a utopia," a notion Latour finds has limited value for social change: The "real" is constructed in a similar manner to ideologies that critical scholar's critique. The premise of a utopian world separates a critical theorist's real from a scientist's real to align both with their appropriate constitutional camps.

The modernist constitution, effectively supported by its contradictions, emerges as a persuasive way of policing acceptable ways of conducting research and is thus "a fully *political* way of distributing power . . . a sort of unwritten compact between what could be and what could not be *discussed*" (Latour, 2010a, p. 476, italics original) as research. Controlled by the modernist constitution, science reduces its discussion to primary qualities, phenomenology "to the residual province of subjective feelings" (Latour, 2004, p. 227), and critical theory to its own utopian reality that does not deal with "real," immanent reality. Clearly dissatisfied with the bifurcated field, Latour developed alternatives to expand scientific discussions. These are crystallized in actor-network theory (ANT; Latour, 2005) and later in "compositionalism."

ANT and Compositionism

Any solution to the modernist constitution, Latour (2005) contends, has to tackle the bifurcation itself to overcome the resulting limiting dualisms. ANT, then, was designed to challenge the division of human action and the agency of objects, social interpretation and the reality of nature, and the idea that science makes objects "natural" reality, whereas social science makes them representational "signs."

Latour's (2004, 2010a) answer to bifurcation is based on the judgment that there is a world waiting to be discovered. But unlike the science of the modernist constitution, he acknowledges that it is an immanent reality of human subjects. Unlike critical theory that dismantles the world they see as constructed by humans in order to reach to utopian reality, the Latourian researcher "composes" the reality of "real"—material and subjective—human elements. These "compositionists" "construct" scientific facts out of matter to politicize their findings. They investigate matters of concern rather than matters of fact. They see immanence (the world constructed by humans) and truth as working together. However, they do not operate within the phenomenological schema that "there is a body, meaning a subject; there is a world, meaning objects; and there is an intermediary, meaning a language, that establishes connections between the world and the subject" (Latour, 2004, p. 208). Instead, they assume "a dynamic definition of the body 'as learning to be affected,'" as actors "*learning to be affected by hitherto unregistered differences through the mediation of an artificially created set-up*" (p. 209, italics original). Such an investigation requires a "slow process of composition and compromise," not "the revelation of the world beyond" (Latour, 2010a, p. 478).

To capture this idea, Latour (2010a) uses the term *compositionism*, which builds a "common world" "from utterly heterogeneous parts that will never make a whole, but at best a fragile, revisable, and diverse composite material" (p. 474). Compositionalists, nevertheless, do not haphazardly merge science and social science or unite physiology and sociology. On the contrary, Latour cautions that simply reattaching the currently separate disciplines conserves the tradition of bifurcation: "We have simply conjoined the worst of politics and the worst of science, that is, the two traditional ways of producing indisputability" (p. 479). Therefore, calling for an uncritical inclusion of the natural sciences to celebrate newly found materiality in social science creates an exclusionary and narrow discussion of the moving body that aligns with, rather than challenges, the modernist constitution. To overcome the limitations of the modernist constitution, compositionalism has to offer an alternative to both "nature" and "society" in the constitution.

Latour (2010a) advocates for the total transformation of the bifurcation to advance social science that includes nonhumans, and science as such accounts for "entangled and controversial and highly disputable matters of concern" (p. 479). The idea of including nonhumans, inanimate entities, or "objects" dominates Latour's transformation of social science. "Objects," like human subjects, compose the "continuity of agents in space and time" (p. 484). They are all actors in the world that a researcher is interested in illuminating, and thus social scientists need a material, mundane, immanent, realistic, embodied definition of the material world "to build a common world" (p. 484). This does not mean simply "talking about" or signifying (analyzing socially constructed meanings) such entities but understanding their active agency, their capacity for action in relationship to humans and their constructions (e.g., organizations, meaning making, the body's movement practices).[7]

Due to their relational nature, the nonhuman entities also create their actors and thus have a force of their own: Far from being socially inert, "they resist, they 'work,' they make things happen, they transform their users" (Hennion & Muecke, 2016, pp. 292–293). Paraphrasing Hennion and Muecke (2016), a dance or sport performance should be understood as "heterogeneous tissue" (or a network) consisting of humans, material entities, and bodies (human and nonhuman), as well as collectives of art or sport organizations and their politics. Such performances assume a life of their own to affect resistance (to meaning making) and create cumulative effects on the involved actors, space, and time. A researcher is then to show how these tissues "hang together" when "mediated" by human and nonhuman entities that actively establish and/or interrupt relations. In other words, how multiple elements "fabricate," or create fabrics of, familiar things should be of major concern to a compositionist researcher.

Latour (2010b) coins the term *factishes* to further explore how objective facts and man-made "fetishes" that are considered to be invested in mythical powers effect and affect our experiences of living in the world.

A compositionist of the body, then, is interested in investigating how artificial and material components affect one's body and how the subject, the individual, learns to be affected by others. When affected, one is exposed to differences in the world and obtains freedom to "articulate" the body in multiple intersections of saying, feeling, and acting. One is "'effectuated,' moved, put into motion by other entities, human or non-human" (Latour, 2004, p. 205). This is the body that "leaves a dynamic trajectory by which we learn to register and become sensitive to what these other elements are" (p. 206). This compositionist looks forward to a prospect of "composing with, compromising, caring, moving slowly, with caution and precaution" (p. 487) to develop research that searches for the common world.[8]

While physical cultural research using Latour's lens is still relatively rare, some scholars (e.g., Kerr, 2014; Weedon, 2015) have examined the agency of nonhuman actors as mediators in physical culture. "Composing" the "common world" of physical culture requires researchers to move slowly to consider their projects "less an object of study, but more an experience to be approached," during which the researchers should allow themselves "to be taken by whatever arises in the midst of experiencing things" (Hennion & Muecke, 2016, p. 304). An ethnographic approach to studying the body in motion thus appears to expose the mediation by humans and nonhumans, as well as the cultural organization of physical activity. To "talk about" the body differently, following Latour (2004),[9] physical cultural studies scholars also need to become sensitive to the "physical" qualities of the entities in question. This requires some familiarity with the scientific constructs of, for example, physiology, biomechanics, and neuroscience when "articulating" the body as a mediator of physical culture. Latour, however, cautions against the uncritical use of science that reinforces the modernist constitution. This means that a constitutionalist has to either learn how to use scientific research differently or collaborate with scientists who do science differently from the premise provided in the modernist constitution. This type of body research, Latour (2004) anticipates, can then act as "a bio-counterpower" (p. 227) that is able to break out of, not align itself with, the effects of bifurcated knowledge production. Such an interdisciplinary engagement, while no doubt productive, can still be difficult to attain. However, to construct his departure from the modernist constitution Latour (2004, 2010a) cites both Foucault's (biopower) and Deleuze's (assemblage) concepts. I also turn to Foucault and Deleuze to discover how they approach the body in motion.

FOUCAULT AND DELEUZE: BODIES WITH/IN POWER

In sport studies, Foucauldian poststructuralist analyses of the body arguably are more prominent than phenomenological approaches or Latourian composition-alism.[10] As its signature, Foucault's analysis places the body directly within power relations: "Power arrangements have an immediate and direct relationships with the body" (Deleuze, 2007, p. 131). An investigation into these relation-ships thus involves the material body in a "concrete" manner, but I want to shed more light on how Foucault's work intertwines the materiality of the body and physical movement into social analysis. To do this, I draw heavily from Deleuze's (1988) reading of Foucault's work that, "in a completely different way" (p. 41), illuminates how knowledge is assembled within power relations.

The Articulable and the Visible

In his discussion of Foucault's work, Deleuze (1988) pictures two types of prac-tical knowledge assembled from visible, "non-discursive," and articulable "dis-cursive" elements (p. 27). The visible elements are comprised of the material bodies, their practices, and their environments, whereas the articulable elements refer to how we talk about, conceptualize, or think about bodies, practices, or environments. Consider, for example, dance, exercise, and sport and their move-ment practices in their environments. For dancing bodies, their specific type of training in a studio and a performance on a theater stage would be comprised of the visible aspects of how we know and understand dance. We also know about bodies exercising in a health club or athletes training or playing in a sta-dium. The material bodies doing certain types of practices in a specific environ-ment thus comprise specific practical knowledge of physical activity: "complexes of actions and passions, actions and reactions, multisensorial complexes" (Deleuze, 1988, p. 50). Foucault (1991) analyzes how these visible elements discipline bodies into docile, useful bodies through the implementation of dis-ciplinary techniques. In the larger schema of Foucault's work, concrete, mate-rial bodies, however, need to be connected with how they are talked about—the articulable elements of how we know about different physical activities.

The "sayable" elements assemble with the visible, material elements to spec-ify "an interpretation" of what we "see" taking place. For example, the naming of movements or exercises in dance (e.g., arabesque), yoga (warrior 3), Pilates (hover), or sport (resistance) training (a squat) distinguishes among different physical activity forms (dance, exercise, sport) and creates training and move-ment practices appropriate to our interpretation or understanding of each movement form. We also know about training physically active bodies in their prospective environments through different "sciences." For example, a combi-nation of sciences (e.g., physiology, psychology, medicine) can be articulable elements for sport, exercise, and dance training and performance. "Science," however, is only one way of knowing. As Deleuze (1988) points out, ethical, aesthetical, and political articulations also shape the interpretations of dance, exercise, and sport. Sport, for example, is discussed in terms of not only scientific aspects of training but also ethics (e.g., unfair advantages through doping, genetics, technology), aesthetics (judging artistic elements in figure skating, diving, ski jumping, rhythmic gymnastics), or how sport is used to promote nationalism, support economics, or advance peace. While the analysis of the "articulable," or what is often referred to broadly as *discourses*, is seen as important in sport studies (King, 2015), there are a few examples of a specific analysis of the articulable elements assembled within specific visible content (e.g., Markula, 2014).

In Deleuze's (1988) reading, the intertwined elements of articulable and visible form "sedimentary beds," the strata "made from things and words, from seeing and speaking, from the visible and sayable, from bands of visibility and fields of readability, from contents and expressions" (p. 43). Here Deleuze employs the terms *contents* (the visible) and *expression* (the articulable). He is careful to note, however, that Foucault's (and Deleuze's) knowledge of an interaction between visible and sayable elements differs from phenomenology and structuralism, which can also be used to examine experience and their interpretation into language. Similar to Merleau-Ponty (1978), Foucault refuses the dualism of "an irreducible difference between two matters" (p. 65), the physiological body and conscious mind. However, his conceptualization of the visible (even if it can be considered close to "experience") does not presume a "natural" or "savage" or "raw" experience unpolluted by knowledge (Deleuze, 1988). Rather, Foucault assumes that there is nothing prior to knowledge because it is a combination of separate visible and articulable elements specific to each historical formation that creates new ways of seeing, displaying, and talking about key issues. For example, physical activity emerges (or remains unsaid and invisible) differently in different eras to become a key concept. Consider a stadium as a way of displaying and competition as a new way of understanding physical activity that then formed a historical formation of sport. Or consider the health club as a new way of displaying the healthy-looking body and commercialism as a new way of understanding physical activity within a historical formation of fitness. Or consider a physiotherapy clinic as a new way of displaying the injury-free body or illness as a new way of understanding physical activity in the historical formation of public health. This understanding differs from Merleau-Ponty's (1978) framework, where knowledge is located in primordial reality underneath human meaning making.

Deleuze's (1988) reading of the historical formations, or the strata, also differs from structuralist theory. Similar to Latour (2004, 2010a), Deleuze and

Foucault reject the notion of material bodies as objects that are assigned meaning by language. Therefore, the sayable elements are not signifiers of meanings (signs) to objects or perceivable "things" (signifieds). The assembling of the visible and articulable elements into content and expression does not mean that the "objects," actions, or other visibilities unite into the sayable and then lose their role as pure elements in knowledge formation. The visible and the sayable do not exist without the other, and thus they comprise equally crucial aspects of the assemblage. Foucault's work (Deleuze, 1988) further emphasizes its difference from structuralism by introducing a further division of form and substance.

Foucault (1991) illustrates how knowledge formations have specific contents and expressions with substance and form. For example, the prisoners are material bodies (the substance) that are further formed into a visible whole through the establishment of a specific space, the prison. The existence of this visible content is "supported" or made understandable by expression, the way we articulate and give meaning to the prison and its prisoners, of certain articulable elements. Statements regarding "delinquency" are formed into a code of "penal law" that articulates who should enter a prison and why. These divisions can be further illustrated by an example of how we divide different physical activities into distinct formations. For example, "exercise" is articulated through such scientific knowledges as medicine, physiology, and psychology (the form) based on a specific substance, health. These forms and substances combine into an expression that then manifests in certain visible content: Physically active bodies (substance) exercise within a specific location of a health club (the form). If we change the articulable substance to "appearance," a different form is needed to capture all the statements around the appearance or the same scientific knowledges (medicine, physiology, psychology) that are now employed in another way to produce visible exercise practices in their locations. We can also use Foucault's idea of historical formation to understand why we interpret some physical activity as sport and others as dance or exercise even if the actual visible movement practices, training, or scientific disciplines informing each practice resemble each other. The articulable substance for sport, for example, can be competition, and thus this physical activity practice employs scientific knowledge differently to different bodies that then train in different locations than those who exercise. Similarly, the articulable substance for dance can be artistic aesthetics, and again this results in assembling of different articulable forms with certain bodies and certain locations (studios or theaters) where the substance of bodies form dance training for performance. The different assemblages can also overlap in such formations as "dance sport" or CrossFit as a "sport of exercise." To analyze how the sayable and the visible interact, the researchers thus must consider their historical conditions, not only to read texts or observe bodies per se. Deleuze (1988), intrigued by how diverse elements assemble, introduces a third agency—power—to account for the formation of the articulable and visible in Foucault's work.

The Diagram of Power

Conceptualizing Foucault's power as a diagram that "organizes" the visible and the articulable elements into formations, Deleuze (1988) defines it in terms familiar to many of Foucault's readers: Power is not a form (e.g., the state) but rather consists of relations between forces; power does not refer (only) to "violence" but rather actions upon actions; power is not repressive but is practiced before possessed; power passes through the hands of both the mastered and the masters. In sum, to engage in power relations is "to affect or to be affected" (p. 60)—an explanation also advanced by Latour (2004). Deleuze (1988) elaborates, "To affect is like a *function* of force," whereas "to be affected is like a *matter* of force" (p. 60, italics original). For example, in Foucault's panoptic diagram, which enabled the disciplinary techniques of the body, the function of force—the anatomopolitics—was to normalize (to affect). The specific categories of people to be normalized were the matter of force (the affected). But within the diagram of biopower, the function—the biopolitics—turned into administering and controlling (affecting) the lives of large and diverse populations (the affected matter). Although Deleuze connected Foucault's diagram of power closely to the historical formations, power is not formalized but "a pure function" (p. 60). Power and knowledge formation exist, however, in "a mutual immanence" in which knowledge actualizes power: How we see and speak is always caught up in power relations that presuppose and actualize knowledge.

Change and Bodies without Organs

Similar to Foucault, Deleuze (2007) is concerned with the "pragmatics" of everyday life and how power arrangements "impose an organization on the bodies" (p. 131). While the major part of Foucault's work focuses on how the body becomes dominated by disciplinary anatomopolitics or biopower,[11] Deleuze is more interested in the body's ability to create change or, in his terms, deterritorialize space from the dominant ways of thinking of the world.

In his work with Guattari, Deleuze (1987) demonstrates how dominant philosophical thought, or what he labeled the "tree-model," sedimented into ways of organizing social fields, knowledge, identity, and the body. This organization, according to Deleuze and Guattari, tends to stratify society based on unified and universalized principles but limits creativity, innovation, and ways of thinking and being. Similar to Foucault, Deleuze (also in his work with Guattari) assigns individuals an important role to play in this change that nevertheless does not take place as a result of "contradiction" between the dominant and dominated groups or is manifest in "resistance" to dominance. Observing that Foucault did not have a chance to fully theorize change, Deleuze (2007) expands the idea about micropolitics to assert that "a social field does not contradict itself, but first and foremost, it leaks out on all sides" (p. 127). These "leaks," or "lines of flight," from the dominant ways of organization have the potential to create change. "Flight lines," he adds, "are not necessarily revolutionary" (p. 127), but they take up new territory—they deterritorialize—from the dominant thought field. According to Deleuze, these lines of flight do not have an equivalent in Foucault's work. They can originate anywhere to initiate different assembling of certain expressions (the sayable) with certain content (the visible material bodies, their practices, and their environment). Living in these "microfields," individuals impact their composition.

As a visible element of assemblages, the doing body has the potential to reveal new ways of thinking, operating, and functioning in the world. Deleuze conceives what he calls the "Body without Organs" (BwO), an unstriated thought plane, to illustrate how the body can act outside of the confines of established, organized thought to engender diverse practices and thoughts. As a smooth plane without organization, "only defined by zones of intensity, thresholds, degrees, and fluxes" (Deleuze, 2007, p. 130), the BwO is not a concrete, material human body per se but a (thought) plane with the potential for multiple ways of thinking and behaving. As such, the BwO does enable a "reconstruction" of the ways we interpret the material body as an "organism": the scientifically defined and organized, universally "true" body. This type of reconstruction is a part of the process of "becoming."

Becoming is not a process of individuals finding their "true" identity, or imitating already-established figures, or changing their behavior or appearance to resemble something closer to an established binary of an identity category (Deleuze & Guattari, 1987). Becoming is a process of loosening the ties to the established knowledges that locate bodies within ready-made, recognizable social identity categories with established behavioral patterns, ideas, or appearances. With a striking resemblance to Merleau-Ponty (1978), Deleuze, in his work with Guattari (1987), defines the moving body as an essential aspect of becoming, because movement, as a pure relation of speed and slowness undefined by perception, escapes dominant expression and interpretation. However, the difference between these two thought systems, similar to Foucault's departure from phenomenology, is that for Deleuze and Guattari, perception and movement do not reveal primordial, universal knowledge but create multiple different ways of thinking that derive from the plane of immanence, the virtual plane where thought has not yet concretized or formalized into knowledge.

Following Deleuze's philosophical path, including his reading of Foucault's work, requires a major departure from the examination of the body as a representation or embodiment of an identity. If the moving body actively challenges the ways we currently know the world and ourselves within the confines of capitalism, it should occupy a central role in sociocultural examinations of physical activity. Everyday movement practices can alter the stratification of "historical formations" if they reveal and then give rise to assemblages of multiple thoughts. However, any assemblage can be reterritorialized (or turn disciplinary and normalizing) without problematization, without the lines of flight that escape the historical formation. The Deleuzean rhizomatic researcher, however, can problematize the organization of an assemblage within specific diagrams of power and, after such problematization, can begin to formulate an alternative practice using the BwO.

CONCLUSION

My necessarily cursory tour of the theoretical frameworks of four French scholars—Merleau-Ponty, Latour, Foucault, and Deleuze—focuses on select readings to illuminate, specifically, their potential for analyses of the body in motion. Drawing from each other's work, often critically, they provide several concepts for productive physical cultural analyses of the moving body. Despite sometimes appearing identical, it is important to observe that these concepts are grounded on fundamentally different assumptions about reality, knowledge, and the body. To conclude, I reflect on the main similarities and differences as they emerge in my discussion.

All four scholars, more or less explicitly, account for the importance of including both the body and its capacity to move into a cultural analysis. In addition, they, like the new materialists, call for a dismantling of the previous divisions of natural science knowledge and social science knowledge. If Merleau-Ponty advocates for the necessity of integrating the physiological and psychological body, Latour goes a step further to critique how both "science" (as a study of objects as "natural" reality) and social science (as a study of objects as representational "signs") need to be revised for meaningful analyses of the moving body. Foucault and Deleuze do not distinguish between the sciences and the body but regard both as elements of knowledge necessary for an understanding of the force of the moving body. Such an "interdisciplinary" examination, however, does not mean simply acknowledging the body as "physiological" and "cultural" or reading it (or any other nonhuman entity) as a social construction, but as Latour specifically emphasizes, both science and social science need to be problematized for the researcher to break out of the confines of the modernist constitution. Deleuze, in his work with Guattari (1987), draws attention to the body's force embedded in its abstract, machinic qualities of connectivity, energy, and momentum that need to be explored if we are to think differently about knowledge, power, and society. A physical cultural scholar of the moving body thus needs to be equipped with an understanding of both how the body moves and how it assumes meanings.

The four scholars are further united in their framing of the body and movement within a version of knowledge as form/substance, expression/content. These divisions have led to separate disciplinary realms: the material body (the substance, the content) is the subject matter of "physiology," while language, meaning making (the expression), is the subject matter of psychology (the consciousness, self that uses language) or sociology. Merleau-Ponty (1978) aims to break out of such a division by claiming a "new meaning for meaning" that gives form to movement through which we know about the world. Foucault and Deleuze discuss the historical formation of knowledge based on the form/ substance and expression/content, but unlike Merleau-Ponty, they explain how power diagrams these formations and consequently how changing these formations is a question of force. This type of inquiry is relatively absent in the current scholarship of physical culture: The investigations into moving bodies and their environments appear somewhat separate from the analyses of "discourses" and identity formation. Therefore, more integrated analyses of the moving body, its practices, and experiences of movement and their expressions is needed for a more complete analysis of the politics of physical culture.

There is a fundamental ontological and epistemological difference between Merleau-Ponty (1978), who assumes "true," unconstructed knowledge to be found outside of power relations, and Latour, Foucault, and Deleuze, who assert that all knowledge is constructed within the historical, political forces. Merleau-Ponty (1978), indeed, accounts for the body or its experiences not as an object of study but rather as a means to knowledge. A physical cultural studies scholar using Merleau-Ponty's phenomenology thus would use physical activity experiences as a starting point for an inquiry of unique knowledge, an essence of a phenomenon, not as the object of inquiry. It is clear, however, that none of the four scholars assign the body as an object of social analysis per se but instead call attention to the body's agency, often explicit in its movement. Latour, in particular, insists that all bodies, including nonhuman bodies, actively mediate strings of actions, the "networks" that constitute the social reality.

Finally, all four scholars acknowledge the centrality of "movement" in social analysis. In my rereading of their works, the moving body emerges as a central element for a new materialist approach that transcends the representational reading of the body as socially constructed. Merleau-Ponty advocates for the physiological and psychological body uniting in movement. Deleuze emphasizes the importance of "what the body does" and highlights the bodily practices as parts of Foucault's historical formation of knowledge. Latour contends that bodies are affected by and affecting the fabrications in which they are involved. Therefore, these scholars clearly establish a support for an analysis of the body in motion.

How do we bring movement, concretely, into an actual analysis of the material and social body? Perhaps this is a task for us, the scholars of physical culture, to accomplish. As experts in the force of the bodies in motion, let us use our unique ability to talk about and see the moving body as a machinic assemblage of metabolism and mechanics, motion and meaning, and force and politics for social change.

NOTES

1. With a substantial existing literature, the following handbooks can be consulted for examples of this type of research: Andrews and Carrington (2013); Coakley and Dunning (2000); Giulianotti (2015); Hargreaves and Anderson (2014); and Silk, Andrews, and Thorpe (2017).

2. Merleau-Ponty (1978) asserts that the phenomenological account should give direct descriptions of our experiences, not causal explanations of psychology, sociology, history, or science. A description requires phenomenological "reduction" (bracketing out of a researcher's own influence as much as is possible) to reveal "a world which it [consciousness] neither embraces nor possess, but towards which it is perpetually directed" (pp. xvii–xviii). Merleau-Ponty, however, views phenomenological inquiry as subjective in a sense that knowledge is gained from one's own particular point of view or experiences of the "pre-existing world."

3. Dance studies scholars have a long history with this type of phenomenological analysis (Franco, 2011). The early work of Sheets-Johnstone (1984), drawing closely from Merleau-Ponty, advocates a description of the "original, pristine body, the preobjective or preobjectivized body" (p. 133) of dance. Merleau-Pontian dance studies continue to highlight the importance of "being-in-the-world" through the dancing body that, as a medium to the world, is open through its innate structures and general basic skills but also cultural skills (such as dance skills; Warburton, 2011). Thus examinations of perception and movement experience and how this solidifies into cultural movement in intersubjective relationships of the self to others is central to phenomenological descriptions of dance (Cooper Albright, 2011). Parviainen (2002), for example, employs insights from several phenomenologists, including Merleau-Ponty, to distinguish between dance skill and dance knowledge generated through "an understanding of the subjective process whereby dancers understand, create, and use knowledge" (p. 22). More recently, Purser (2017), who interviewed 16 professional contemporary dancers, employs Merleau-Ponty's nondualist conceptualization of "body-subjectivity" to suggest that the dancers' lived experiences allowed them to be directly conscious of their bodies and "the here-and-now" (n.p.). In her study, Rouhiainen (2008) emphasizes how intersubjectivity, the interaction with others in her dance space, enables a dancer to discover and respect individual differences and thus teach ethical relationships to others in the world.

4. Some sport philosophers have embraced Merleau-Ponty's work on intentionality. Breivik's (2017) phenomenological description, for example, uses Merleau-Ponty's concept of "motor-intentionality" (body knowledge used to act intentionally without "consciousness") to illustrate how good athletes use bodily knowledge to enhance sport performance. Aggerholm (2013) notes how aesthetic expression—a type of theatrical attitude that "bears players into another world"—enables the players to express new meanings through their physical game performance (p. 220).

5. Through phenomenological descriptions of one golfer's self-consciousness, Ravn and Christensen (2014), for example, demonstrate how prereflective, performative body

awareness enabled the athlete to internalize and feel good about her performance as a successive flow of movement skills. Such an awareness, as "a third" dimension of "self-consciousness" (in addition to consciousness and unconsciousness), functions as an important aspect of athletes' learning and development of skills and thus enhances their performance.

6. In sport studies, phenomenology, including Merleau-Ponty's work, has been established as a distinctive qualitative methodology to study subjective body experiences (Allen-Collinson, 2009; Kerry & Armour, 2000; Ravn, 2013, 2017). As Merleau-Ponty's approach advances the understanding of a phenomenon by doing it (e.g., Kerry & Armour, 2000), some (e.g., Hockey & Allen-Collinson, 2007) favor first-person experiences (i.e., descriptions of the researcher's own experiences), while others argue for secondhand descriptions collected through participant observation or interviews (Clegg & Butryn, 2012; Ravn & Christensen, 2014; Ravn & Høffding, 2017). Nevertheless, the emphasis is on the importance of studying lived bodily experiences beyond social and cultural meanings through detailed descriptions of bodily experiences directed to reveal a dimension of knowledge unreachable by other approaches to the body's performance of physicality. The body experiences highlighted by these phenomenologists lie underneath psychological or sociological influences, and thus qualitative first- or second-person descriptions are required for "authentic phenomenological" description. While these descriptions are to reveal new ways of knowing about the "being-in-the-world" and help fine-tune dance or sport performance and practice, they exclude cultural analysis.

7. Henning, in particular, tends to understand "objects" very broadly. For example, he identifies art, religion, morality, ethics, truth, and culture as sociological objects. His own area of study is music, which he treats as a type of object while recognizing "an art work" (e.g., a musical composition and performance) as an object with its own agency.

8. For a more detailed discussion of Latour's understanding of a network as a string of actions assembled together by a collective of mediators, both human and nonhuman actors with agency, see Latour (2005). He further advocates for a three-part succession for studying such networks: (1) deployment of "controversies" (understanding groups as continually forming through a focus on action, an account of the agency for the nonhuman actors through a focus on matters of concern) to study possible actors in future assemblages; (2) the study of how these actors stabilize their assemblages; and finally, (3) the study of how assemblages can renew the sense of being in the same collective.

9. Latour (2004) provides a detailed critique of the science of the body in his "How to Talk about the Body?"

10. There is indeed a significant amount of Foucauldian sport research including analyses of a docile body normalized through disciplinary techniques (e.g., Barker-Ruchti & Tinning, 2010; Denison, 2007; Denison, Mills, & Jones, 2013; Markula & Pringle, 2006); "discourse" as a device to frame the normalization of sporting and exercising bodies (e.g., Chapman, 1997; MacKay & Dallaire, 2013; Markula & Pringle, 2006; Thorpe, 2008); and the dominant sciences (e.g., medicine, physiology, psychology) that underline sport or fitness practices (Markula & Kennedy, 2011; Markula & Pringle, 2006; Mills & Denison, 2016). Foucauldian researchers have also identified how bodies invested with an embodied subjectivity actively negotiate their physical activity experiences within relations of dominance (e.g., Chase, 2006; Cox & Thompson, 2000; Markula, 1995; Wesely, 2001).

11. Deleuze (1988) identifies how Foucault, in his later work, developed ways to understand how change might take place at the microlevel of everyday life when he became increasingly interested in how individuals actively construct a self through problematization of the discursive formations and the power relations.

REFERENCES

Aggerholm, K. (2013). Express yourself: The value of theatricality in soccer. *Journal of the Philosophy of Sport, 40*(2), 205–224.

Allen-Collinson, J. (2009). Sporting embodiment: Sports studies and the (continuing) promise of phenomenology. *Qualitative Research in Sport and Exercise, 1*(3), 279–296.

Andrews, D. L., & Carrington, B. (2013). *A companion to sport.* Oxford, England: John Wiley & Sons.

Barker-Ruchti, N., & Tinning, R. (2010). Foucault in leotards: Corporeal discipline in women's artistic gymnastics. *Sociology of Sport Journal, 27*(3), 229–250.

Booth, D. (2012). *Australian beach cultures: The history of sun, sand and surf.* London, England: Routledge.

Breivik, G. (2017). Searle, Merleau-Ponty, Rizzolatti—three perspectives on Intentionality and action in sport. *Journal of the Philosophy of Sport, 44*(2), 199–212.

Chapman, G. E. (1997). Making weight: Lightweight rowing, technologies of power, and technologies of the self. *Sociology of Sport Journal, 14*(3), 205–223.

Chase, L. F. (2006). (Un)disciplined bodies: A Foucauldian analysis of women's rugby. *Sociology of Sport Journal, 23*, 229–247.

Clegg, J. L., & Butryn, T. M. (2012). An existential phenomenological examination of parkour and freerunning. *Qualitative Research in Sport, Exercise and Health, 4*(3), 320–340.

Coakley, J., & Dunning, E. (2000). *Handbook of sports studies.* Thousand Oaks, CA: Sage.

Cooper Albright, A. (2011). Situated dancing: Notes from three decades in contact with phenomenology. *Dance Research Journal, 43*(2), 7–18.

Cox, B., & Thompson, S. (2000). Multiple bodies: Sportswomen, soccer and sexuality. *International Review for the Sociology of Sport, 35*, 5–20.

Deleuze, G. (1988). *Foucault.* New York, NY: Continuum.

Deleuze, G. (2007). *Two regimes of madness: Texts and interviews 1975–1995.* New York, NY: Semiotext(e).

Deleuze, G., & Guattari, F. (1987). *A thousand plateaus.* London, England: Athlone.

Denison, J. (2007). Social theory for coaches: A Foucauldian reading of one athlete's poor performance. *International Journal of Sports Science & Coaching, 2*, 369–383.

Denison, J., Mills, J., & Jones, L. (2013). Effective coaching as a modernist formation: A Foucauldian critique. In P. Potrac, W. Gilbert, & J. Denison (eds.), *The Routledge Handbook of Sport Coaching* (pp. 388–399). London, England: Routledge.

Foucault, M. (1991). *Discipline and punish: The birth of the prison.* London, England: Penguin Books.

Franco, M. (2011). Editor's note: What is dead and what is alive in dance phenomenology. *Dance Research Journal, 43*(2), 1–4.

Fullagar, S. (2017). Post-qualitative inquiry and the new materialist turn: Implications for sport, health and physical culture research. *Qualitative Research in Sport, Exercise and Health, 9*(2), 247–257.

Giardina, M. D., & Newman, J. I. (2011a). Physical cultural studies and embodied research acts. *Cultural Studies ↔ Critical Methodologies, 11*(6), 523–534.

Giardina, M. D., & Newman, J. I. (2011b). What is this "physical" in physical cultural studies? *Sociology of Sport Journal, 28*, 36–63.

Giulianotti, R. (ed.). (2015). *Routledge handbook of the sociology of sport.* Abingdon, England: Routledge.

Hargreaves, J., & Anderson, E. (2014). *Routledge handbook of sport, gender and sexuality.* Abingdon, England: Routledge.

Hennion, A., & Muecke, S. (2016). From ANT to pragmatism: A journey with Bruno Latour at the CSI. *New Literary History, 47*(2 & 3), 289–308.

Hockey, J., & Allen-Collinson, J. (2007). Grasping the phenomenology of sporting bodies. *International Review for the Sociology of Sport, 42*(2), 115–131.

Kerr, R. (2014). From Foucault to Latour: Gymnastics training as a socio-technical network. *Sociology of Sport Journal, 31*, 85–101.

Kerry, D. A., & Armour, K. M. (2000). Sport sciences and the promise of phenomenology. Philosophy, method and insight. *Quest, 52*, 1–17.

King, S. (2015). Poststructuralism and the sociology of sport. In R. Giulianotti (ed.), *Routledge handbook of the sociology of sport* (pp. 94–114). Abingdon, England: Routledge.

Larsson, H. (2014). Materialising bodies: There is nothing more material than a socially constructed body. *Sport, Education and Society, 19*(5), 637–651.

Larsson, H., & Quennerstedt, M. (2012). Understanding movement: A socio-cultural approach to exploring moving humans. *Quest, 64*(4), 283–298.

Latour, B. (1993). *We have never been modern*. Cambridge, MA: Harvard University Press.

Latour, B. (2004). How to talk about the body? Normative dimension of science studies. *Body and Society, 19*, 205–229.

Latour, B. (2005). *Reassembling the social: An introduction to actor-network theory*. New York, NY: Oxford University Press.

Latour, B. (2010a). Compositionist manifesto. *New Literary History, 41*, 471–490.

Latour, B. (2010b). *On the modern cult of the factish gods*. Durham, NC: Duke University Press.

MacKay, S., & Dallaire, C. (2013). Skirtboarders.com: Skateboarding women and self-formation as ethical subjects. *Sociology of Sport Journal, 30*, 173–196.

Markula, P. (1995). Firm but shapely, fit but sexy, strong but thin: The postmodern aerobizing female bodies. *Sociology of Sport Journal, 12*, 424–453.

Markula, P. (2014). Reading yoga: Changing discourses of postural yoga on the *Yoga Journal* covers. *Communication & Sport, 2*, 143–171.

Markula, P., & Kennedy, E. (2011). Introduction: Beyond binaries: Contemporary approaches to women and exercise. In E. Kennedy & P. Markula (eds.), *Women and exercise: The body, health and consumerism* (pp. 1–26). New York, NY: Routledge.

Markula, P., & Pringle, R. (2006). *Foucault, sport and exercise: Power, knowledge and transforming the self*. London, England: Routledge.

Merleau-Ponty, M. (1978). *Phenomenology of perception*. London, England: Routledge & Kegan Paul.

Mills, J. P., & Denison, J. (2016). Charting the development of contemporary endurance-running training theory. In W. Bridel, P. Markula, & J. Denison (eds.), *Endurance running: A socio-cultural examination* (pp. 50–60). Abingdon, England: Routledge.

Parviainen, J. (2002). Bodily knowledge: Epistemological reflection on dance. *Dance Research Journal, 34*(1), 11–26.

Purser, A. C. E. (2017). Dancing like a girl: Physical competence and emotional vulnerability in professional contemporary dance. *Women in Sport & Physical Activity Journal*, Advance online publication.

Ravn, S. (2013). Interacting spaces in Argentinean tango. In S. Ravn & L. Rouhiainen (eds.), *Dance spaces: Practices of movement* (pp. 99–115). Odense, Denmark: University Press of Southern Denmark.

Ravn, S. (2017). Dancing practices: Seeing and sensing the moving body. *Body & Society, 23*(2), 57–82.

Ravn, S., & Høffding, S. (2017). The promise of "sporting bodies" in phenomenological thinking—how exceptional cases of practice can contribute to develop foundational phenomenological concepts. *Qualitative Research in Sport, Exercise and Health, 9*(1), 56–68.

Ravn, S., & Krogh Christensen, M. (2014). Listening to the body? How phenomenological insights can be used to explore a golfer's experience of the physicality of her body. *Qualitative Research in Sport, Exercise and Health, 6*(4), 462–477.

Rouhiainen, L. (2008). Somatic dance as a means of cultivating ethically embodied subjects. *Research in Dance Education, 9*(3), 241–256.

Sheets-Johnstone, M. (1984). Phenomenology as a way of illuminating dance. In M. Sheets-Johnstone (ed.), *Illuminating dance: Philosophical explorations* (pp. 124–145). Cranbury, NJ: Associated University Presses.

Silk, M. L., Andrews, D. L., & Thorpe, H. (2017). *Routledge handbook of physical cultural studies*. Abingdon, England: Routledge.

Thorpe, H. (2008). Foucault, technologies of self, and the media discourses of femininity in snowboarding culture. *Journal of Sport & Social Issues, 32*(2), 199–229.

Thorpe, H. (2014). Moving bodies beyond the social/biological divide: Toward theoretical and transdisciplinary adventures. *Sport, Education and Society, 19*(5), 666–686.

Thorpe, H. (2016). Athletic women's experiences of amenorrhea: Biomedical technologies, somatic ethics and embodied subjectivities. *Sociology of Sport Journal, 33*(1), 1–13.

Warburton, E. C. (2011). Of meanings and movements: Re-languaging embodiment in dance phenomenology and cognition. *Dance Research Journal, 43*(2), 65–83.

Weedon, G. (2015). Camaraderie reincorporated: Tough Mudder and the extended distribution of the social. *Journal of Sport & Social Issues, 39*(6), 431–454.

Wesely, J. (2001). Negotiating gender: Bodybuilding and the natural/unnatural continuum. *Sociology of Sport Journal, 18*, 162–180.

2 · OBJECTIFIED BODIES AND INSTRUMENTAL MOVEMENT

What Might Merleau-Ponty Say about Fitness Tracking?

MARY LOUISE ADAMS

This chapter adds to critiques of contemporary fitness cultures by looking at digital fitness tracking technologies and their relationship to instrumental approaches to the body, physical activity, and human movement. Devices like the Fitbit or the Apple Watch manifest the long-standing Western proclivity to treat the body as some sort of machine and to frame the relationship between self and body as an instrumental one. The machine metaphor goes back at least as far as Descartes, who, in elevating the mind as that which defines humanness, distanced it from everything else—"the earth, animals, the human body"—all of which he classified as "merely mechanically interacting matter" (Bordo, 1999, p. 63). The separation of mind from matter is what philosophers refer to when they speak of Cartesian dualism, or the mind/body split, or the distinction between reason and nature. Cartesian modes of thinking are evident all around us: in the capitalist exploitation of the natural world, in colonization and the centuries-long aftereffects of conquest, and in biomedical approaches to the body as an amalgam of separate systems and parts. In relation to human movement and physical cultural forms, this dualism has contributed to the technological and instrumental view of the body that prevails in mainstream kinesiology and physical education, in health promotion of physical activity, and in the everyday practices of both commercial and casual fitness cultures, such as the increasingly popular practices of step-counting and fitness-related data collecting.

In the discussion that follows, I discuss how the materialist views of the body put forward by French phenomenologist Maurice Merleau-Ponty have helped me think about the possible consequences of fitness tracking devices for our embodied subjectivities. How do such technologies transform the reality of the body? What other corporeal realities might be lost as they do so? While this chapter is primarily theoretical, it is informed by critical reflection on my own experience wearing a Fitbit wristband. Like many users, I was, at first, diligent about "getting my steps." But over the course of about a year, the Fitbit started to spend more time on my dresser than on my wrist. Eventually the Fitbit stopped working, and I didn't bother to fix it or replace it. I was tired of wearing a black plastic bracelet and trying to stay ahead of the Fitbit's constantly draining battery. More important, I was troubled by the way step-counting had come to insinuate itself into my days and how it seemed to be shifting my relationship with my body. Simply put, my Fitbit was continually bringing my body to my awareness—sometimes because I was thinking about the steps the device was or could be counting and sometimes because the lights or vibrations of the device were trying to capture my attention. With the Fitbit on my wrist, I found myself paying attention to step-counting—and by association to my health, my fitness, my body—not just when I was explicitly trying to get exercise but throughout the day. Lecturing, walking to the mailbox, cooking—my Fitbit turned the usually unnoticed walking that facilitates these activities into a fitness practice; it "fitness-ized" the most banal movements of my daily life.

In the literature on fitness tracking, scholars have framed the technology in myriad ways—as discipline and surveillance (Sanders, 2017), workplace regulation (Moore & Robinson, 2016), monetization of exercise (Lupton, 2013), gamification of health (Maturo & Setiffi, 2016), labor (Till, 2014), or ubiquitous health care (Gilmore, 2016). My concern is with the way tracking devices rely on a binary relationship between self and body, fostering an instrumental orientation toward movement and reinforcing notions (and thus experiences) of the mind as separate from and superior to the body. To help make sense of this process, I have turned to Merleau-Ponty, whose philosophy of embodiment is, at its root, a critique of the view that human subjects exist in an objective and directive (administrative) relationship with their bodies and the world around them.

While he is not a "new" materialist, Merleau-Ponty is keenly attentive to matter as productive of the social, its inequalities and its potential for freedoms. What makes his work especially useful for this chapter is his focus on the material capacities of the body itself—such as its capacity for perception and movement—and his understanding that, while situated in and shaped by culture and history, bodies are more than cultural and historical effects. To oversimplify, while Merleau-Ponty shares poststructuralist and Foucauldian concerns with the body and bodily experience as constructed through discourse, he also asks questions about the precultural capacities of the body, and he takes these into account in his explanations of how bodies are positioned or experienced in their social milieu. As he writes in *Phenomenology of Perception*,

> For man, everything is constructed and everything is natural, in the sense that there is no single word or behavior that does not owe something to mere biological being—and, at the same time, there is no word or behavior that does not break free from animal life, that does not deflect vital behaviors from their direction [sens] through a sort of *escape* and a genius for ambiguity that might well serve to define man. (Merleau-Ponty, 2012/1945, p. 194)

This focus on the entanglement of the material and the cultural has much to offer analyses of the moving body, and yet the materialist perspectives of Merleau-Ponty are only infrequently brought to bear on sport sociology or critical studies of movement and fitness cultures (for examples, see Liu & Howe, 2012; Grimshaw, 1999; Hockey & Allen-Collinson, 2007). Merleau-Ponty's work is somewhat better known in the philosophy of physical education, where it is the ground for Margaret Whitehead's (2001) influential concept of physical literacy and where it supports arguments about the importance of embodied learning (Stolz, 2014) and the value of the moving body as a site of meaning and source of knowledge (Standal, 2015). In this chapter, I hope to add to this literature by pointing to the more political aspects of Merleau-Ponty's work and showing the utility of his ideas on embodiment for critical critiques of contemporary fitness cultures. At the root of these ideas lies a rejection of mind/body dualism and an understanding of the centrality of the body to our experience and to our conceptualizations of ourselves and the world around us.

THE PERSISTENCE OF MIND/BODY DUALISM, OR WHY WE NEED TO THEORIZE EMBODIMENT

The problems stemming from Cartesian rationalities have occupied generations of critics—philosophers, feminists, environmentalists, critical theorists, anticolonial theorists, and activists. In her book *Feminism and the Mastery of Nature*, Australian feminist and environmentalist Val Plumwood (1993) makes clear the consequences of dualistic modes of thinking, identifying their role in global economic exploitation, oppression and injustice, and widespread and multiplying ecological crises. For Plumwood, the subordination of the body and the disavowal and degradation of the environment are part of the same general failure in Western knowledges to recognize fully the materiality of human existence in terms of the body and its dependence on and embeddedness in the natural world. Western reason has failed to recognize its own "enabling conditions—the body, ecology and non-human nature"—because these have been labeled as inferior and seen merely as background to human achievement (Plumwood, 2002, p. 17). Rational dualisms justify and naturalize inequality, and their binary logic supports relationships of domination (e.g., racism, colonialism, sexism, ableism) across local and global social systems. Plumwood argues that we need to, and can, develop "antidualist" remedies via new forms of thought and broad cultural transformations that would resolve "the nature/culture and reason/nature dualisms that split mind from body, reason from emotion, across their many domains of cultural influence" (p. 4). Central to any antidualist project would be the expansion of modes of knowing that recognize a more bodily mind and a more "mindlike" and agential body. We could, Plumwood (1993) writes, "conceive of their relationship in friendlier terms" (p. 124).

While he might not have used the term *friendly*, the entanglement of mind and body is one of the principle themes of Merleau-Ponty's philosophy and is key to what makes his work a valuable political resource for those of us concerned with human movement and physical cultures. Diana Coole (2010) writes that Merleau-Ponty aimed "to explain a generative, self-transformative, and creative materiality without relying on any metaphysical invocation of mysterious, immaterial forces or agencies" (p. 93). In other words, he wanted to explain the capacity to transform the world as a capacity of a material body. His understanding of the fundamental materiality of human existence provides conceptual support for efforts to articulate the place of humans in a broader ecology. It also helps make the case that how we view and treat bodies, our own and others, matters. In the context of contemporary fitness cultures, Merleau-Ponty might lead us to ask questions about discourses and practices, such as fitness tracking, that encourage us to relate instrumentally to our bodies as health-producing objects that need to be managed—perhaps by pushing them to get more steps or more "active minutes." Merleau-Ponty's concept of the lived body, which I explain below, can help us see what's at stake when we experience our physically active bodies as mechanical or machine-like. It can also help us see what might be gained by living our embodied selves differently.

Following Merleau-Ponty, I argue that in the current context, in which physical activity is so often understood as a health intervention, fitness tracking, like many other health-promoting practices, encourages us to take an instrumental view of movement. In so doing, it encourages us to relate to our bodies as objects. By distancing us from our bodies, tracking interrupts our embodied outward orientation toward the world of which we are a part. Put simply, instrumental forms of embodiment can make our worlds smaller, as I explain below.

In the next section, I briefly contextualize the practice of activity tracking. I then offer an overview of the political roots of Merleau-Ponty's project before presenting some of his specific insights about human embodiment. Finally, I take up elaborations of Merleau-Ponty's work by medical philosopher Drew Leder and feminist phenomenologist Iris Marion Young to show how his ideas on embodiment might provide useful tools for understanding the implications of tracking.

FITNESS TRACKING AS RESPONSE TO HEALTH ANXIETY

Digital tracking devices, like other forms of technology, have no single role to play for the people who use them. People use fitness trackers in a variety of personal and institutional contexts and make sense of tracking and the data it produces through a range of discursive and material contexts. Here I am primarily interested in the most basic functions of digital fitness tracking technology—that is, in their ability to count steps and to capture our attention by way of vibrations, flashing lights, or other haptic or visual notifications.

Data tracking devices have emerged in a neoliberal environment in which the pursuit of health is a moral obligation and, for many, a marker of identity (Crawford, 1980; Kickbusch, 2007). Part of what motivates the cultural obsession with "healthy living" is what Robert Crawford (2004) calls an "escalating spiral of control and anxiety" in response to the ubiquity of health information: "The continuing expansion of knowledge about threats to health, the prolific communication and insatiable consumption of that knowledge and the professional and lay mandate to protect and improve health together aggravate the very insecurities they are designed to quell" (p. 506). Health consciousness is a response to information about the dangers to our health, and it affords "a mandate to undertake instrumental actions for protection" (p. 507). Tracking technology responds to the desire for information and the desire to act on it. As health promotion devices, fitness trackers collect masses of personal data on our bodies to inspire us to change our behaviors.

Digital fitness tracking technology is a broad category that includes a range of devices: wearables like the Apple Watch or the Samsung Gear S3 watch, smart phone applications like MyFitnessPal or the iPhone Health app, and fitness bands like those made by Garmin or Fitbit. The capabilities of these apps and devices vary. Most count steps; some measure heart rate and/or calculate calories burned. Some claim to measure sleep quality. Others remind users to stand up and move around. Marketed as a means of motivating and measuring physical activity, fitness trackers are framed as health-enhancing tools that people can use to increase their physical activity levels and diminish their risk of future illness. Whereas earlier generations of activity-measuring devices—Garmin watches, for instance—were marketed to people already heavily invested in fitness and looking to improve athletic performance, devices like the Fitbit wristband tend to be marketed to people as part of a "healthy lifestyle": "See how Fitbit can help you exercise, eat, sleep, & live better" (Fitbit, n.d.).

The tracking of personal health information is not new—think of bathroom scales and measuring tapes. Nor is the tracking of physical activity new. Pedometer-like objects have been around since the 1600s. What *is* new is the tremendous reach of the new technologies—their constant proximity to our bodies, their ability to collect massive amounts of data, their public sharing and social media components, and their ability to deliver users' data back to corporate manufacturers who then sell it to third parties. What is also new is their pervasiveness. Between 2010 and 2016, Fitbit sales went from 58,000 units to almost 2.3 million units per year (Statista, n.d.). And yet Fitbit sales are now surpassed by sales of smart watches that incorporate fitness tracking capability, a market that is expected to continue to expand (Dudley-Nicholson, 2017). Beyond the consumer market, there is also a growing interest in wearable fitness trackers as tools that can be used in institutional contexts, such as in schools (Williamson, 2015) and workplaces (Moore & Robinson, 2016). In 2015, for instance, Fitbit announced it would be providing devices for 335,000 employees of Target, the U.S. discount store (Cipriani, 2015). Corporate wellness programs—especially in the United States, where health insurance is often linked to employment—may push the use of fitness trackers beyond the health-conscious sectors of the middle class, which typically furnish the biggest market for commercial self-improvement initiatives.

Social scientists have raised myriad issues related to tracking technologies, among them the commercialization of the personal information they gather, privacy and surveillance, the quantification of everyday life, and the translation of complex human activities into numbers. Researchers have identified tracking as a kind of technological biopedagogy that normalizes and disciplines the body (Rich & Miah, 2014; Lupton, 2013) and directs us toward self-optimization through self-governance. The self-tracking body is governed by reason and rendered as a "knowable, calculable and administrable object" (Pantzer & Ruckenstein, 2015, p. 97). In this chapter, I hope to add to these discussions by addressing two features of fitness tracking that distinguish it from some of the other practices—such as running on a treadmill or taking a boot-camp exercise class—that we might associate with instrumental views of physical activity and the body. First, tracking devices such as the Fitbit wristband turn all the movements we make into fitness. As the ads for the Apple Watch say, "Every move counts, so count every move." It is this capacity of fitness trackers to quantify physical activity that makes them different from other aspects of fitness culture that take up a designated time and space in someone's day. Second, devices like the Fitbit are constantly communicating with their users, physically trying to pull users' awareness toward their bodies. This particular aspect of tracking technology is key to the argument that follows.

In the next section, I introduce Merleau-Ponty's broader project. After that, I present an overview of two key aspects of his philosophy—the lived body and intentionality—that have been especially useful for thinking about the type of embodiment produced through tracking technology and practices.

THE POLITICAL PROJECT OF MERLEAU-PONTY'S PHILOSOPHY

Merleau-Ponty was a phenomenologist. By his definition, phenomenology is "the attempt to provide a direct description of our experience such as it is, and without any consideration of its psychological genesis or of the causal explanations that the scientist, historian, or sociologist might offer of that experience" (Merleau-Ponty, 2012/1945, p. lxx). While sometimes presented in the social sciences simply as a "first-person" method in which the aim is to find out how people make sense of their experiences, phenomenology is also fundamentally a philosophical approach to the study of subjectivity and consciousness and an account of the conditions necessary to make experience—meaningful life—possible. The body is central to such an account.

Founded by Edmund Husserl in the early 20th century, the phenomenological movement evolved as a critique of Cartesian mind/body dualism and the objectifying, mechanistic view of the world at work in the natural sciences. Husserl wanted to develop a new approach to science that would not depend on the God's-eye view that distances scientists from the world of which they are a part. He argued for a reflective scientific practice firmly grounded in the "life-world" (Husserl, 1970/1954, p. 104). Husserl's rejection of the distanced, "objective" view of science was both methodological and political. In Husserl's final work, *The Crisis of European Sciences and Transcendental Phenomenology* (1970/1954), a series of lectures published after his death, he argues that "positive sciences," in their lack of self-reflection, had turned "indifferent" to questions that were "decisive for humanity" (p. 6). He was concerned with the hegemony of scientific ways of knowing and the ethical and moral vacuity of an "objective" science that had little to offer to everyday life. "Merely fact-minded sciences make merely fact-minded people," Husserl writes.

> In our vital need . . . this science has nothing to say to us. It excludes in principle precisely the questions which man, given over in our unhappy times to the most portentous upheavals, find the most burning: questions of the meaning or meaninglessness of the whole of this human existence. (p. 6)

Husserl was a Jew writing in the mid-1930s after having been forced from his university position in Germany. The "crisis of European sciences" about which he writes was the crisis of fascism, and pre–World War II Europe; it was the crisis of modernity, in which the elevation of the "merely" factual as the valued goal of scientific inquiry had left pressing questions of ethics and politics in shadow (Coole, 2001). Husserl's text anticipates the scathing critique of science and rationalism that appears in Max Horkheimer and Theodor Adorno's *Dialectic of Enlightenment*, first published in 1947 in the aftermath of the Holocaust and the atomic nightmare unleashed by U.S. bombs in Hiroshima and Nagasaki. In the preface to the book, Horkheimer and Adorno (2002) write, "What we had set out to do was nothing less than to explain why humanity, instead of a truly human state, is sinking into a new kind of barbarism" (p. xiv). The well-known opening line of the text is as follows: "Enlightenment, understood in the widest sense as the advance of thought, has always aimed at liberating human beings from fear and installing them as masters. Yet the wholly enlightened earth is radiant with triumphant calamity" (p. 1). The main target of their critique was instrumental reason, the growing distance between science and the natural world, and the domination of nature and human collective life inherent to capitalism. In the face of "enlightened reason," Horkheimer and Adorno called for more deeply critical reflexive thought with the aim, they said in a later edition, "to preserve and disseminate freedom, rather than to accelerate, however indirectly, the advance toward the administered world" (p. xii).

The renewal of reason was the task Husserl had set for phenomenology (Coole, 2001), which, by the end of his life (he died just before the war, in 1938), he had recast as a deeply political project. Merleau-Ponty was greatly influenced by Husserl's later works, and he had a similarly political orientation toward phenomenology. Merleau-Ponty embraced phenomenology out of frustration with the "abstractions" of Kantian philosophy, which had proved to be of little relevance during the war (Coole, 2001, p. 17). In phenomenology, Merleau-Ponty (2012/1945) saw the potential for a form of "radical reflection" applicable to the political questions of his time (p. lxxxv). A contemporary of De Beauvoir and Sartre, he was, like both of them, an existentialist thinker who found phenomenology well suited to the existential project of trying "to find a way of thinking about our condition" (as cited by Busch, 2008, p. 30). It was a task that needed new conceptual tools, and Merleau-Ponty drafted his version of these in *Phenomenology of Perception* (2012/1945), a massive work that offers a compelling philosophy of embodiment.

"MY BODY IS THE PIVOT OF THE WORLD"

At its most fundamental level, Merleau-Ponty's work, like Husserl's, is a challenge to dualistic modes of thought that perpetuate the subject/object and mind/body binaries. Merleau-Ponty is critical of forms of knowledge in which living beings are abstracted from their context and objectified. His notion of radical reflection involves moving beyond the dogmatisms of tradition, common sense, and certain forms of science to see the life-world from a different angle, "to rediscover the world in which we live, yet which we are always prone to forget" (Merleau-Ponty, 2008/1948, p. 32). While he does not reject science per se, he advocates for forms of inquiry that address the world "revealed to us by our senses" and that "do not start out from measurement and comparisons" (p. 34) or conclude with causal formulas and laws. Where certain forms of science aim at mastery of the world, phenomenological reflection, Merleau-Ponty says, quoting Eugene Fink, involves a stance of "wonder" (p. lxxvii).

At the center of *Phenomenology of Perception* lies an extended argument that presents the body as active and agentic and the ground of human subjectivity. The focus on perception provides Merleau-Ponty an opportunity to take up scientific and philosophical debates about human consciousness and existence. His objective is to convince us that perception is not simply the mental effect of a mechanistic physiological cause or the product of representational processes or judgments in the mind. His account of perception is an ontology of human existence (Käufer & Chemero, 2015); it is also, and fundamentally, a story about embodiment.

Perception is the body's relationship to the world. This idea is key to the analysis of fitness tracking devices that I develop in the next sections. Perception is the most basic expression of our human being and involves our sensory, affective, and motor capacities. It is "the background from which all acts stand out, and is thus presupposed by them" (Merleau-Ponty, 2012/1945, p. lxxiv). David Abram (1997) calls it the "reciprocity" between the world and the self, "a sort of silent conversation that I carry on with things, a continuous dialogue that unfolds far below my verbal awareness" (p. 52). Perception is the connection between our sensing, moving bodies and our environment. It's the variability of my steps and the shifts in my posture as I walk over uneven terrain; it's the way my feet and my legs measure the rise of the steps as I walk up the stairs. Such actions demonstrate a kind of outside-of-consciousness awareness, a bodily knowledge or skill in negotiating the environment. Perception, in this sense, is a tacit, pre-conceptual "knowing" that is prior to and necessary for conscious reflection and experience; it is the first order of expression. As Nick Crossley (2012) puts it,

Phenomenological reflection unearths within perception a form of embodied agency which ordinarily escapes reflection and remains unaware of itself as such;

an agency which subtends conscious awareness.... We are involved in the world, in an embodied and practical way, prior to any conscious sense that we might enjoy of either self or world. This agency is not yet a subject, in the usual sense.... It is also more than an object. It is a "third" term between the two: body-subject. (p. 135)

Merleau-Ponty argues that the world becomes meaningful for us *only* through our bodies, their capacities, orientations, movements, and practical engagements. It is not my thinking that makes the world appear to me, as a Cartesian would say; it is my body's perceptual involvement with the world, its preconceptual response to the world's solicitation, that manifests through my senses and my movements:

If it is true that I am conscious of my body through the world and if my body is the unperceived term at the centre of the world toward which every object turns its face, then it is true for the same reason that my body is the pivot of the world. (Merleau-Ponty, 2012/1945, p. 84)

Merleau-Ponty conveys the interrelation of the body-subject and the world with the concept of the "lived body." The lived body is material and experiential. It involves our senses and our motor and cognitive capacities. It is "an intending entity" (Leder, 2001, p. 123) directed out toward the environment, objects, and other people. In philosophy, intentionality is a concept that refers to the directedness of consciousness—a recognition that consciousness is always *about* or *of* something. Consistent with his general project, Merleau-Ponty says that intentionality is not just a mental phenomenon but accomplished through the body. Motricity, he says, is the "original intentionality" (Merleau-Ponty, 2012/1945, p. 139). He talks of "motor intentionality," which is evident as our bodies respond, through action, to things in the world: "The gesture of reaching one's hand out toward an object contains a reference to the object, not as a representation [in the mind], but as this highly determinate thing toward which we are thrown" (p. 140). My phone rings and I reach across the table to pick it up. I don't think about how to pick it up; I just reach out to answer it or to make the ringing stop. This goal is already evident in the series of movements my body makes—the way it leans across the table, the way my arm moves out, the shape my hand takes. The phone, as an object present to me, is intended by my actions. And these actions are possible because over a lifetime my body has learned how to address objects around it:

A movement is learned when the body has understood it, that is, when it has incorporated it into its "world," and to move one's body is to aim at things through it, or to allow one's body to respond to their solicitation, which is exerted upon the body without any representation. Motricity is not, as it were, a servant of consciousness. (p. 140)

This is what Merleau-Ponty means when he tells us that our bodies, at a level prior to reason, make it possible for us to have a world. He is not talking here of the body as it is reflected in the phrase "healthy mind in a healthy body" that has been woven through the history of instrumental physical education and fitness culture. Rather, he is referring to a body-subject that is "the mediator of a world" (p. 146) with the potential, by way of an evolving, embodied understanding, to respond to the possibilities of that world.

In the next section, I continue the discussion of intentionality by looking at the work of phenomenologists Drew Leder and Iris Marion Young, whose extensions of Merleau-Ponty's work make evident the usefulness of his ideas for thinking about the implications of fitness tracking technology for our experiences of embodiment.

THE CONDITIONS OF TRACKING:
INTENTIONALITY AND THE NO-LONGER-ABSENT BODY

In his book *The Absent Body*, Drew Leder (1990) writes that one of the "essential structures of embodiment" is the tendency of our bodies to "recede from direct experience" (p. 1) as we go about our daily activities. Leder calls this phenomenon "the absent body" and describes it is a product of the intentionality or outward orientation of the lived body. All day my body does things without my conscious involvement. I do not, for instance, think about chewing when I eat. I do not tell my fingers how to hold my pen or how to make letters when I write. I do not tell my breathing to slow down as I am falling asleep. I do not direct the various movements that enable me to stand up from my chair. I just stand up, focused on the task I want to accomplish: going to the kitchen to make tea, letting the dog out, greeting someone at the door. "While in one sense," Leder (1990) writes, "the body is the most abiding and inescapable presence in our lives, it is also essentially characterized by absence. That is, one's own body is rarely the thematic object of experience" (p. 1). Our ordinarily absent bodies, he says, free us up to respond to the invitations of the world; they allow us to accomplish our aims.

By contrast, when we are sick or in pain, our bodies move from the background to the foreground of our experience to become "an active presence whose call we must resist" (Leder, 1990, p. 76). As an unavoidable and undesirable presence, the body that is ill or in pain shifts the typical structures of our embodied subjectivity:

> Whenever our body becomes an object of perception, even though it perceives itself, an element of distance is introduced. I no longer simply "am" my body, the set of unthematized powers from which I exist. Now I "have" a body, a perceived object in the world. (p. 77)

My body is no longer a "'*from*' structure" that orients me out to the world but "that *to* which" I attend (p. 74). Leder uses the term "*dys*-appearance" to refer to the process that brings the body into consciousness, with the prefix "dys-" conveying a sense of difficulty (p. 84).

Medical phenomenologists have used Leder's work to talk about how pain and illness shift one's orientation from outward to inward, diminishing one's "possibilities to act in the world" and "frustrating one's bodily intentionality, that is, one's possibility to endow meaning to one's life and world" (Slatman, 2014, p. 550). The body in pain is intrusive, and activities that had once been in the background—for instance, the mundane, unreflected-upon walking that takes a person around the kitchen as they cook—move to the foreground of experience when that person is hurting. In a study of chronic pain, Jennifer Bullington (2009) writes that "when the purely instrumental (going up the stairs) becomes the focus for time, energy and thought, I experience a shrinking of my world, a shrinkage which fetters me to the here and now in an unnatural way" (p. 105). She argues that the body in pain becomes an objective body "standing in the way of my dealings with the world" (p. 102).

Pain and illness are not the only experiences that bring our bodies to the foreground of consciousness or that lead us to treat our bodies as objects (Zeiler, 2010). Sport, dancing, sex, pregnancy, and the learning of new physical skills are experiences that interrupt the body's typical absence. When we learn new skills, for instance, we may well have to relate to our bodies as objects, directing their unfamiliar movements. Physical pleasures like sex or eating and other sensory experiences can also make our bodies intensely present to us. But in these situations, the bodily presence supports rather than challenges our intentional links with the world, including those that connect us to other people (Leder, 1990, p. 75). The experience of pain and illness, by contrast, makes it hard to focus on anything beyond the body, as it consumes our attention and becomes the focus of our actions as we try to make ourselves better, hence *dys*-appearance.

Leder's arguments about the ordinary absence of the body are key to my analysis of the implications of fitness tracking technologies like the Fitbit, which have been designed to keep our bodies present in our awareness. With its lights and vibrations, my Fitbit constantly tells me to notice my body, precluding my body's ordinary absence from my consciousness. In calling our attention to our bodies, tracking devices engender a type of embodiment that mimics that of a person who is ill or in pain. I am, of course, absolutely not arguing here that the experiences of data tracking and being ill or in pain are the same; rather, tracking devices, just as pain and illness, interrupt the outward orientation of the body and, in so doing, pull us into a process that, Bullington (2009) would say, shrinks our worlds.

Digital tracking has emerged at a historical moment in which "exercise is medicine" and "sitting is the new smoking" and in which public health approaches have long since shifted from structural to individual explanations for ill health. In this context, fitness trackers are promoted as tools to help us reduce our risk of future disease. The ever-present tracking bracelet means that it's not just the physical activity we decide to do as "exercise" that gets tracked but every movement we make. Worn on the body, fitness bracelets insinuate themselves— and the health-related risk discourses that motivate their use—into all we do. And thus walking to the photocopier down the hall from my office becomes an opportunity for me to attend to my body and the project of step-counting that is meant to serve my future health.

Fitness trackers ask us to treat our bodies as objects that might constantly be improved; they remind us that no matter how good we might be feeling now, we nevertheless need to work at achieving health for later. The bodies that these technologies invite to activity are not generative bodies open to the world but mechanical bodies that require vigilance and attention. In this situation, my body ceases to be the "anchorage of [my] world" (Merleau-Ponty, 2012/1945, p. 146) and becomes that to which I (as the governing mind) attend. It is not just that self-tracking makes us focus too much on ourselves, although that is a concern, and one that could be leveled at fitness cultures generally. The issue here is that in a context in which physical activity is most often promoted as a response to the risk of future dysfunction, fitness trackers ask us to engage with our bodies as problems to be solved, and they recruit our movement to that end. Thus trackers foster an instrumental relationship between body and self, interrupting the body-subject relation. By bringing our bodies to the foreground of our awareness, they mimic (albeit to a much lesser extent) the process of *dys*-appearance that Leder sees in actual debilitating situations such as pain or illness. What I want to suggest is that these devices encourage a form of embodiment that constantly draws us to our bodies, which is not how we tend to experience our bodies when we are feeling capable and strong.

What might the *dys*-appearing body mean for the embodied subjects who use fitness tracking devices? Merleau-Ponty would say that when we act instrumentally toward our own bodies, we disturb the interweaving of mind/body/ world. Leder talks of a disrupted intentionality. Feminist phenomenologist Iris Marion Young (2005) talks about "inhibited intentionality." In her classic article "Throwing like a Girl," she uses the term to describe a motor intentionality in which the body is not fully open to possibility in the world (p. 38). Young was writing about how gender norms constrain the physicality of girls and women. While her examples feel outdated (she wrote originally in 1977), the fact of differently gendered physicalities is not. She argues that what may disrupt a girl's ability to put her full body into an action such as throwing is her lack of confidence (fostered by her cultural and historical context) in the capacities of her prereflective body and the fact that she comes to gain experience of her own body, via the gaze of others, as a thing distanced from herself as a subject. As Young puts it, "We feel as though we must have our attention directed upon our bodies to make sure they are doing what we wish them to do" (p. 34). Inhibited intentionality is a consequence of a girl's inability to express the full potential of her physical capacity to act in the world. When she throws "like a girl," the "I can" of motor intentionality (Merleau-Ponty, 2012/1945, p. 139) is overlaid with the "I cannot" of gender norms. Remember that for Merleau-Ponty intentionality is not the conscious intentions of the subject; it is the preconceptual outward orientation of the body, the relationship between the body-subject and its environment. Inhibited intentionality occurs, therefore, when the capacity of the body is disconnected (by cultural ideologies about gender in the case of the girl who throws) from possibility in the world (Young, 2005, p. 38).

Young's work provides a useful account of the relationship between culture and the styles of embodiment that emerge in specific cultural contexts. While the parallel is not exact, her notion of inhibited intentionality has helped me think about tracking practices in the context of health-related risk discourses and present-day health promotion. Health-related risk discourses and the intense and worried health consciousness that they provoke lead us, like the girl of Young's (2005) account, to focus on our bodies "to make sure they are doing what we wish them to do" (p. 34). Contemporary health discourses and the practices they inspire—not just fitness tracking but other approaches to physical activity that have instrumental aims—also make us familiar with the idea and the experience of our bodies as objects. The quantification involved in tracking and the way it abstracts our movement from the life-world simply reinforces this objectification. When we move simply to get more steps, the body refers back on itself, and the unity of the lived body, or body-subject, is disrupted. The (too sedentary, too fat, or likely to get sick) body becomes that which needs to be improved, while at the same time the physically active body becomes that which will do the improving. The body-subject's motor intentionality—that is, its capacity to respond to the world—is not gone, but it is inhibited by culture.

CONCLUSION

My aim in this chapter was to consider the materiality of the body upon which fitness tracking devices work. I turned to Merleau-Ponty's philosophy of the body to illustrate some of the implications of fitness tracking for how we relate to our bodies and to movement. Merleau-Ponty's ideas encourage us to look at the body as we experience and live it and to think about how the body feels and how it makes possible our engagements with the world. In this way, he offers a perspective that is still not well represented in the literature on tracking. Existing literature on digital tracking technology has taken up the body from various directions—for instance, as positioned in social and economic relations (Till, 2014; Moore & Robinson, 2016); as subject to biopower and constructed by discursive formations (e.g., those around gender, health and appearance; Sanders, 2017); or as quantified, datafied, and analyzable (Pantzar & Ruckenstein, 2015). In new materialist analyses of tracking, the body is explicitly decentered, appearing as just one component of a tracking assemblage or network that might also involve any number of other things, such as devices, data, sales people, algorithms, sidewalks, or trails (Fox, 2017). Here the body remains the focus, as I have considered how tracking devices and practices shift the way the body/subject engages with the world. I address the body as the locus of perception and thus the basis of knowledge and experience. It is understood as the vehicle through which we are able to encounter a world of openings and possibilities. Movement is one of the means by which we and the world become open to each other.

In the current cultural context, in which physical activity is often understood as a health intervention, fitness tracking, like other practices intended to produce our future good health, encourages us to take an instrumental view of our bodies and movement. In light of widespread health consciousness, tracking technology reminds us that our bodies are always potentially ill and thus in need of attention and regulation. Physical activity in this equation is seen not for its inherent value as a ground of experience and knowledge for the body-subject but as a means of securing the subject's health.

Of course, fitness tracking does not preclude the range of reasons people move and the meanings they might make of their moving bodies. Whether I am wearing a Fitbit or not, I walk my dog because I like hanging out with her and because we both like being in the woods. But while she is completely engaged in her tasks and fully immersed in her world, I cannot help but experience my own walking not just as a form of moving that engages me in my environment but as something that is also producing my health. This point of view produces a bodily self-awareness that situates the thinking/reflecting me as separate from the body that makes my reflecting possible. Through such an attitude, I experience my body not as the "pivot of the world" but as an object that I can direct to take more steps.

What might a nonobjectified body look like? What might it be like to experience the moving body (the moving self) in noninstrumental ways? Maybe it is something like this:

After hours of steady walking, with the long rhythm of motion sustained until motion is felt, not merely known by the brain, as the "still centre" of being. . . . Walking thus, hour after hour, the senses keyed, one walks the flesh transparent. But no metaphor, *transparent*, or *light as air*, is adequate. The body is not made negligible, but paramount. Flesh is not annihilated but fulfilled. One is not bodiless, but essential body. (Shepherd, 2011/1977, p. 106)

Here [in the mountains] there may be lived a life of the senses so pure, so untouched by any mode of apprehension but their own, that *the body may be said to think* [emphasis added]. (Shepherd, 2011/1977, p. 105)

This is not the body addressed by my Fitbit or the type of walking motivated by step-counting technology. The quote comes from a small book called *The Living Mountain* (2011/1977) by Scottish writer Nan Shepherd. Completed in 1944 but not published until 1977, the book is her account of coming to know the Cairngorm Mountains that sit to the west of her home in Aberdeen. Lauded by reviewers as a masterpiece of nature writing, *The Living Mountain* presents a world view that is remarkably similar to the phenomenology of Merleau-Ponty. As Robert Macfarlane (2011) notes in his introduction to a new edition of Shepherd's book, it is quite unlikely that the authors would have been known to each other. And yet *The Living Mountain* makes concrete and evocative some of the most important ideas that have come to be associated with the *Phenomenology of Perception*.

The body in Shepherd's work is no detached mechanical object; it is not bossed around by a superordinate mind; it is a fundamental source of knowledge tangled up with the world around it. Shepherd (2011/1977) helps us grasp the body/subject/world relationship that is the lived body:

It is therefore when the body is keyed to its highest potential and controlled to a profound harmony deepening into something that resembles trance, that I discover most nearly what it is *to be*. I have walked out of the body and into the mountain. (p. 106)

We are not minds with bodies, as Cartesian thought would have it; we are our bodies, and our embodied subjectivities (we) are coextensive with the world. Our bodies have, therefore, the capacity to do much more than simply keeping us from getting sick.

REFERENCES

Abram, D. (1997). *The spell of the sensuous.* New York, NY: Vintage.

Bordo, S. (1999). Excerpts from "The flight to objectivity." In S. Bordo (ed.), *Feminist interpretations of René Descartes* (pp. 48–69). University Park: Pennsylvania State University Press.

Bullington, J. (2009). Embodiment and chronic pain: Implications for rehabilitation practice. *Health Care Analysis, 17,* 100–109.

Busch, T. (2008). Existentialism: The new philosophy. In R. Diprose & J. Reynolds (eds.), *Merleau-Ponty: Key concepts* (pp. 30–43). Stocksfield, England: Acumen.

Cipriani, J. (2015, October 20). Fitbit beats back competition with wellness program. *Fortune.* Retrieved August 29, 2017, from http://fortune.com/2015/10/20/fitbit-wellness -program/

Coole, D. (2001). Thinking politically with Merleau-Ponty. *Radical Philosophy, 108,* 17–28.

Coole, D. (2010). The inertia of matter and the generativity of flesh. In D. Coole & S. Frost (eds.), *New materialisms: Ontology, agency, and politics* (pp. 92–115). Durham, NC: Duke University Press.

Crawford, R. (2004). Risk ritual and the management of control and anxiety in medical culture. *Health, 8*(4), 505–528.

Crawford, R. (1980). Healthism and the medicalization of everyday life. *International Journal of Health Services, 10*(3), 365–388.

Crossley, N. (2012). Phenomenology and the body. In B. S. Turner (ed.), *Routledge handbook of body studies* (pp. 130–143). London, England: Routledge.

Dudley-Nicholson, J. (2017, May 28). Fitbit fumbles: Apple Watch overtakes fitness bands as consumers embrace smartwatches. Retrieved August 29, 2017, from http://www.news.com .au/technology/gadgets/wearables/fitbit-fumbles-apple-watch-overtakes-fitness-bands-as -consumers-embrace-smartwatches/news-story/d78f99b8d58b08a31c7e8f68293726ea

Fitbit. (n.d.). Fitbit homepage. Retrieved from https://www.fitbit.com/en-ca/home

Fox, N. J. (2017). Personal health technologies, micropolitics and resistance: A new materialist analysis. *Health, 21*(2), 136–153.

Gilmore, J. N. (2016). Everywear: The quantified self and wearable fitness technologies. *New Media & Society, 18*(11), 2524–2539.

Grimshaw, J. (1999). Working out with Merleau-Ponty. In J. Arthurs & J. Grimshaw (eds.), *Women's bodies: Discipline and transgression* (pp. 91–116). London, England: Cassell.

Horkheimer, M., & Adorno, T. W. (2002). *Dialectic of enlightenment: Philosophical fragments.* Stanford, CA: Stanford University Press. (Original work published 1947)

Husserl, E. (1970). *The crisis of European sciences and transcendental phenomenology: An introduction to phenomenological philosophy* (D. Carr, trans.). Evanston, IL: Northwestern University Press. (Original work published 1954)

Käufer, S., & Chemero, A. (2015). *Phenomenology: An introduction.* Cambridge, MA: Polity Press.

Kickbusch, I. (2007). Responding to the health society. *Health Promotion International, 22*(2), 89–91.

Hockey, J., & Allen-Collinson, J. (2007). Grasping the phenomenology of sporting bodies. *International Review for the Sociology of Sport, 42*(2), 115–131.

Leder, D. (1990). *The absent body.* Chicago, IL: University of Chicago Press.

Leder, D. (2001). A tale of two bodies: The Cartesian corpse and the lived body. In D. Welton (ed.), *Body and flesh: A philosophical reader* (pp. 117–129). Oxford, England: Blackwell.

Liu, L., & Howe, P. D. (2012). Phenomenology and embodiment in cross-cultural sporting contexts: A case of Chinese female students. *Asia Pacific Journal of Sport and Social Science,* *1*(2–3), 169–185.

Lupton, D. (2013). Quantifying the body: Monitoring and measuring health in the age of mHealth technologies. *Critical Public Health, 23*(4), 393–403.

Macfarlane, R. (2011). Introduction. In N. Shepherd (ed.), *The living mountain* (pp. ix–xxxvii). Edinburgh, Scotland: Canongate.

Maturo, A., & Setiffi, F. (2016). The gamification of risk: How health apps foster self-confidence and why this is not enough. *Health, Risk & Society, 17*(7–3), 477–494.

Merleau-Ponty, M. (2012). *Phenomenology of perception* (D. Landes, trans.). London, England: Routledge. (Original work published 1945)

Merleau-Ponty, M. (2008). *The world of perception* (O. Davis, trans.). London, England: Routledge. (Original work published 1948)

Moore, P., & Robinson, A. (2016). The quantified self: What counts in the neoliberal workplace. *New Media & Society, 18*(11), 2774–2792.

Pantzar, M., & Ruckenstein, M. (2015). The heart of everyday analytics: Emotional, material and practical extensions in self-tracking market. *Consumption Markets & Culture, 18*(1), 92–109.

Plumwood, V. (1993). *Feminism and the mastery of nature*. London, England: Routledge.

Plumwood, V. (2002). *Environmental culture: The ecological crisis of reason*. London, England: Routledge.

Rich, E., & Miah, A. (2014). Understanding digital health as public pedagogy: A critical framework. *Societies, 4*, 296–315.

Sanders, R. (2017). Self-tracking in the digital era: Biopower, patriarchy, and the new biometric body projects. *Body and Society, 23*(1), 36–63.

Shepherd, N. (2011). *The living mountain*. Edinburgh, Scotland: Canongate. (Original work published 1977)

Slatman, J. (2014). Multiple dimensions of embodiment in medical practices. *Medicine, Health Care and Philosophy, 17*, 549–557.

Standal, O. F. (2015). *Phenomenology and pedagogy in physical education*. London, England: Routledge.

Statista. (n.d.). Number of Fitbit devices sold worldwide from 2010 to 2016. Retrieved August 29, 2017, from https://www.statista.com/statistics/472591/fitbit-devices-sold/

Stolz, S. A. (2014). *The philosophy of physical education: A new perspective*. London, England: Routledge.

Till, C. (2014). Exercise as labour: Quantified self and the transformation of exercise into labour. *Societies, 4*, 446–462.

Whitehead, M. (2001). The concept of physical literacy. *European Journal of Physical Education, 6*(2), 127–138.

Williamson, B. (2015). Algorithmic skin: Health-tracking technologies, personal analytics and the biopedagogies of digitized health and physical education. *Sport, Education and Society, 20*(1), 133–151.

Young, I. M. (2005). *On female body experience: Throwing like a girl and other essays*. New York, NY: Oxford University Press.

Zeiler, K. (2010). A phenomenological analysis of bodily self-awareness in the experience of pain and pleasure: On dys-appearance and eu-appearance. *Medicine, Health Care and Philosophy, 13*, 333–342.

3 · BODY OBJECTS, POLITICAL PHYSICS, AND INCORPORATION

Object-Oriented Ontology for Sport and Physical Culture

CHRISTOPHER M. MCLEOD AND MATTHEW G. HAWZEN

Sociologists of sport and physical cultural studies (PCS) scholars have long wrestled with the physically active body and its relation to culture, society, and politics. They have studied how active bodies—"and the subjectivities they inhabit, perform, and embody" (Andrews & Silk, 2011, p. 7)—provide the keys to understanding physical culture, power, and power relations. In this work, two propositions about the body and its significance guide inquiry: First, the body is "always dialectic," which is to say its meanings are produced by, and productive of, wider contexts and social structures (Giardina & Newman, 2011, p. 40). Second, the body is political and politicizing, which means it actively contributes to systems of power rather than passively receiving their affects (Silk & Andrews, 2011). With this understanding of the active body as both dialectical and political, sociologists and PCS scholars use Birmingham School of Cultural Studies and feminist-inspired methodologies of *articulation* and *embodiment* to study the lived realities of active bodies in society.[1] As Giardina and Newman (2011) explain, "Meaningful, textual, sensual, lived, performative, fleshed bodies can only exist within and through articulations of culture(s)," and, as they go on to note, "the physical ... articulates the body to structures of power, ideology, and determinism" (pp. 39–40). If bodies are articulated, then embodiment captures the way that bodies are uniquely capable of articulation compared with other more mundane and less active objects: Bodies of social and cultural theory articulate by bearing and projecting, or embodying, certain meanings, power relations, histories, and identities.

With articulation and embodiment as our points of departure, we invite readers to explore a different philosophy for studying bodies in sport and physical culture. In this chapter, we entertain the speculative realism of object-oriented ontology (OOO) to see what it can tell us about the physically active body and contemporary body politics.

Object-oriented philosophers maintain that all things count as objects, including bodies, abstractions, material entities, sentient beings, conglomerates, signs, or anything else whatsoever. OOO is important for readers of this volume because it offers new and surprising trajectories for theories of the body, embodiment, and body politics. Specifically, it paves the way for an *irreductionist* theory of bodies (Latour, 1988), which is to say OOO avoids undercutting bodies toward their constituent cells, hormones, and contexts at the same time as it forbids proliferating bodies upward to their affects, meanings, and politicizations. Compared with theories from sociology, cultural studies, and new materialism, OOO establishes an autonomous, incommensurable body as the basis of cultural and political theory.

In this chapter, we introduce OOO to scholars of sport and physical culture. More specifically, we build on the work of object-oriented philosophers such as Graham Harman (2002, 2005, 2009, 2011, 2016), Timothy Morton (2013a, 2013b), Ian Bogost (2012), Levi R. Bryant (2011), and Bruno Latour (1988, 2007) to develop a theory of the body, embodiment, and body politics. If we are successful, this chapter will provide an alternative approach to studying active bodies and offer novel insights into topics scholars of sport and physical culture hold dear: namely, the body and its political constitution.

This chapter proceeds with three sequential arguments. First, we argue that objects must be the targets of inquiry rather than their material ingredients or their enveloping contexts. Second, we develop the premises of OOO to predict how object-oriented philosophers might conceive of bodies. One implication of this analysis is that a given body cannot be "for" another object. As a consequence of this finding, and in contrast to the axiological positions of much work in the sociology of sport and PCS, we contend that bodies are not inherently political. Therefore, we push OOO to determine when and how an object can be political. Using a brief case study of the National Football League's (NFL) concussion crisis, we develop the concept of *incorporation* in order to show that bodies are frequently, but not uniquely, political.

OBJECT OBJECTIONS

Object-oriented ontology is a metaphysics built on the premise that objects are the stuff that make up the world. According to OOO, objects are the only topic worthy of philosophy. Luckily, everything qualifies as an object so long as it is autonomous; an object is autonomous if it is a thing unto itself, irrespective of its relations with anything else, especially the human mind. For example, a rock is an object. A pair of pajamas is also an object. These are easy examples. It gets harder when we consider compound objects, such as the Ladies Professional Golf Association (LPGA); fleeting objects, such as a backflip; past objects, such as early 20th-century Russian factory workers; intangible objects, such as an elusive digital Pikachu; sentient objects, such as Tim Tebow; and taboo objects, such as subjects. But all these things are equally objects so long as they are irreducible to their parts, their effects, their relations, other objects, or our perceptions.

This is the starting point of OOO—autonomous objects everywhere. Before showing what Harman and his colleagues have discovered about the nature of these autonomous objects, we need to convince readers that objects are the best starting point for philosophy. We argue sociologists of sport and PCS scholars must study things rather than the physical culture and contexts that surround them, the language that refers to them, the matter that constitutes them, or the perceptions that refer to them. There are only two alternatives to studying objects, and we must dispel both.

Alternative One: Undermining. If you decide not to study objects such as rocks, pajamas, or the LPGA, then there are two philosophical alternatives. Harman (2011, 2016) calls the first *undermining*. Undermining arguments hold that objects are built of something more basic, and this more basic thing is the true topic of philosophy: "An object is undermined if we explain it in terms of its smaller constituents, by way of downward reduction" (Harman, 2016, p. 8). According to an underminer, rocks are not really rocks but granite; granite is not really granite but molecules; molecules are not really molecules but atoms; atoms are not really atoms but intensities; intensities are not really intensities . . . and so on. Undermining is common in positivist sciences, but it is also common in materialism and cultural theory because humans and things are said to be dependent on their background (or underground) conditions. According to Harman (2016), these approaches are untenable because an object is not equal to its parts, its base material, or its environmental influences. For example, the LPGA is still the LPGA when it gets new members and when its members retire. It is still the LPGA when it gains and loses major sponsors. And it is still the LPGA when Donald Trump is elected president of the United States.

Alternative Two: Overmining. The second alternative asserts that objects are less important than their effects or how they are perceived (Harman, 2011, p. 12). Harman (2011, 2016) calls this *overmining*. One type of overmining can be seen in actor-network theory and Spinozian-influenced philosophies, in which objects are replaced by actants or affects. According to Harman (2016), the problem with this type of overmining is that "it allows objects no surplus of reality beyond whatever they modify, transform, perturb, or create." This is a problem because "if objects were nothing more than their current expression in the world, they could not do anything differently in the time that follows" (pp. 10–11). Thus, according to overminers, there is no reason anything should change. To the contrary, Harman (2016) argues a thing must first exist in order to act: "Instead of replacing objects with a description of what they do (as in ANT) or what they are made of (as in traditional materialism), OOO uses the term 'object' to refer to any entity that cannot be paraphrased in terms of either its components or its effects" (p. 3).

Another type of overmining is any approach that grants some things—such as human consciousness or language—more reality than other things. According to Morton (2013b), this type of overmining "decides that [objects] are only granted realness status by somehow coming into the purview of the more real entity" (p. 45). This second type of overmining is a symptom of what Meillassoux (2008) called "correlationist" philosophies. Correlationists argue that humans are incapable of thinking reality outside of thought because in thinking reality, humans always think it in thought. In other words, correlationists say epistemology is always first and ontology is only after. The correlationist position is inadequate because it is rare for humans to apprehend objects in thought; usually we only think about things when they go wrong. Correlationism is equivalent to saying your gas tank only exists when you notice the fuel light flashing. Each day, only a few things present themselves to a mere seven billion humans. A philosophy limited to this sprinkling of events can hardly claim the title.[2]

Most approaches use a combination of undermining and overmining, which Harman (2007) calls *duomining*. In contrast to these approaches, OOO calls for a return "to the 'things themselves'" (Husserl, 2001, p. 168). To study things in themselves, OOO must establish the autonomy of objects without undermining or overmining (Harman, 2016, p. 41). Starting with objects only, and nothing else, means these objects must be separate from their contexts, their own pieces, and any other objects they relate with. Ultimately, OOO must be able to say the same thing about rocks as it says about the LPGA and Pikachu. Since our goal is body theory, our OOO must be able to say the same thing about corpses as it says about factory workers and Tim Tebow. Moreover, because bodies are only one type of object, our body theory must have good reason to say something about these objects that is different from 20-pound dumbbells, Orangetheory

Fitness, a fantasy football profile, or a stirring rendition of "You'll Never Walk Alone."

THE QUADRUPLE OBJECT

Every object has four components: a real withdrawn reality, a perceived sensual identity, a set of real qualities that distinguish it, and a sensual parade of accidents that skip across its surface. Every one of these components is at tension with each of the others. We will find that these tensions are a rich trove of metaphysical contradiction; they are the bounty of OOO. However, before we meet these tensions, we must start with the four components of every object, which means beginning with Harman and his first book, *Tool-Being* (2002). Here Harman extends Heidegger's famous tool analysis (found in *Being and Time* [1927/2008]). In doing, he discovered the first of four components: the real withdrawn reality of an object.

In Heidegger's tool analysis, he describes how a piece of equipment such as a hammer can be approached in two ways. We can pick it up and use it, or we can contemplate it. When we pick it up and use it to hammer a nail, we encounter the hammer as ready-to-hand. We ignore it and attend to the nail, the board, or the emerging piece of furniture. On the other hand, if we contemplate the hammer, it changes from ready-to-hand to present-at-hand. A present-at-hand hammer is palpable, weighable, and describable; it has a wooden handle, which is smooth and dark from many years of use, and it has a pitted head and claw from striking and ripping out nails.

When we contemplate the hammer and make it present-at-hand, parts of the hammer present themselves to us that were not there when we were using it. However, when contemplated, it loses its use for hammering. The same happens for all sorts of things we contemplate. Try walking by purposefully contracting your own leg muscles and you lose the ability to walk. This difference between ready-to-hand and present-to-hand shows that a hammer has at least two faces of existence: one when we use it and one when we contemplate it. As soon as we try to access an object, its reality seems to hide from us, as though it keeps running to the other side of a coin.

Harman (2002) extends the tool analysis by arguing that we always reduce the hammer to present-at-hand, including when we use it. This is because the hammer exceeds its uses just as it exceeds our efforts to contemplate it: "Insofar as a tool is 'used,' it is no less present-at-hand than an image in consciousness. But a tool is not 'used'; *it is*. And insofar as it is, the tool is not exhausted by its relations with human theory *or* human practice" (Harman, 2011, p. 44). Moreover, Harman argues that our relationship to the hammer (a selective encounter with the hammer's use qualities or image) is the same as the nail's relationship with the hammer. The nail responds to the hammer's weight and hardness, but it is quite unperturbed by its smooth wooden handle or its color. We can imagine a third object, an iron filing or a bacterium, that relates to the hammer in a way that humans are not privy to. Importantly, the hammer remains the same hammer for me in use, for me in contemplation, for the nail, for the iron filing, and for the bacterium. Therefore, this hammer must hold some subterranean reality in reserve. It must refuse to share this reality not just with me, or me plus the nail, but any combination of objects imaginable, *including the hammer's own parts.* Harman (2011) concludes, "The withdrawal of objects is not some cognitive trauma that afflicts only humans and a few smart animals, but expresses the permanent inadequacy of any relation at all" (p. 44). Morton (2013b) elaborates, "Withdrawal means that at this very moment, this very object, as an intrinsic aspect of its being, is incapable of being anything else: my poem about it, its atomic structure, its function, its relations with other things" (p. 17). Withdrawal is the first premise of OOO and must be maintained at the center of any subsequent theory of bodies. Rather than underpinning a flat ontology of planes or intensities, Bogost (2012) calls this a tiny ontology of points, analogous to a universe cluttered with vacuum-sealed entities or black holes.

Heidegger's tool analysis describes the fundamental characteristic of real objects: their withdrawal. But it remains the case that we interact with objects as objects. Last week, Chris bought a new pair of shoes. At home, he fondles and ogles his shoes, and when he is at school, he fantasizes about wearing them. When he looks at his new shoes, he does not see a copy, or an abstraction, or a set of nerve firings, or three hours of socially necessary labor time. He perceives *this* pair of shoes. Therefore, along with a real withdrawn object, there must also be a present, sensual object of perception. Harman (2005) developed his philosophy of sensual objects with reference to the "carnal phenomenologists." Later, he attributed the major insights into sensual objects to Husserl (Harman, 2011). Husserl's enduring observation was that a single intentional object of perception (which is synonymous with a "sensual object" in OOO—e.g., Chris's shoe) is always at tension with the accidents of its appearance as well as its real qualities.

As an example of the tension between a sensual object and its accidents, imagine circling Doak Campbell Stadium on foot during a single day in May, from dawn to dusk. In the crisp Tallahassee morning, you see rich, warm brick glinting in the early light. As the thunder claps arrive in the afternoon, the brick becomes dull and foreboding. Finally, the storm hits, and the bricks start dripping bloody streaks. Doak Campbell is the same object all day, yet it is encrusted with an incredible diversity of sensual qualities. We can remove these accidents—warm brick, dull brick, and bloody brick—from Doak without changing its identity for us.

Doak Campbell, however, also relies on real qualities, which, unlike its accidents, cannot be removed from the stadium without destroying it. For instance, Doak is red. It would not be the same stadium if it were blue. Nevertheless, we cannot recreate the stadium simply by adding redness to other qualities (like cylindrical, solid, and hollow) because Doak deploys redness in its own specific way. In other words, real qualities "are imbued from the start with the reality or style of the object to which they belong" (Harman, 2011, p. 28).

We are now in a position to repeat the major components that define any object, including bodies. Objects have a fourfold structure that consists of a real object (withdrawn hammer), a sensual object (Doak Campbell for us), real qualities (hardness, redness), and accidents or sensual qualities (warmth, dullness, blood). Rather than specifying four types of objects, this fourfold structure designates the tensions at play within a single object. In other words, we could have used the hammer throughout this section because like Doak Campbell, it also has a sensual object and accidents. We do not have room to consider all the tensions and their implications here, but two points are worth noting.

First, in OOO, each object is riddled with internal contradictions. An object appears to us yet withdraws from access. It contains things that are not it. Its qualities are not the object, but the object is impossible without its qualities. Morton (2013b) writes, "Objects present us with the following paradox: objects are both objects and non-objects" (p. 27). He continues, "An object is a non-object not because it is 'really' something else, a void or some featureless lump or a moment in my reflective process—but because an object isn't something else" (p. 30, emphasis added). According to Morton, this contradictory being at the heart of an object is like an engine that accounts for the birth, life, change, and death of all objects.

Second, if real objects are withdrawn, then the question of how one object interacts with another becomes a fantastic impossibility.[3] This is further complicated by the fact that object-oriented philosophers demote relations by taking away their status as basic ontology. Object-oriented philosophers cannot stand objects *and* relations being the two basic types of things that make up the world; to elevate objects, they reject relations. That being said, if relations are not on the same footing as objects, then what are they? Harman's (2011) ingenious solution is to say relations are objects too: "Notice that our relation with an entity can itself become a unified object that withdraws from the scrutiny of all other entities, including we ourselves: as when we form marriages and business partnerships, or join the Foreign Legion" (p. 113). He goes on to formulate this proposition in more detail: "If certain components are arranged in such a manner as to give rise to anything that exceeds them, in such a way that it can withstand certain changes in these components, then they have entered a genuine relation with each other as real objects rather than merely stoking one another's sensual facades" (p. 117). Any relation that cannot be reduced to its components satisfies the basic principle of autonomy and is therefore an object. The result is a universe of cascading objects, vacuous objects within vacuous objects, objects all the way down.

THIS BODY IS A QUADRUPLE OBJECT

According to OOO, bodies are objects, which makes them ontologically equivalent to turtles, turnips, toothbrushes, and tourist attractions. In this section, we will give examples that refer to this set of bodies: a corpse, Russian factory workers in 1917, Tim Tebow, Simone Biles, and a decathlete. Given that every object is autonomous, each of these bodies is unique—from other objects and from other bodies. Whenever possible, scholars should retain this autonomy by describing specific bodies. A specific predicate such as the name "Tim Tebow" is no more or less closer to describing the real Tim Tebow than a vague predicate such as "The Body," but vague predicates hide the fact that we rarely see vague bodies (except, of course, in physiology textbooks, where it makes sense to inquire into their vagueness). Vague predicates encourage vague analyses of bodies that are not vague but real and unique.

A given body (instead of "the" body) is real, which means it is utterly withdrawn. For example, a corpse is always only *this* corpse and nothing else. A prosecutor could photograph the corpse, slice it open with a scalpel, present it to a jury, and cremate it in flames, but the prosecutor is unable to exhaust the corpse in these actions because it keeps an inaccessible reality to itself (e.g., even the flames of a crematorium fail to exhaust corpses because there is always enough left to put in an urn). Instead, cameras, scalpels, juries, and flames each encounter a sensual corpse. Paraphrasing Bogost (2012), a scalpel and a jury have an idea of the corpse, and this idea really is present, but the corpse itself withdraws infinitely. Moreover, it withdraws from the scalpel in exactly the same way as it withdraws from the jury. This is the first bodily tension: between a real body that withdraws and a sensual body that appears.

The second tension manifests an "and" structure that refers to a body's set of real qualities. For instance, we could theorize a certain number of qualities that characterize Tim Tebow—let's say he digests *and* has motility *and* ages *and* perceives. Without these real qualities, there is no way to distinguish Tebow from a pyramid or the Fourth of July. Nevertheless, there is still a tension here in that it is impossible to exhaust his body by describing its qualities. Even if we made a machine capable of putting Tebow through every trial imaginable, even if we deconstructed his body from a God's-eye view and rebuilt it with parts created in meticulous detail on a 3-D printer, we would never attain, deconstruct, or build Tim Tebow's body.

A given body also has an "or" structure that refers to its sensual qualities or accidents. For example, a decathlete might be clothed *or* naked, jumping *or* sprinting, under daylight in Olympic Stadium *or* under halogen lights at IMG Academy. Whereas a body depends on its real qualities, these accidents are inessential. In fact, an object is characterized by the fact that it can have any number of accidents and still be the same object. A decathlete who moves from the 100-meter sprint to the discus is still the same athlete. Even though the decathlete displays a bewildering mess of limbs, colors, and surfaces as she spins in the middle of her discus throw, the officials never confuse her for another body. This is the third tension: The accidents of movement are inessential but remain inextricably anchored to bodies.

A given body can also have a "part" structure. Simone Biles can be a part of the Olympic team, part of the United States, or part of a social movement, and she herself can be made of organs, cells, and degrees of freedom. In each case, Biles is never fully exhausted by her memberships or by her components. This fourth tension is present in the fact that we can describe, count, or collect every part of Simone Biles's body, and yet her body always emerges as something over and above its parts.

The most shocking implication of OOO, however, concerns what a body is not. A given body cannot be "for" something else. For example, Russian factory workers in 1917 could not have been instruments for capital accumulation because they did not fundamentally change after the February or October Revolutions. In other words, because they did not change when capitalism was removed, they could not have been for capitalism. By analogy, this hammer remains as *this* hammer whether we use it to hammer a nail or to rip down old weatherboard. Similarly, Tim Tebow cannot be a symbol of paleo- or neoconservatism (cf. Hawzen & Newman, 2017) because he refuses to disappear even as his popularity fades. If Tebow and factory workers were given to their relations, then we would be required to conclude that they are killed and then remade, from moment to moment, as their context changes. Their poor bodies would be drowned in the storm of their relations; they would be no more or less than termination points for other things.

This is important. In order to be an object, a body must withdraw, so it cannot be "for" another object. In fact, according to OOO, it is unclear how bodies, as objects, relate to anything at all! This provocative impossibility of relating seems to undercut many of the deeply held assumptions in sociology of sport and PCS—namely, dialectical relations, radical contextualization, and the political given-ness of bodies. What can scholars of sport and physical culture do with a body that withdraws?

We can do a lot. Take, for example, the problem of embodiment, which the editors of this volume identified in their introduction. There is only one thing we can say for sure about embodiment. It is this: *To embody, you must not be the thing you embody*. To embody conservatism, Tim Tebow cannot be conservatism. Put differently, Tim Tebow matters because he is *not actually* conservatism. If he became conservatism, he would cease to be important. He would cease to matter to neoconservatives, he would cease to matter to professional sports teams, and he would cease to matter to scholars. He is important precisely because he is Tim Tebow, only Tim Tebow, and no one and no-thing else. This is what OOO shows. Paraphrasing Morton (2013b), a body is *this* body and is able to embody *because it is not something else* (p. 30). In our opinion, this evident and productive contradiction is the most interesting insight of OOO and is a deserving topic of analysis for PCS scholars and sociologists of sport.

This contradiction of embodiment might be interesting, but is it feasible? After all, if OOO has merit, this means researchers must forgo relations in favor of real withdrawn objects. Rather than reject this conclusion offhand, we will now trial a theory of real withdrawn bodies. Having established the OOO foundation for embodiment, we will develop a theory of embodiment and body politics based on an ontology that loves things but demotes interactions.

To see why an object-oriented theory of body politics is important, consider how these scholars within and outside of sociology and PCS have ruminated on "embodiment" and "body politics":

- "The production of bodies is a means to the constitution of social beings and social relations" (Therberge, 1991, p. 126).
- "The body is cultural, and *culture embodied* is *political* . . . [because] the body is . . . given to cultural systems and political structures" (Newman, Giardina, & McLeod, 2016, p. 592).
- "The body is less an entity than a relation, and it cannot be fully dissociated from the infrastructural and environmental conditions of its living" (Butler, 2014, p. 8).
- "The body is a particularly useful instrument for concealing public motives as private ones because the body *seems* irreducibly private and individuated" (Duncan, 1994, p. 62, emphasis added).
- "An anatomo-politics of the human body" is "centered on the body as a machine: its disciplining, the optimization of its capability . . . the parallel increases of its usefulness and its docility, [and] its integration into systems of efficient and economic control" (Foucault, 1990, p. 139).

Each of these discussions of embodiment and body politics begins with an assumption that the body is related, has *potential* to relate, *becomes* related, or is "for" another object (a cultural system, a political structure, an economy, or an apparatus of control). Politics would seem to be underpinned by a relational metaphysics. What happens if we substitute this for an object-oriented metaphysics?

INCORPORATED OBJECTS

The phrase *incorporated objects* is borrowed from Drew Leder's (1990) phenomenological inquiry *The Absent Body* and Gavin Weedon's (2015) essay "Camaraderie Reincorporated." In *The Absent Body*, Leder uses the etymological combination of *in* and *corpus* to posit incorporation as the process by which movements are internalized into bodies. Weedon alludes to Latour's use of "reassembly" and deploys "reincorporation" to consider the place of the body alongside an eclectic group of objects that produced camaraderie as a "more-than-human" phenomenon in mud-running events. We initially define *incorporate* as an object joined with another object to create a new object (or body). Incorporation is a type of fusion (Harman, 2011). The opposite of incorporate is *excorporate*, which we define as an object split into its constituent parts. Excorporation is a type of fission (Harman, 2011). Incorporated and excorporated objects do not relate—they are either entirely separate, completely conjoined, or paradoxically, both. The paradoxical state of being at once separate and conjoined can be called *embodiment*.

In the previous section, we argued that a given body is simply another object, which means it is not ontologically distinct from rocks, toothbrushes, Nike Swooshes, or zone defenses. It is just as real and withdrawn as any other real object. It is also just as intentional and immanent as any other sensual object. This means that, against many assumptions of sociology of sport and PCS, the body is not inherently more or less political than any other object. Moreover, neither context nor relations can account for any special quality of bodies—political or otherwise—because objects are not relational.

We argue that if bodies are indeed especially political (but not uniquely so), then this is because of the frequency with which they incorporate. In other words, although bodies are no more or less political than potatoes that incorporate pig DNA, dirt that incorporates rain, or T-Mobile commercials that incorporate Justin Bieber, bodies habitually incorporate objects and are therefore commonly political.

To demonstrate incorporation, we will briefly describe the life cycles of a selection of objects pertaining to the NFL's concussion crisis:

- athletes who played football during the 1900s and the early 2000s
- head-to-head collisions
- Mike Webster's brain
- tau proteins

- the NFL
- chronic traumatic encephalopathy (CTE)

Rather than providing an exhaustive analysis (which would require a longer list), our aim is to demonstrate the theoretical value of incorporation. By dealing with these objects in sets of two or more, we aim to show how the question of politics is about objects that are not yet fully incorporated, meaning they are both distinct parts and new objects at once. By analogy, a person standing in a doorway is not yet fully incorporated because he or she is at once inside the room and out.

As should be clear by now, none of these objects (football players, collisions, Webster's brain, tau proteins, the NFL, or CTE) are political by themselves. Instead, we must look to their incorporations and excorporations.

Incorporation 1. A head-to-head collision is an object made when two football players collide headfirst. The collision object is at once greater than the two football player parts but also ceases to exist as soon as the players disentangle themselves. Football administrators have attempted, at various times in history, to prevent these collisions from happening. In the early 1900s, rule changes (including the introduction of the forward pass) were implemented to stem football-related deaths. During the mid-1900s, increasingly sophisticated helmet technology reintroduced collisions. In 2013, the NFL banned leading with the head to mitigate the risk of concussion.

What is most interesting about head-to-head collisions in the late 1900s is how collisions became autonomous objects, wholly distinct from the player parts that make them. The first technology to grasp collisions and grant them autonomy was video replay, which made collision objects that could be bought and sold to audiences, sponsors, and broadcasters. These replays could be exchanged irrespective of whose heads constituted them in the first place. The second technology to help collisions gain permanency was coaching: specifically, the strategy of leading with the head (Keilman, 2017). Former players such as Dave Meggyesy (1971) and Gary Shaw (1972) recollect training camps where collisions became a self-fulfilling unit, existing regardless of exactly which body was implicated. Ryan Riddle (2012) describes a similar experience when recounting a 2006 training camp: "A massive collision ensued right at the line of scrimmage between me and the fullback" (para. 32), "both of us were fighting to make this team and giving equal effort to do so" (para. 34), and "I was beginning to make a good impression on the coaches—unfortunately at the cost of some valuable brain cells, for sure" (para. 35). From the perspective of OOO, the compound object of head-to-head collisions achieved autonomy from its parts. Head-to-head collisions subsequently became something that had effects and could be targeted in the efforts to remove concussions from American football.

Incorporation 2. Mike Webster was born on March 18, 1952. He played for 17 years in the NFL and retired in 1990. Webster died on September 24, 2002, at the age of 50. Dr. Bennet Omalu conducted the autopsy of Webster's body and brain. After ordering a range of special tests, Omalu discovered large accumulations of tau proteins in Webster's brain tissue. Tau proteins are "a substance which serves to stabilize cellular structure in the neurons but which may become defective and subsequently may cause major interference with the function of the neurons" (BIRI, 2017, p. 1). The tau proteins in Webster's brain appeared as "brown and red splotches. All over the place" (Laskas & Veasay, 2009, p. 1).

Tau proteins already exist in brain tissue. However, in Webster's case, they had accumulated around blood vessels, interrupting normal functioning and killing nerve cells. Over time, tau proteins further accumulated in Webster's frontal and temporal lobes and eventually overwhelmed his brain (McKee et al., 2013).

Incorporation 3. When tau proteins accumulate in the brain, persist over a period of years, and are coupled with brain atrophy and when this accumulation is the result of traumatic impacts to the cranium, it is called CTE (BIRI, 2017). CTE and the NFL were both topics in Omalu's 2005 *Neurosurgery* publication (Omalu et al., 2005). Soon after the article was published, three scientists on the NFL's payroll wrote a letter to the editor requesting Omalu's article be retracted (Laskas & Veasay, 2009). It was not until March 2016 that an NFL administrator, Jeff Miller (the senior vice president for health and safety), first acknowledged a relationship between football and neurodegenerative diseases such as CTE (Fainaru, 2016). This was 15 years after the NFL used incomplete data to obscure the danger of concussions (Schwartz, Bogdanich, & Williams, 2016), 4 years after they reportedly tried to influence the NIH's brain injury program (Pallone, 2016), and 3 years after settling a lawsuit brought by 5,000 former players. Bein (2016) argues this settlement was reached on the basis that CTE was not serious, meaning "the NFL can now acknowledge CTE without having to act or assume liability" (p. 1).

The NFL's concussion crisis is about incorporating objects. In the subcases reviewed above, the matters of concern revolve around two different states of being that can be described as (1a) separate athletes playing football versus (1b) athletes in head-to-head collisions; (2a) Mike Webster's brain separate from tau proteins versus (2b) Mike Webster's brain with accumulated tau proteins; and (3a) the NFL separate from CTE versus (3b) the NFL with CTE. These objects oscillate from excorporated to fully incorporated and back over their lifespan. For example, rule changes separated athletes' heads in the early 1900s, whereas helmets, video replays, and coaching strategies fused heads in the late 1900s, and finally, more rule changes in the 2000s endeavored to separate heads once again and remove head-to-head collisions (see figure 3.1). Webster's brain

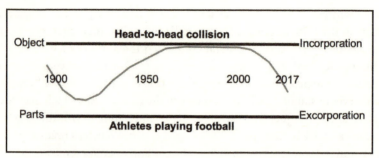

FIGURE 3.1. The life cycle of head-to-head collisions and their constituent parts. Full incorporation occurs when the collision object is autonomous from, and can dictate, its player parts. Excorporation occurs when the parts gain autonomy from the object. In the case of collisions, excorporation will destroy the object because collisions require players.

fully incorporated tau proteins throughout his life, but Dr. Omalu separated them postmortem and then spent significant resources trying to prove that they really had overwhelmed Webster's brain (see figure 3.2). NFL officials resisted CTE for 14 years and then accepted it within their governing body after settling with former players (see figure 3.3). To be clear, these objects were not relating to one another. Rather, they were riven with contradictions as to whether they were really separate objects, full objects, or parts of other objects. It is this contradiction that ebbed and flowed, leaving controversial objects in its wake.

We must make one more observation. These objects seemed most divisive when they were partially, rather than fully, incorporated. In figures 3.1–3.3, the objects were most contentious when they occupied the middle ground between parts and objects. Head-to-head collisions were a matter of public debate in the early 1900s and in the 2000s, but not in the late 1900s, when they took on a life

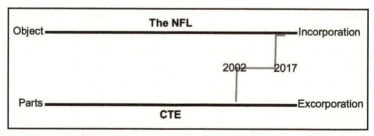

FIGURE 3.2. The life cycle of Mike Webster's brain and the tau proteins that accumulated in it and eventually became the object precipitating the NFL's concussion crisis.

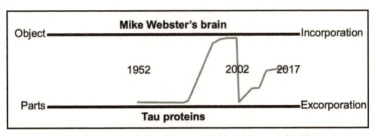

FIGURE 3.3. The life cycle of the NFL and CTE, which was finally incorporated in 2016.

of their own in video replays and training camps. Webster's brain was an issue only after Dr. Omalu separated tau from it in an attempt to show the proteins could be found in other football players. CTE was debated in the media, in academic journals, and at conferences for 14 years. When the NFL finally accepted CTE, it became one more thing players consent to when they "choose" to play. In other words, these objects seemed more political when they were most contradictory and less political when they were unified and simple.

A POLITICAL PHYSICS OF CTE

None of the object-oriented ontologists write political theory. However, in 2014, Harman published a second book on Bruno Latour devoted solely to Latour's political philosophy. This book, *Reassembling the Political*, is brimming with clues about object-oriented politics (Harman, 2014). Harman thinks of Latour as an immanent object-oriented philosopher, making Latour's work an important part of the OOO canon. This is especially the case for Latour's object-oriented democracy, or *Dingpolitik* (inspired by Noortje Marres's interpretation of Lippmann's and Dewey's political pragmatism), which has parallels with OOO.

In 2007, Latour described the political life cycle of an object. He followed Marres (2005) to define politics as neither a procedure, a domain, nor an essence but "a type of situation": "It is something that moves; it is something that has a trajectory" (Latour, 2007, p. 814). Latour suggested using different meanings of the adjective "political" to qualify "certain moments, stages or segments" in the trajectory of an issue (p. 814). He demonstrated this by describing the political life cycle of a blood screening procedure for Dutch women as a sequence of stages cycling from "political-1" through "political-5" (see de Vries, 2007). Let us apply these political stages directly to the NFL concussion crisis.

An object is *political-1* when it produces new associations between humans and nonhumans. This is the traditional domain of ANT, science and technology studies, and new materialism. It describes the plethora of objects that are being discovered and created by scientists and engineers (or removed from corpses by

Dr. Omalu) on a day-to-day basis. In the NFL concussion crisis, the excorporation of tau proteins from Webster's brain and the creation of CTE are examples
of political-1.

An object is *political-2* whenever it "generates a concerned and unsettled public" (Latour, 2007, p. 816). This is the domain of Lippmann and Dewey and can
be summarized by Marres's (2005) dissertation (with the catchy title "No Issue;
No Public"), which explains that public assemblies are presupposed by issues
or, in our case, objects. Political communities do not create political issues but
assemble around ones that already have a bearing on their lives. In our case, players, fans, scientists, and administrators gathered around collisions, tau proteins,
and CTE. It should also be noted that most objects do not pass from political-1
to political-2.

An object is *political-3* when "the machinery of government tries to turn the
problem of the public into a clearly articulated question of common good and
general will" (Latour, 2007, p. 816). This stage of politics is most closely associated with the political philosophy of Carl Schmitt and his duel concepts of
sovereignty and the state of exception. In the present case study, in the 1900s
and again in the 2000s, presidents of the United States commented on football
deaths. In 2016, Jeff Miller from the NFL first acknowledged that a relationship
existed between football and neurodegenerative diseases such as CTE during a
roundtable discussion on concussions convened by the U.S. House of Representatives' Committee on Energy and Commerce.

An object is *political-4* when it is included in the traditional procedures of
deliberative democracy. As with all the other political adjectives, Latour (2007)
cautions that an object need not reach political-4, nor should it. Citizens should
not be required to reach a consensus on all objects. In our case study, there
were no apparent instances where collisions, tau proteins, or CTE were pushed
through the democratic process. However, we might consider alternative processes in place of deliberative democracy here. For instance, consumers of the
NFL may have consented or dissented through the market system, and parents
may have acted by removing their children from tackle football.

Finally, an object is *political-5* when it has "become part of the daily routine
of administration and management" (Latour, 2007, p. 817). Latour (2007) gives
the example of sewers in Paris, which are no longer political. His next remark is
worth quoting at length:

> Should we abstain from calling [sewers] political in another sense of the adjec-
> tive? Of course not, because not only did they used to be loudly disputed con-
> troversies (historians of science and technology, feminist scholars, do nothing
> else but "repoliticizing" them through a kind of historical reverse engineering),
> but also because they might reopen at any moment. . . . This is the stage that

fascinated Michel Foucault as suggested by this much-abused expression of "governmentality": all those institutions appear on the surface to be absolutely apolitical, and yet in their silent, ordinary, fully routinized ways they are perversely the most important aspects of what we mean by living together—even though no one raises hell about them and they hardly stir congressmen out of their parliamentary somnolence. (p. 817)

For our purposes, political-5 refers to fully incorporated objects such as head-to-head collisions in the late 1900s and the NFL's acknowledgment of CTE in 2016. These objects have given up blurry boundaries to become clearly defined matters of fact. Their life of controversy is put on hold. But, Latour reminds us, their parts can explode back out at any moment.

Adding Latour's (2007) political stages to our existing analysis of incorporation depicts the political life of things (see figure 3.4). Each political stage corresponds with a more or less incorporated object. New objects such as tau proteins and CTE are differently political when they are discovered compared with when a concerned public assembles around them and again when they are totally subsumed within objects such as Webster's living brain and the NFL's concussion policy.

In considering this model of political physics, we do not need to be perfectly true to Latour's stages because his primary insights are these: an object has a political life course; an object is differently political throughout this life course; and an object can become entirely incorporated within another, such as a body or a government, and cease being imminently political (although it can always be pulled out from this hibernation). A good example of a fully incorporated body is a child who has been vaccinated against the measles. A vaccinated body can incorporate a disease the same way a modern city can incorporate sewers, subways, and stadiums after protracted public debate. Indeed, there are some things we might want to advance to political-5 and leave there for a while. We might want to incorporate nutritious breakfasts, automated cars, or physical

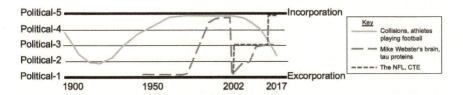

FIGURE 3.4. Combining object incorporation with Latour's (2007) five political adjectives. The life cycle of an object, which oscillates between incorporation and excorporation, corresponds with different political stages.

education curriculums by making them unquestioned parts of our daily routines. Similarly, there are many objects we might want to excorporate to political-3 by unsettling the matter and pushing it into public debate.

When it comes to human bodies, we argue they do not differ ontologically from governing bodies, bodies of water, or baboon bodies (see Strum & Latour, 1987). What makes a human body political is not its unique capacity to do anything, and certainly not its unique ontological status, but simply the prevalence of its incorporation. Harman (2005) confirms we are on the right track with this interpretation when he argues humans are the most object-oriented of objects. It is only humans' love of objects that explains any difference between humans and animals, plants, and rocks. PCS scholars and sociologists of sport must decide whether our analysis of collisions, tau proteins, and CTE can also be said of identities, movements, performance-enhancing drugs, brands, subjectivities, statements, ideologies, and capital.

Finally, what we call *embodiment* must happen somewhere between political-1 and political-5. An object only embodies another object when it has not yet incorporated it, which is to say that embodiment is contradictory and embodiment is productive because it is contradictory. Tebow is not conservatism, but he also is. He embodies it because he is not it. Webster's brain embodies tau proteins only after Dr. Omalu differentiates between tau and nerve cells. The NFL no longer embodies CTE because CTE has been fully incorporated in concussion tests and player safety programs like Heads Up Football. According to our political physics, one way to define embodiment is as a contradiction in being that assembles a public. Tim Tebow is a contradiction in being that assembles a public. He is the same as Mike Webster's brain in 2002 and the NFL until 2016. One day, like Webster and the NFL, Tebow may not embody.

CONCLUSION

In this chapter, we have made three central arguments. First, objects must be at the center of ontology and, consequently, social and political theory. Second, bodies are objects, meaning any theory of the body or body politics must begin with OOO and its first premise of withdrawal. This means taking the impossibility of relation, context, and being-for seriously. Third, if bodies are ontologically equivalent to rocks, toothbrushes, Nike Swooshes, or zone defenses, a theory of body politics cannot be grounded in any ontological divide between, for example, the subject and the object. Instead, bodies can only be said to be commonly political because they incorporate other objects (and are incorporated within other objects) more frequently than most things.

Incorporation reveals a political physics of objects circling, staving off, swallowing, and then spitting each other out. OOO is a rich, if strange, philosophy.

It draws scholars of sport and physical culture toward autonomous bodies and away from relations and holds fascinating possibilities for studying embodiment and politics.

NOTES

Thanks to the editors and to Eleanor MacDonald for useful feedback.

1. It is worth noting that PCS as a subdiscipline is not necessarily unique or novel in its appropriation of these methodologies (see Adams et al., 2016). Adams et al. (2016) argue that the methodological approaches of PCS are grounded in feminisms, which is not always made clear by authors of PCS. In addition, Giardina and Newman (2011) point to the wider "corporeal turn" in other disciplines that they seek to follow in their work and that make the method/theories of articulation and embodiment appropriate strategies for scholars of sport and physical culture to understand active bodies in context and society. While PCS is not the focus of this chapter, it serves as our conversation partner. We return to PCS and sociology of sport throughout the chapter to highlight interesting differences between our reading of OOO and other approaches to studying the body.

2. See also Bryant (2011), who defended ontology-first inquiry using a transcendental argument.

3. All the major OOO philosophers have devoted portions of their major works to exploring this conundrum: Harman in *Guerrilla Metaphysics* (2005), Morton in *Realist Magic* (2013b), Bryant in *Democracy of Objects* (2011), and Bogost in *Alien Phenomenology* (2012).

REFERENCES

Adams, M. L., Davidson, J., Helstein, M. T., Jamieson, K. M., Kim, K. Y., King, S. . . . Rail, G. (2016). Feminist cultural studies: Uncertainties and possibilities. *Sociology of Sport Journal, 33*(1), 75–91.

Andrews, D. L., & Silk, M. L. (2011). Physical cultural studies: Engendering a productive dialogue. *Sociology of Sport Journal, 28*(1), 1–3.

Bein, L. (2016). Why the NFL can finally admit football causes CTE and not owe players anything: The NFL is finally copping to CTE now that it's not liable for it. *SB Nation*. Retrieved from http://www.sbnation.com/2016/4/18/11451036/nfl-concussion-settlement-cte

BIRI. (2017). What is CTE? *Brain Injury Research Institute*. Retrieved from http://www.protectthebrain.org/Brain-Injury-Research/What-is-CTE-.aspx

Bogost, I. (2012). *Alien phenomenology, or, what it's like to be a thing*. Minneapolis: University of Minnesota Press.

Bryant, L. R. (2011). *The democracy of objects*. Ann Arbor, MI: Open Humanities Press.

Butler, J. (2014). Rethinking vulnerability and resistance. Madrid. Retrieved from http://bibacc.org/wp-content/uploads/2016/07/Rethinking-Vulnerability-and-Resistance-Judith-Butler.pdf

de Vries, G. (2007). What is political in sub-politics? How Aristotle might help STS. *Social Studies of Science, 37*(5), 781–809.

Duncan, M. C. (1994). The politics of women's body images and practices: Foucault, the panopticon, and shape magazine. *Journal of Sport & Social Issues, 18*(1), 48–65.

Fainaru, S. (2016, March 15). NFL acknowledges, for first time, link between football, brain disease. *ESPN*. Retrieved from http://www.espn.com/espn/otl/story/_/id/14972296/top-nfl-official-acknowledges-link-football-related-head-trauma-cte-first

Foucault, M. (1990). *The history of sexuality, volume I: An introduction*. New York, NY: Vintage.

Giardina, M. D., & Newman, J. I. (2011). What is this "physical" in physical cultural studies? *Sociology of Sport Journal, 28*(1), 36–63.

Harman, G. (2002). *Tool-being: Heidegger and the metaphysics of objects*. Chicago, IL: Open Court.

Harman, G. (2005). *Guerrilla metaphysics: Phenomenology and the carpentry of things*. Chicago, IL: Open Court.

Harman, G. (2009). *Prince of networks: Bruno Latour and metaphysics*. Melbourne, Australia: Re.Press.

Harman, G. (2011). *The quadruple object*. Winchester, England: Zero Books.

Harman, G. (2014). *Bruno Latour: Reassembling the political*. London, England: Pluto Press.

Harman, G. (2016). *Immaterialism: Objects and social theory*. Cambridge, MA: Polity Press.

Hawzen, M. G., & Newman, J. I. (2017). The gospel according to Tim Tebow: Sporting celebrity, whiteness, and the cultural politics of Christian fundamentalism in America. *Sociology of Sport Journal, 34*(1), 12–24.

Heidegger, M. (2008). *Being and time*. New York, NY: Harper. (Original work published 1927)

Husserl, E. (2001). *Logical investigations*. London, England: Routledge. (Original work published 1900)

Keilman, J. (2017, February 15). New tackling methods aim to make football safer, but proof still lacking. *Chicago Tribune*. Retrieved February 2017 from http://www.chicagotribune .com/sports/highschool/ct-football-tackling-safety-met-20150821-story.html

Laskas, L. M, & Veasay, N. (2009, September 14). Bennet Omalu, concussions, and the NFL: How one doctor changed football forever. *GQ*. Retrieved from http://www.gq.com/ story/nfl-players-brain-dementia-study-memory-concussions

Latour, B. (1988). *The Pasteurization of France*. Cambridge, MA: Harvard University Press.

Latour, B. (2007). Turning around politics: A note on Gerard de Vries' paper. *Social Studies of Science, 37*(5), 811–820.

Leder, D. (1990). *The absent body*. Chicago, IL: University of Chicago Press.

Marres, N. (2005). *No issue, no public: Democratic deficits after the displacement of politics* (Unpublished doctoral dissertation). University of Amsterdam, The Netherlands. Retrieved from http://dare.uva.nl/search?arno.record.id=165542

McKee, A. C., Stein, T. D., Nowinski, C. J., Stern, R. A., Daneshvar, D. H., Alvarez, V. E., ... Cantu, R. C. (2013). The spectrum of disease in chronic traumatic encephalopathy. *Brain: A Journal of Neurology, 136*, 43–64.

Meggyesy, D. (1971). *Out of their league*. New York, NY: Paperback Library.

Meillassoux, Q. (2008). *After finitude: An essay on the necessity of contingency*. London, England: Continuum.

Messner, M. (2002). *Taking the field: Women, men, and sports*. Minneapolis: University of Minnesota Press.

Morton, T. (2013a). *Hyperobjects: Philosophy and ecology after the end of the world*. Minneapolis: University of Minnesota Press.

Morton, T. (2013b). *Realist magic: Objects, ontology, causality*. Ann Arbor, MI: Open Humanities Press.

Newman, J. I., Giardina, M. D., & McLeod, C. M. (2016). Embodiment and reflexive body politics. In M. L. Silk, D. L. Andrews, & H. Thorpe (eds.), *Routledge handbook of physical cultural studies* (pp. 589–598). London, England: Routledge.

Omalu, B. I., DeKosky, S. T., Minster, R. L., Kamboh, M. I., Hamilton, R. L., & Wecht, C. H. (2005). Chronic traumatic encephalopathy in a National Football League player. *Neurosurgery, 57*, 128–134.

Pallone, F. (2016, May). Democratic staff report: The National Football League's attempt to influence funding decisions at the National Institutes of Health. *U.S. House of Representatives Committee on Energy and Commerce.* Retrieved from https://democrats-energycommerce .house.gov/sites/democrats.energycommerce.house.gov/files/Democratic%20Staff %20Report%20on%20NFL%20NIH%20Investigation%205.23.2016.pdf

Riddle, R. (2012, July 19). Memoirs from an NFL training camp. *Los Angeles Times.* Retrieved from http://bleacherreport.com/articles/1264662-memoirs-of-an-nfl-training-camp

Schwartz, A., Bogdanich, W., & Williams, J. (2016, March 24). N.F.L.'s flawed concussion research and ties to tobacco industry. *New York Times.* Retrieved from https://www .nytimes.com/2016/03/25/sports/football/nfl-concussion-research-tobacco.html?_r=0

Shaw, G. (1972). *Meat on the hoof: The hidden world of Texas football.* New York, NY: St. Martin's Press.

Silk, M. L., & Andrews, D. L. (2011). Toward a physical cultural studies. *Sociology of Sport Journal, 28*(1), 4–35.

Strum, S. S., & Latour, B. (1987). Redefining the social link: From baboons to humans. *Social Science Information, 26*(4), 783–802.

Theberge, N. (1991). Reflections on the body in the sociology of sport. *Quest, 43*(2), 123–134.

Weedon, G. (2015). Camaraderie reincorporated: Tough Mudder and the extended distribution of the social. *Journal of Sport & Social Issues, 39*(6), 431–454.

4 · TELOMERE BIOLOGY IN AN AGE OF PRECARITY

A "New" Materialist Experiment in a More-Than-Human Kinetics

GAVIN WEEDON

> What happens when the best biologies of the twenty-first century cannot do their job with bounded individuals plus contexts, when organisms plus environments, or genes plus whatever they need, no longer sustain the overflowing richness of biological knowledges, if they ever did? What happens when organisms plus environments can hardly be remembered for the same reasons that even Western-indebted people can no longer figure themselves as individuals and societies of individuals in human-only histories? (Haraway, 2016, pp. 30–31)

> The future health of our society is being shaped right now, and we can measure part of that future in telomere base pairs. (Blackburn & Epel, 2017, p. 327)

INTRODUCTION

This chapter explores how "new" materialist philosophy matters for scholars of the moving body. It is an experiment in asking what happens when biology and society are not together brought to bear on the body, as in the great fantasies of interdisciplinary integration, but when bodies themselves are shown to materialize the interface that makes those two worlds thinkable. What happens when moving bodies can no longer be pursued only as so many objects compelled to motion by cultural forms, biopolitical imperatives, and the desires they elicit and oblige or as biological vessels knowable only by a law unto themselves? When some of our most cherished ideas about bodies and societies, nature and culture, history and humanity are put at risk and put us at risk, do we have the analytical tools to respond?

The "new" materialisms are philosophies and research practices that respond to these questions and the trouble they invoke. Like all materialisms, these "new" ideas and approaches have gained traction in response to the idealisms and upheavals of their time. Just as Baruch Spinoza would not accept Descartes's privileging of the mind, positing instead a monistic mind-body-brain substance, an affective force always in motion, so the "new" materialisms will not countenance reductions to the human social: subjectivity, rationality, and language. In their place are relational, processual categories intended to exceed all reductions to nature or society. Just as Marx and Engels were animated by the rise of industrial capitalism in 19th-century Europe and concerned that the ideational pontifications of their "Young Hegelian" peers were not grounded enough in material social life to be worthy of revolutionary times, the "new" materialisms refuse the category of "the human" as the conceptual unit for making habitable futures in the 21st century. Attempts at expressing relationships among people, the planet, technology, and other animals follow from this conviction. My invocation of these "older" materialisms is intended to nuance the regrettable use of the word *new*, which I nonetheless retain, scare quotes always in attendance. The "new" moniker is best understood as denoting the contemporary coming-together of insights from feminist theory, eco-philosophy, poststructuralism, posthumanism, "older" materialisms, genetics, quantum physics, life sciences and neurosciences, Indigenous knowledges, Buddhism, queer theory, science studies, and cultural studies. Many of these insights are inflected with a renewed significance or at least brought together for new audiences, especially insofar as they are articulated as responses to biotechnological advances, environmental crises, animal welfare movements, and other threats and possibilities for (more-than-)human life. Rather than inventions of and for the present, the constellation in which the "new" materialisms appear in relation to these ideas is what might be said to make them "new."[1]

The "new" materialisms have grown in influence as ecological crises have become ordinary in our present. This is not hyperbole, yet hyperbole is surely part of the problem. The human impact on earth is so profound that in August 2016, the International Geological Congress moved to declare ours the age of the Anthropocene: an age in which the human-as-species wields a force equivalent to the earth's own atmosphere and geophysical systems. An "unintended experiment of humankind on its own life support systems" (Steffen, Crutzen, & McNeill, 2007, p. 614), the Anthropocene demarcates the age in which human activity has altered the planet by chemical, climate, and socioeconomic measures. Most histories tie its birth to the emergence of industrial capitalism, with the mid-20th century cited as the "great acceleration" that distinguishes this era from the Holocene (Steffen, Crutzen, & McNeill, 2007); other periodizations cite the agricultural revolution over ten thousand years ago as the inception of the changes that the Anthropocene names.

Conceptualizing the Anthropocene is contested terrain that helps map the stakes of the "new" materialisms. Yet simply distinguishing between those who are somehow "for or against" does not adequately express the "new" materialist sensibility. For example, there are those for whom the Anthropocene affirms a triumph of humanism, for whom such an age is a victory of the technocrats, another step forward in the march of progress propelled by the efficacious equation of human ingenuity plus technological development.[2] Inversely, there are those for whom that same process is seen as an unprecedented disaster for life on earth borne of human hands and brains that should alert us to the mortal danger of anthropogenic climate change. Rather than take sides, the key point to highlight is the convergence of these perspectives: Whether triumphant or foreboding, each is faithful to the anthropocentric, anthropogenic premises of the Anthropocene story. Put another way, in heralding us, the *anthropos*, as the engines of change, these perspectives are in many ways two sides of the same humanistic coin.

The "new" materialisms can be distinguished in part from these twinned perspectives by taking issue with the Anthropocene concept itself (though in this it is not alone; see Grusin, 2017). Rather than proceed from the premise of environmental triumph or catastrophe, "new" materialist philosophy seeks out the ways in which "human activity" and "human impact" are in fact the process and outcome of complex entanglements between human and nonhuman life and matter. Put another way, an explosion of interest in nonhuman life across the social sciences and humanities illustrates the embeddedness of humans-in-nature at a time when ecological problems cannot be comprehended as external to the rest. It follows that the Anthropocene must be confronted as *more than human* in both conceptualization and scale of effects for life forms of all kinds, and responses to the Anthropocene must not stoke the hubris of human exceptionalism in order to mitigate its legacy. Transcendental humanism is not the answer.[3] The "new" materialisms confront this paradox: to expose the conceited fantasy of human control over a nature that it has transcended and somehow respond ethically to the seismic task befalling humanity-as-species. As we will see, this does not mean or require a suspension of politics—a common concern with the "new" materialisms—but instead broaches politics as an inclusive, expansive question toward habitable multispecies futures.

It is therefore as a *foundational, nonanthropocentric, ecological* premise that the "new" materialisms challenge the singularity and boundedness of "the human."

For example, noting that "the notion of the human has come to be perceived as an index of a historically specific fantasy of mastery over the self, the earth, and all its many creatures," Samantha Frost (2016) embarks on a "new" materialist project to develop a politically viable theory of the human as a "biocultural creature" composed of proteins, oxygen, ideas, and other worldly substances and forces (p. 1). Donna Haraway has written extensively of becoming-with the multiplicity of critters, from bacteria and primates to mice, pigeons, and barking companions, that co-constitute and share in her own biology and help make us who "we" are. Her visions have helped manifest multispecies entanglements that continue to shape the social sciences and (post)humanities; they glimpse habitable futures in the ruins of the Anthropocene. Karen Barad (2003), inspired by Haraway, feminist theory, and quantum physicist Niels Bohr, invites those inclined toward social and political analysis to "meet the universe halfway" in trying to understand how life materializes:

> Crucial to understanding the workings of power is an understanding of the nature of power in the fullness of its materiality. To restrict power's productivity to the limited domain of the "social," for example, or to figure matter as merely an end product rather than an active factor in further materializations, is to cheat matter out of the fullness of its capacity. (p. 810)

Cheating matter, including the materiality of the human body, out of the "fullness of its capacity" by omitting that capacity from our accounts of the world risks ironically reinscribing the kinds of human exceptionalism—claims to the superiority and distinctiveness of human bodies, cultures, and labor—that characterize the Anthropocene. Apprehending the vibrant materiality of bodies, heeding their lively entanglements with other forms and forces, and considering the productive power of nonhuman matter: These are the tenets by which "new" materialisms set out to better understand the embeddedness, as well as the precarity, of life.

Returning to the experimental objective of this chapter, I have sought an object of inquiry that articulates the lessons of "new" materialist philosophy to the interests of sport, physical culture, and human movement studies. Telomeres, chromosomal caps that play a crucial role in the vitality and vulnerability of bodies, are already a hot topic in kinesiology and sport, health, and exercise science due to revelations about the effects of physical activity on telomere length (Puterman et al., 2010) and positive implications for health (offsetting disease) and longevity (cellular aging). Yet knowledge of the fascinating activity of telomeres has mostly led to orthodox prescriptions to sleep more, stay positive, exercise, and be self-aware (Blackburn & Epel, 2017) and other truisms that appeal to resilient individualism. I think there is more to learn from telomeres than well-rehearsed mantras for individual biological health. They are, as Elizabeth Blackburn suggests in my epigraph, metrics of biological vulnerability at a time when the cost of sustaining individual health is a critical social issue. What's more, they are sources of untold authority on the condition of precarity shared across *all* biological life in a moment of ecological precarity, and they tell stories that cannot be confined to biology or society, nature or culture. Such stories are already being told and heeded in the social sciences in the form of the "new biology" (Blackman, 2016) and related ventures into the complexity, plasticity, and indeterminacy of bodies that shape and are shaped by their environments in a material, biological sense (Mansfield & Guthman, 2015; Senanayake & King, 2019).

In what follows, I take telomeres seriously as the kind of dynamic, (re)generative, intra-active, material-discursive phenomena that "new" materialisms name. Following a snapshot of telomeres, telomere science, and the telomere-related insights of sport and exercise science, I set about reading telomeres *through* "new" materialist philosophy in order to better understand both. I focus first on intra-activity, or the entanglement of nature and culture, as a means of understanding the relationality of bodies as the condition of their becoming. Telomeres, I argue, illustrate the entwined concerns of biologists and sociologists as immanent to the body in motion. Critique's limits and generative possibilities are then discussed in relation to telomere science, "new" materialism, and ways of knowing that are prevalent for understanding physical culture. By way of conclusion, I speculate on a "more-than-human kinetics" informed by the lessons of telomeres and "new" materialist philosophy amid the uncertainty and precarity of the Anthropocene.

WHAT ARE TELOMERES AND WHAT DO THEY DO?

Telomeres are caps found at either end of chromosomes: the "ends of our DNA." Every human cell is composed of 23 pairs of chromosomes, and each of the total of 46 chromosomes is capped by a telomere (see figure 4.1).

Through the reiterative process of cell division, DNA sequences repeat to seek out binding proteins. The extent to which these sequences are communicated efficaciously is a complex matter that requires understanding the interplay of bodies, genes, and what for brevity is probably best termed their *ecologies*: the worlds in which telomeres—and the creatures they inhabit—flourish and perish. Suffice it to say for now that in the postgenomic era, genes open up more questions than they provide answers, as genetic scripts are influenced by many actors through the process of cell division, and not all of them play their parts obediently, reliably, or *within* the biological body (Fausto-Sterling, 2004;

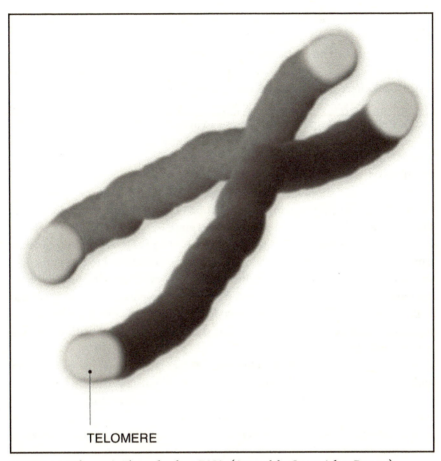

TELOMERE

FIGURE 4.1. Telomeres: The ends of our DNA. (Artwork by Stuart Adam Brown.)

Meloni, 2014). Despite its heuristic value, figure 4.1 therefore cannot capture the dynamic relationality of telomere activity. But we have to start somewhere.

Until recently, it was thought that telomeres were "junk DNA." Scientists had hypothesized that they might play a protective role for chromosomes, but the details and implications were a mystery. Elizabeth Blackburn's intervention, for which she won the Nobel Prize for Medicine in 2009, affirmed that telomeres do indeed protect chromosomes during the process of cell division. In a 1975 biochemical study of single-cell pond-dwelling creatures called *tetrahymena*, Blackburn found that telomeres are composed of repeated DNA patterns that vary significantly in length. In her 2017 book *The Telomere Effect*, coauthored with Elissa Epel, Blackburn recalls her excitement at this discovery: "The telomeres of pond scum were sending a message: There is something special here at the ends of chromosomes. Something that would turn out to be vital for the health of human cells" (pp. 46–47).[4]

Those variations in length are, it transpired, indicators of health and longevity. The longer the telomere, the greater the protective barrier to the chromosome. Once a telomere has shortened enough to expose the chromosome, the cell enters a state of senescence, and the physical and physiological signs of aging, as well as their health implications, become manifest. The body's vital systems thus become vulnerable to a host of cardiovascular and neurological diseases. Hence telomeres are at once a *protector* of the body's immune, autonomic, and cardio-vascular systems and a *metric* of healthy aging.[5]

In subsequent studies of the DNA structure of telomeres in *tetrahymena*, Blackburn observed a striking phenomenon. Under certain conditions, telo-meres will actually grow in length. This bears emphasis: Understandings of DNA hitherto held that such growth would be impossible, as DNA cannot regenerate from without. The key to this regenerative process was found to be telomerase, an enzyme that repairs telomeres through the process of cell division. Telomer-ase replicate biochemical information from DNA sequences and reproduces it as brand-new DNA. This sequence replication attracts the proteins needed for the reproduction of particular cells as well as stimulating further growth or mainte-nance of telomerase.

Far from junk DNA, then, telomeres protect the chromosomes they cap, and telomerase enzymes lengthen—and thus defend against the shortening of—telomeres. Telomeres, in turn, protect chromosomes and the creatures they inhabit from cellular senescence and the associated onset of disease. Telo-meres and telomerase, as the Latin root of their names indicate, do in fact serve a purpose.

Owing to these insights, telomeres have become objects of inquiry across the health and exercise sciences and industries for those interested in the effects of stress under the skin. Studies increasingly link telomere length to a whole host of "stressors," trauma, or experiences of adversity, variably defined. An abridged list ranges from chemical exposure, diet, sleep patterns, depression, and smoking to discrimination, disability, unemployment, physical and emotional abuse, and caregiving (cf. Blackburn & Epel, 2017; Shalev et al., 2013). With remarkable consistency, diverse forms of stress are associated with attenuated telomeres and the onset of morbidity and mortality. While the links between psychological stress and mortality are well established (Russ et al., 2012), telomeres are pro-viding researchers with a verifiable measure of the extent of one's exposure to stress and therefore an index of the multitude of stressors that affect health over the lifespan. What's more, researchers do not have exclusive access to this knowl-edge: Individuals can now pay biotechnology companies to noninvasively mea-sure their telomeres through mailing services—indeed, without leaving their homes—and purchase supplements that claim to boost telomerase (see Dumit, 2003, p. 38, for a relevant analysis of "objective self-fashioning" through biomedical knowledge). The selling point here is not just that consumers can "measure their aging" relative to chronological age but that they can respond to this information with behaviors and products intended to fuel telomerase, lengthen telomeres, and therefore resist or defer the aging process. Biomedical consumerism notwithstanding, telomeres shorten with each replication as we age—that is, each time a cell divides—and so this is not a process that can be arrested forever. Our bodies are not composed of, nor are they able to produce, an infinitesimal number of these enzymes to stave off aging indefinitely. Attempts to supplement telomerase might even have an inverse effect, resulting in the unregulated cell division associated with cancer (Zhu et al., 2016). Nonetheless, in a very real sense, telomerase have been shown to attenuate cellular aging by protecting the proclivity of the chromosome to divide and replicate.

On this basis, researchers in sport and exercise science have taken an interest in telomere biology. Exercise has been shown to help lengthen telomeres by stimulating telomerase without the need for potentially carcinogenic supplementation. For example, in studies that measure self-reported physical activity and perceived stress in humans, a positive correlation between exercise and telomere health has been identified (Cherkas et al., 2008; Werner et al., 2009). One study of the effectiveness of physical activity for mitigating against the implications of stress for disease and aging is titled and extols the "power of exercise" (Puterman et al., 2010). By measuring telomere length in "sedentary" and "physically active" individuals—conceptions of movement to which I will return—Eli Puterman and colleagues (2010) produced evidence for the "buffering potential of physical activity on the detrimental effects of stress on cellular longevity" (p. 4). LaRocca, Seals, and Pierce (2010) observed a link between leukocyte telomere length and maximal aerobic capacity in healthy human adults, suggesting that this "may well represent a novel molecular mechanism linking vigorous habitual exercise/aerobic fitness to reduced cellular senescence and preserved physiological function with aging" (p. 167). Bucking the trend, Laine and colleagues (2015) measured the telomere length of male former professional athletes and found no difference between the telomere length of these individuals and their peers in a control group in later life. Such findings jar with the assumption that participation in high-performance sport has a causal correlation with good health in later life and will thus not surprise sociologists of sport and exercise (see Theberge, 2008; Young, 1993).

Nevertheless, the message emerging from telomere science conducted by health and exercise scientists is one of empowerment: A combination of physical activity, awareness, and resilience can mitigate against the deteriorating effects of stress and life trauma on telomere length. What's more, it can even aid in the process of regenerating telomeres by stimulating telomerase enzymes. "Lengthen your telomeres, strengthen your telomerase!" might be the less-than-catchy mantra among 21st-century health promoters, for whom telomeres offer the latest link between exercise, vitality, and (im)mortality.[6]

Amalgamating contemporary research for a wider audience, Blackburn and Epel (2017) offer the *Telomere Manifesto*, comprising "the elements of telomere maintenance that we believe to be most crucial for a healthier world" (p. 327). Here are some of those elements:

- Evaluate sources of persistent, intense stress. What can you change?
- Transform a threat into a challenge appraisal.
- Be active.
- Develop a sleep ritual for more restorative and longer sleep.
- Cultivate a few good, close relationships.
- Reduce inequality.

Skeptics, we will return to these shortly.

INTRA-ACTIVITY: THE ENTANGLEMENT OF NATURE AND CULTURE

What can telomere biology tell us about "new" materialisms and the material bodies, human and otherwise, that "new" materialist philosophy foregrounds? And what can reading telomere biology and "new" materialist philosophy through each other do for understanding bodies in motion?

For emphasis, telomere research puts forward striking evidence of an interface between biology and society. It is now widely accepted that in addition to hereditary factors, "a wide range of environmental and behavioral factors also appear to affect [telomere length] and telomerase" (Shalev et al., 2013, pp. 1836–1837). Once thought to be immutable, DNA can regenerate and, under certain conditions, be regenerated by enzymes that are not cut from its cloth. Anatomy emerges neither as destiny nor as the immaculate conception of one's environment. This is a hugely significant insight in itself. The regenerative capacity of telomeres and telomerase, as well as their susceptibility to "external stressors" from chemical exposure and nutritional intake to unemployment and discrimination, challenges erstwhile assumptions about the immutability of genome functioning and indeed of what has long been bordered off as bodily nature (Alaimo, 2010). The susceptibility of telomeres to worlds beyond the biological body evidences the permeability of these borders (see Mansfield & Guthman, 2015, and Meloni, 2014, on epigenetics and the plasticity of biology in the postgenomic era).

This *entanglement of nature and culture* signals the shared ground of telomere science and "new" materialist philosophy as well as a crucial distinction between them. The distinction is this: Research on telomeres has mostly pursued the *interaction* between the preexisting ontological realms of biology and society. The variables tend to be telomere length, on the one hand, and aging, morbidity, mortality, and their known causes and conditions (such as stress), on the other. Telomere research then indexes those sources of trauma that might affect telomere length and, by implication, our vital systems. For example, unemployment is indexed as a stressor associated with shortened telomeres. This model of interaction expresses the vulnerability of our (internal) chromosomes to (external) stressors and emphasizes our ability to act on this knowledge through means of prevention and resilience.

In contrast to this interactive model, "new" materialisms refuse a distinction between nature and culture, foregrounding instead the inseparability and intrinsic relationality of naturecultures (a monistic moniker coined by Haraway). For "new" materialists, the world is not the sum of nature and culture interacting. Rather, worlds emerge from *ongoing intra-activity* (Barad, 2003, 2007). Intra-activity is the premise that individuated things do not precede their becomings: Difference is not the outcome of an encounter between two things that were hitherto alone and external to each other. Such ontological separations bequeath the tasks of either representing the material world from the outside or mining its truths through rational detachment, both of which require the knowing subject to crack the puzzle.[7] In robbing matter of its own capacity to divide and conquer, each of these perspectives—objectivism and representationalism—harbor humanistic underpinnings. By contrast, intra-activity is the ongoing patterning of difference, in which biological, evolutionary, and human histories are always already entwined. A generative tool for refusing dualisms and thinking their relations anew, intra-activity is perhaps the touchstone concept for "new" materialisms.

Intra-activity takes us down the rabbit (worm?) hole of quantum physics to places we cannot go in this chapter with any hope of return. Especially relevant, though, is quantum theory's reworking of how space, time, and matter are configured as empirically observable phenomena that implicate the observer not deterministically or incidentally but as part of the ongoing configuring of the world under study. This in itself is not a radically "new" insight: Poststructuralism, several schools of feminism and queer theory, and those working in science studies have demonstrated many times over that the apparatus assembled to elicit knowledge—including its author—is implicated in that knowledge (Latour & Woolger, 1979; Law, 2004). The difference cited by "new" materialists is that in many such approaches, it remains humans, especially those endowed with the authority to make knowledge (such as scientists), who encode the world with meanings. Through this epistemological alchemy, matter is endowed with life by objectivists and constructivists alike.

In a posthumanist twist, matter in Barad's "new" materialism produces its own partiality, its own separations or "cuts" in the world, without the necessity of a knowing subject to enact the crucial divisions or to stitch them back together. All matter, including the biological matter of the body, is the outcome of relational materializations, of intra-activity (Barad, 2003, 2007). And by this performative logic, all relata (singular phenomena) conceal the relations through which they are co-constituted. *The matter of the body is always moving; movement is not the effect of the world on the body's biology but the condition of body-in-world.* On this, Haraway (2004) is as eloquent as she has been prescient:

> Biology is relentlessly historical, all the way down. There is no border where evolution ends, where genes stop and environments take up, where culture rules and nature submits, or vice versa. Instead, there are turtles upon turtles of naturecultures, all the way down. (p. 2)[8]

Understanding biology as "relentlessly historical" and moving bodies through the monistic neologism "naturecultures" signals the intra-active complexity of telomeres. Far from little bits of biological nature—inert, passive, and awaiting human representation—telomeres are phenomena that attenuate with the world's ongoing intra-activity. All bodies emerge intra-actively, not as individual units that we must will to know and reunite, but as flows within flows, worlds within worlds (see Moore, 2015). Like the cells they are constantly becoming-with, telomeres do not comprise an inside or outside, despite the persistence of our biological imaginations (Keller, 2010). Frost (2016) moves to correct this atomistic imaginary by describing the cell as less like an atom or molecular substance than a roiling cityscape of constant activity that undoes and remakes itself. Where does a city stop and start? Do its borders define and contain its qualities? The point is that *an intra-active understanding of telomeres recognizes their capacity to lengthen and shorten not as an aberration of nature brought about by events in society but as their precarious condition of becoming.*

The plasticity of telomeres, while a crucial insight, is therefore better apprehended as part of a dynamic, historical, biocultural condition: a "new" materialist understanding of bodies in motion informed by developments in the life sciences, most notably epigenetics. In a sense, then, telomeres at once inform and perform "new" materialist philosophy. The knowable interface of biology and society is materialized by telomeres. That is their ontological force.

CRITIQUE, TELOMERE SCIENCE, AND PRECARITY

Perhaps, then, telomere science need not be teleological. Rather than necessarily culminate in ethopolitical tactics to mitigate the wearing down of individual human bodies (cf. Rose, 2006), apprehending telomeres as dynamic constellations of bodies in motion might sensitize us to our shared condition of precarity. Anna Tsing (2015) identifies this condition as one of economic and ecological ruination. More than merely representing those ruins, however, she sets about asking how life is even possible amid them—specifically, how mushrooms are able to flourish in inhospitable terrain. Here she is on the precarity of life in the Anthropocene:

> Precarity is the condition of being vulnerable to others. Unpredictable encounters transform us; we are not in control, even of ourselves. Unable to rely on a stable structure of community, we are thrown into shifting assemblages, which remake us as well as our others. We can't rely on the status quo; everything is in flux, including our ability to survive. Thinking through precarity changes social analysis. A precarious world is a world without teleology. Indeterminacy, the unplanned nature of time, is frightening, but thinking through precarity makes it evident that indeterminacy also makes life possible. (Tsing, 2015, p. 20)

Instead of understanding telomere biology as providing techniques for individual telomere maintenance and telomerase stimulation, telomeres make evident that the flourishing and compromising of all bodies is the precarious condition of life: They make our shared vulnerability an observable, biophysical phenomenon. While historical materialisms since Engels's study of Victorian Manchester have observed the enervating effects of industrial capitalism on the health of its subjects, telomeres tell stories of precarity that get under the skin. To hear them, a "new" materialist apprehension of matter must be complemented by an affirmative sensibility toward critique.

The "new" materialisms are envisioned as "creative alternatives to critique" (Van der Tuin & Dolphijn, 2012, p. 14), or a combination of "critique with creativity" (p. 22).[9] These alternatives take varied forms that thread through transversal genealogies of "new" materialist philosophy and other bodies of knowledge, substance, and force. The aim is to avoid the limits of critique as well as the misplaced confidence of teleological triumph or defeat and come to terms with the indeterminacy of living in the Anthropocene, not knowing if there will be a happy ending (cf. Tsing, 2015).

What are the limits of critique? According to Bruno Latour (2005), whose actor-network theory remains influential and controversial in relation to the "new" materialisms, critique has a fatalistic propensity in too much of the social sciences:

> To insist that behind all the various issues there exists the overarching presence of the same system, the same empire, the same totality, has always struck me as an extreme case of masochism, a perverted way to look for a sure defeat while enjoying the bittersweet feeling of superior political correctness. (p. 252)

Let's experiment. Hands up those readers who scanned through the indexes of stressors and life traumas shown to affect telomere length and were quick to aggregate this disaggregated list into systems, structures, and the like. For those committed to critical, contextual, and predominantly cultural approaches for understanding human movement, the interventions prescribed by telomere scientists surely warrant skepticism. The jump from telomere biology to individualized telomere maintenance habits omits the structures, systems, and patterns of health inequality that make many of these habits necessary and their causes perpetual. Individualized recommendations for healthy telomeres such as those listed in Blackburn and Epel's (2017) *Telomere Manifesto* recapitulate the shortcomings of all such individualized prescriptions for health and longevity. They deal in disaggregated symptoms and mostly in individualized solutions.

However, there are problems with the swift move to structures and systems. While poverty and unemployment, for example, are easily aggregated as symptoms of capitalism, aggregation threatens to ossify the dynamism of actually existing capitalism (Mann, 2013) into a "social structure" whereby explanation risks becoming a shorthand substitute for analysis. More pressing for living in an age of precarity is critique's political expediency or lack thereof. For Latour (2004), the deployment of skepticism by climate change deniers and their ilk rings the death knell for critique as an efficacious strategy for political action. Casting doubt on matters of fact is too easy, too easily weaponized, and too easily serves to affirm and defend one's preexisting beliefs and interests. Contemporary instances of the weaponizing of critique for political gain hardly need mention.

Critique is not only flawed on account of its being hijacked for egregious ends, though. The problem named by Latour, who is not alone in taking this perspective, is that critique as a default position of the social sciences and humanities indulges a kind of masochistic humanism, whereby the culpability for all the world's problems can be traced back to human hands and brains. The structures we have created are revealed as insurmountable—unless "we" overcome them. Ironically, the prisons we humans make for ourselves and our habitat are also evidence of our scheming brilliance—self-flagellation and congratulation, all in one gesture. Humanism lurks behind this fatalistic logic. The problem of the dominant mode of critique as an appeal to preexisting social structure is not that it fails to name a truth or correctly identify a villain but its humanistic fatalism and political inexpediency.

Actor-network theory, Latour's own weapon of choice, nonetheless stands accused of perpetuating the limits of interactive thinking in its pursuit of a world of objects in action. Object-oriented ontology (Harman, 2009), for example, has been mobilized as an attack on postmodernism and critical theory that empties out the generative promise of the "new" materialisms. There is also skepticism surrounding Latour's persistent disengagement with matters of subjectivity and his suspicion of critical theory as if it had played no part in the genealogy of "new" materialist philosophy and posthumanist theory—which of course it has (Braidotti, 2013).

A further critique of critique moves from concerns with the limits of identity politics into some of the most cherished concepts in the social sciences and humanities. Jasbir Puar's writing on intersectionality and assemblage theory is exemplary. Of critical responses to Deleuze and Guattari's notion of the assemblage, Puar (2012) has described how such a diffuse concept "seems to inspire doubt about its political 'applicability,' while intersectionality seems to hold fast as a successful tool for political and scholarly transformation" (p. 50). "Part of the assumption," she continues, "is that representation, and its recognized subjects, is the dominant, primary, or most efficacious platform of political intervention, while a Deleuzian nonrepresentational, non-subject-oriented politics is deemed impossible" (p. 50). This same assumption informs suspicions that the "new" materialisms are "a-political," which is often another way of saying that they offer no guarantee that a preexisting object of critique—whiteness, sexism, corruption—will still be the villain at the end of the analysis.

Telomere science has suggested that human individuals who inhabit iniquitous categories of identity are especially (and for anyone who has been paying attention, predictably) susceptible to telomere erosion (Russ et al., 2012). In light of the longstanding failure of biology and sociology to embrace their shared terrain, this is a victory, a great leap forward. Nevertheless, Puar's critique is powerful. When we operationalize the intersections of race, gender, class, and nation in order to oppose them, are we and others more able to escape them—or do we become objects of our critique? In my own work, I have found that critiques of representation, for instance, seldom take me where I want to go; more frequently they confine me to the grids of intelligibility—categories of identity, systems of oppression—that animate trouble in the first instance. Intersectionality weaves a sticky web. Barad (in Van der Tuin and Dolphijn, 2012) is bold on this: "I am not interested in critique. In my opinion, critique is over-rated, over-emphasized, and over-utilized, to the detriment of feminism" (p. 49). Puar and Barad differ

from Latour in their efforts to retain a focus on subjectivity and discursive matters, though they seek to do so without deploying subject positions as primary
or given analytic frames.

The price of generativity—of seeing how things might be otherwise—might
be to put at risk the explanations we hold dear: the matters and structures that
much of social and cultural theory has identified as culpable. As Elizabeth
St. Pierre (2016) puts it, we must "postpone the leap to application" (p. 113) so
as to remain curious and hopeful as to how things could be otherwise. Recalling Tsing (2015), precarity and indeterminacy are not aberrations of life in the
Anthropocene. They are its conditions of becoming.

Thus a "new" materialist response to critical readings of telomere science
might be a request by way of a question: Can we not start there? Can our orientation to novelty not be one of skepticism, even if we are right in our mis-
givings? Can we resist the temptation to rush to explanation, even when our
critical faculties are flaring? For if the concern is with health and fitness in the
most encompassing, historical sense—a planetary understanding of the vitality
and vulnerability of the earth and all its critters, people included—then there
are surely more far-reaching and consequential lessons to learn from bodies of
research such as telomere science and the bodies from which they emerge. How
might apprehending telomeres as "onto-epistemological" phenomena (Barad,
2007)—matter that is immanently entangled with the meanings and knowledges made with and about it—broaden our focus beyond projects of individual
biomedical enhancement and health promotion without taking up residence in
the fatalistic sanctity of "structural" explanations (cf. Massumi, 2002)? I think
this is a foremost challenge for studying human movement practices.

A MORE-THAN-HUMAN KINETICS?

> Humanism takes for granted much of what needs to be investigated.
>
> (Barad, 2007, p. 27)

If the human can no longer be accepted by biologists or sociologists as a primary or coherent conceptual unit, if it ever could, then the proclaimed art and
science of the moving body has much to ponder, many movements to follow.[10]
"We have never been individuals" proclaim Gilbert, Sapp, and Tauber (2012)
in their article on the symbiotic condition of all biological life, and sociologists
surely concur. And yet many of those committed to the sociocultural study of
the moving body in sports and physical cultural practices have delimited the
human body in motion as a shared empirical focus at the very moment that
the *problem* of human centrality has resumed its place atop the agenda of the

social sciences and humanities (Silk & Andrews, 2011).[11] What is the fate of this corporeal focus in the age of the Anthropocene?

That anthropocentrism is among the touchstones that unite the struggles of our times casts an uneasy light on all of the human sciences. Disciplines and fields of study that center the human in their approaches to understanding biology, history, and society are found complicit in the crises of the Anthropocene. Some have gone so far as to contend that fields that limit their analyses to the human bios, psyche, and social are unfit for the purpose: ethically comprised by their humanism even in otherwise critical, emancipatory projects (see Braidotti, 2013). As such, the study of sport, physical culture, and the moving body—each of which center the human and human movement in their own ways—is far from alone in these challenges.

Surely interdisciplinarity is not the answer. The harmonious unification of (sub)disciplinary knowledge is the institutional dream of the university to pool its epistemological resources in a triumphant liberal consensus. Insofar as this is limited to an epistemological exercise, it represents the apogee of human exceptionalism, the most hubristic attempt at transcendental humanism this author can envisage. Insofar as it is limited to an ontological exercise, insights across the sciences affirming the entanglement of matter and meaning—from Foucault to quantum physics—are neglected. Alan Ingham (1997) warned that such subdisciplinary integrations would be insufficient substitutes for a philosophy of human movement that emerges from a common empirical focus in kinesiology and sport and exercise science. And yet that common empirical focus—the human body in motion—also needs rethinking lest it take for granted what needs to be investigated (cf. Barad, 2007).

The "new" materialisms open up the key concepts of physical culture. The "physical" is expansively conceived to include the materiality of the world of and beyond biology. Culture is appreciated as something held in common across all life, from plants to ant colonies—something that can never omit or externalize nature. The "new" materialisms apprehend the body as an ontological force that will not be contained by any disciplinary canon or by any attempts to draw boundaries that distinguish it as human. These same ideas trouble the boundedness of the body foregrounded by kinesiology and, I would add, refocus it on the *kinetics* of human movement. The dynamism of telomeres is, after all, a kind of movement, a kinetics that is at once genetic and more than human, environmental and corporeal, inside and outside of the body. Hence my calling on Barad's (2003, 2007) notion of "onto-epistemology" to denote that it is the boundaries of the cell and the body and the knowledge-making boundaries of human movement studies that telomeres refuse and remake. Movement in this context is not just something bodies do or do not do, a matter of sedentary or active physicality. It is their condition of intra-active materialization in constituent, constitutive ecologies.

A final word for telomere science, then, which presents a connective unity of health, physical activity, and long life in a 21st-century world of genetics: In the age of the Anthropocene, it comes up short. So do critiques of these innovations that themselves rush to critique and its humanisms that tie us in misleadingly neat and tidy knots. The knowledge that we can lengthen our telomeres does not render telomere maintenance tactics the solution. "We" are many (Gilbert, Sapp, & Tauber, 2012). Surely the ethical question is not, *How can I live longer?* Nor, of course, can it be, How might I appear younger? Might the ethical question be, *What is worth shortening our telomeres for?* This question does not indulge the quest for immortality and more importantly implies that individual health enhancement or longevity always comes at a cost. Better still, might a corollary be, *Whose telomeres are shortened on account of my persistence and flourishing?* This question conveys that the proteins that form a protective sheath for telomeres and protect us from cellular aging and disease *come from somewhere*: a multispecies sensibility to health and movement as more than human that taps an ecological imperative (see Samantha King, chapter 8 in this volume).

Telomeres materialize the vulnerability of bodies as a shared condition, as precarity held in common. We are not in control even as the translation of telomere science into the popular press puts the health-conscious reader front and center. Far from an aberration, the fluctuation of telomeres is the condition of our flourishing, "the centre of the systematicity we seek" (Tsing, 2015, p. 20). Beyond fantasies of individual biological sovereignty and resilience as the basis of our collective persistence, there might be something in the state of precarity conveyed by telomeres, shared among and beyond our species' kin, that can help in the making of habitable futures.

NOTES

Thank you to Paige Patchin, Sammi King, and Gui Artioli for their lively conversations and editorial guidance in the making of this chapter.

1. There is no definitive genealogy of "new" materialist thought. Citing Spinoza and Marx makes this cursory timeline susceptible to accusations of Eurocentrism if read as portraying some definitive timeline. As well as familiarity, I cite these materialisms on the basis of the responsiveness to the hegemonic intellectual and sociopolitical trends of their historical moments rather than as irrefutable touchstones in any "new" materialist tradition.

2. In the social sciences, this view is often represented through ecological modernization theory which argues for the compatibility of economic development and environmental policy, or "green capitalism." See Wilson and Millington (2013) for a discussion of how theory manifests in sport.

3. There are versions of posthumanist theory, research and art that celebrate the augmentation of the human body, implying its preexisting purity and thus—ironically or not—exacerbating the problems of anthropocentrism. This is the narrative of transcendence, of triumph over the real and threatened contamination of what makes us human. For "new" materialists and related fields, this notion of the body augmented belies the insights of life and social scientists that "we have never been human" (Haraway, 2008), "never been individuals" (Gilbert, Sapp, & Tauber, 2012), and so must do better when thinking the embeddedness and connectivity of life and matter.

4. There is scope here for a discussion—indeed, a study—of Blackburn and telomeres as engaged in a dance of coproduction, akin to Bruno Latour's (1988) account of Louis Pasteur and his discovery of microbes. I note this to guard against the advancement of a narrative that emphasizes Blackburn's singular genius—one that, in doing so, might reduce telomeres to the role of passive matter waiting to be discovered (see also Mialet, 2012). "New" materialisms are skeptical of any narrative that foregrounds human mastery over a world of things awaiting discovery. Any such story would need to include the critters, the kin, and the rest of the cast that have made Blackburn and telomeres famous. *Tetrahymena*, single-cell organisms that live in freshwaters on a diet of food and sex, offered a simplistic biology for Blackburn to study their telomeres in 1975. Mice enlisted in telomere science have coproduced knowledge about aging and cellular health under experimental conditions that are not possible—or not ethical—to replicate in humans. Agency, in materialist philosophies old and new, is never the attribute of an individual, human or otherwise.

5. "Telomere syndromes" are instructive for understanding the functions of telomeres. These syndromes designate inherent disorders characterized by extremely short telomeres. Signs of aging, such as gray hairs, fragile skin, and wrinkles around the eyes and mouth manifest earlier in the lifespans of those born with telomere disorders. In the bloodstreams of these individuals, stem cells cannot effectively repair telomeres, which erode away. Exposed by the shortened telomeres, chromosomes then become susceptible to signs and effects of aging and disease. Stem cell transplants are now possible for the one in a million people born with telomere syndromes. The insight here is that telomere syndrome is not a radical difference from the expected function of telomeres. It is an accelerated version, a sped-up aging process that makes lucid the role of telomeres in that process (Armanios & Blackburn, 2012). Moreover, it indicates a link between telomere biology and ageing.

6. To put this another way, telomere research extends the vitality attributed to physical activity by health promoters of centuries past—the doctors, healers, educators, and political figures who have espoused the many virtues of physical activity since the early 19th century (Whorton, 1982; Vertinsky, 2017)—into the 21st-century world of genetics.

7. The habit of beginning with separate things and comprehending their coming-together is historically sedimented in natural and social science and has at its heart a thinking human—detached somehow such that reflection on the world's interactions is possible. Barad (2007) follows this habit to Democritus, who, in positing a world of singular phenomena, comprehensible by tracing each "thing" down to its smallest unit, bequeathed modern Western thought an "atomist physics" that manifests in Descartes, in Newton, and in the worlds made in the image of their thought (Barad, 2007; see also Deleuze & Guattari, 1987).

8. I came across this passage when reading Katie King's blog (http://playingcatscradle .blogspot.co.uk) on cat's cradle and other games for knowledge making, to whom and to which I am grateful for inspiration in writing this chapter.

9. At the same time, they are themselves objects of criticism. These approaches are accused of being "a-political" insofar as they do not explicitly engage and support the most visible,

established sites of struggle for critical social scientists and social justice activists. They are accused of not adequately representing the schools of theory and philosophy that preceded and informed them—hence the scare quotes that flank the word "new." And as theories affiliated with science studies have become accustomed to, they are accused of admitting politics into every crevice of life such that discernment of abuses of power and means of intervention are inhibited.

10. "Along with a de-exceptionalizing of human bodies, multiple forms of matter can be bodies—bodies of water, cities, institutions, and so on" (Puar, 2012, p. 57).

11. The challenges to the prevailing orthodoxies for studying the moving body are well known: Kinesiology and sport and exercise science departments are largely divided into subdisciplinary areas of study that more or less emphasize the inside (nature) or outside (society) of human movement (cf. Andrews, 2008; Vertinsky, 2009). These specialized fields of expertise therein gravitate toward the human body, which in turn serves as an empirical touchstone for otherwise eclectic, somewhat disparate bodies of knowledge.

REFERENCES

Alaimo, S. (2010). *Bodily natures: Science, environment, and the material self*. Bloomington: Indiana University Press.

Andrews, D. (2008). Kinesiology's inconvenient truth and the physical cultural studies imperative. *Quest, 60*, 46–63.

Armanios, M., & Blackburn, E. H. (2012). The telomere syndromes. *Nature Reviews Genetics, 13*, 693–704.

Barad, K. (2003). Posthumanist performativity: Toward an understanding of how matter comes to matter. *Signs: Journal of Women in Culture and Society, 28*(3), 801–831.

Barad, K. (2007). *Meeting the universe halfway: Quantum physics and the entanglement of matter and meaning*. Durham, NC: Duke University Press.

Blackburn, E., & Epel, E. (2017). *The telomere effect: A revolutionary approach to living younger, healthier, longer*. London, England: Orion Spring.

Blackman, L. (2016). The new biologies: Epigenetics, the microbiome and immunities. *Body and Society, 22*(4), 3–18.

Braidotti, R. (2013). *The posthuman*. Cambridge, MA: Polity Press.

Cherkas, L. F., Hunkin, J. L., Kato, B. S., Richards, J. B., Gardner, J. P., Surdulescu, G. L., . . . Aviv, A. (2008). The association between physical activity in leisure time and leukocyte telomere length. *Archives of Internal Medicine, 168*, 154–158.

Deleuze, G., & Guattari, F. (1987). *A thousand plateaus: Capitalism and schizophrenia* (B. Massumi, trans.). Minneapolis: University of Minnesota Press.

Dumit, J. (2003). Is it me or my brain? Depression and neuroscientific facts. *Journal of Medical Humanities, 24*(1–2), 35–47.

Fausto-Sterling, A. (2004). Refashioning race: DNA and the politics of health care. *Differences: A Journal of Feminist Cultural Studies, 15*(3), 1–37.

Frost, S. (2016). *Biocultural creatures: Towards a new theory of the human*. Durham, NC: Duke University Press.

Gilbert, S. F., Sapp, A., & Tauer, A. I. (2012). A symbiotic view of life: We have never been individuals. *Quarterly Review of Biology, 87*(4), 325–341.

Grusin, R. (ed.). (2017). *Anthropocene feminism*. Minneapolis: University of Minnesota Press.

Haraway, D. (2004). Introduction, a kinship of feminist figurations. In *The Haraway reader* (pp. 1–6). New York, NY: Routledge.

Haraway, D. (2008). *When species meet*. Minneapolis: University of Minnesota Press.

Haraway, D. (2016). *Staying with the trouble: Making kin in the Chthulucene*. Durham, NC: Duke University Press.

Harman, G. (2009). *Prince of networks: Bruno Latour and metaphysics*. Melbourne, Australia: Re.Press.

Ingham, A. G. (1997). Toward a department of physical cultural studies and an end to tribal warfare. In J. M. Fernandez-Balboa (ed.), *Critical postmodernism in human movement, physical education, and sport* (pp. 157–182). Albany: State University of New York Press.

Keller, E. F. (2010). *The mirage of a space between nature and nurture*. Durham, NC: Duke University Press.

Laine, M. K., Eriksson, J. G., Kujala, U. M., Raj, R., Kaprio, J., Bäckmand, M., . . . Sarna, S. (2015). Effect of intensive exercise in early adult life on telomere length in later life in men. *Journal of Sports Science and Medicine, 14*, 239–245.

LaRocca, T. J., Seals, D. R., & Pierce, G. L. (2010). Leukocyte telomere length is preserved with aging in endurance exercise-trained adults and related to maximal aerobic capacity. *Mechanisms of Aging and Development, 131*, 165–167.

Latour, B. (2004). Why has critique run out of steam? From matters of fact to matters of concern. *Critical Inquiry, 30*(2), 225–248.

Latour, B. (2005). *Reassembling the social: An introduction to actor-network theory*. Oxford, England: Oxford University Press.

Latour, B., & Woolgar, S. (1979). *Laboratory life: The construction of scientific facts*. London, England: Sage.

Law, J. (2004). *After method: Mess in social science research*. London, England: Routledge.

Mann, G. (2013). *Disassembly required: A field guide to actually existing capitalism*. Oakland, CA: AK Press.

Mansfield, B., & Guthman, J. (2015). Epigenetic life: Biological plasticity, abnormality, and new configurations of race and reproduction. *Cultural Geographies, 22*(1), 3–20.

Massumi, B. (2002). *Parables for the virtual: Movement, affect, sensation*. Durham, NC: Duke University Press.

Meloni, M. (2014). Biology without biologism: Social theory in a postgenomic age. *Sociology, 48*(4), 731–746.

Mialet, M. (2012). *Hawking incorporated: Stephen Hawking and the anthropology of the knowing subject*. Chicago, IL: University of Chicago Press.

Moore, J. (2015). *Capitalism in the web of life: Ecology and the accumulation of capital*. London, England: Verso.

Puar, J. (2012). I'd rather be a cyborg than a goddess: Becoming-intersectional in assemblage theory. *PhiloSOPHIA, 2*(1), 49–66.

Puterman, E., Lin, J., Blackburn, E. H., O'Donovan, A., Adler, N. E., & Epel, E. S. (2010). The power of exercise: Buffering the effect of chronic stress on telomere length. *PLoS One, 5*, e10837.

Rose, N. (2006). *The politics of life itself: Biomedicine, power and subjectivity in the twenty-first century*. Princeton, NJ: Princeton University Press.

Russ, T. C., Stamatakis, E., Hamer, M., Starr, J. M., Kivimaki, M., & Batty, G. D. (2012). Association between psychological distress and mortality: Individual participant pooled analysis of 10 prospective cohort studies. *BMJ, 345*, e4933.

Senanayake, N., & King, B. (2019). Health-environment futures: Complexity, uncertainty and bodies. *Progress in Human Geography, 43*(4), 711–728.

Shalev, I., Entringer, S., Wadhwa, P. D., Wolkowitz, O. W., Puterman, E., Lin, J., & Epel, E. S. (2013). Stress and telomere biology: A lifespan perspective. *Psychoneuroendocrinology, 38,* 1835–1842.

Silk, M. L., & Andrews, D. L. (2011). Towards a physical cultural studies. *Sociology of Sport Journal, 28*(1), 4–35.

St. Pierre, E. A. (2016). The empirical and the new empiricisms. *Cultural Studies ↔ Critical Methodologies, 16*(2), 111–124.

Steffen, W., Crutzen, P. J., & McNeill, J. R. (2007). The Anthropocene: Are humans now overwhelming the great forces of nature? *Ambio, 36*(8), 614–621.

Theberge, N. (2008). "Just a normal bad part of what I do": Elite athletes' accounts of the relationship between health and sport. *Sociology of Sport Journal, 25,* 206–222.

Tsing, A. L. (2015). *The mushroom at the end of the world: On the possibility of life in capitalist ruins.* Princeton, NJ: Princeton University Press.

Van der Tuin, I., & Dolphijn, R. (2012). *New materialism: Interview and cartographies.* London, England: Open Humanities Press.

Vertinsky, P. (2009). Mind the gap (or mending it): Qualitative research and interdisciplinarity in kinesiology. *Quest, 61,* 39–51.

Vertinsky, P. (2017). A question of the head and the heart: From physical education to kinesiology in the gymnasium and the laboratory. *Kinesiology Review, 6*(2), 140–152.

Werner, C., Furster, T., Widmann, T., Poss, J., Roggia, C., Hanhoun, M., . . . Laufs, U. (2009). Physical exercise prevents cellular senescence in circulating leukocytes and in the vessel wall. *Circulation, 120,* 2438–2447.

Whorton, J. (1982). *Crusaders for fitness: The history of American health reformers.* Princeton, NJ: Princeton University Press.

Wilson, B., & Millington, B. (2013). Sport, ecological modernization, and the environment. In D. L. Andrews & B. Carrington (eds.), *A companion to sport* (pp. 129–142). Oxford, England: John Wiley & Sons.

Young, K. (1993). Violence, risk, and liability in male sports culture. *Sociology of Sport Journal, 10,* 373–396.

Zhu, X., Han, W., Xue, W., Zou, Y., Xie, C., Du, J., & Jin, G. (2016). The association between telomere length and cancer risk in population studies. *Scientific Reports, 6,* 22243.

PART II BODY TECHNOLOGIES

5 · BIG BODIES, BIG DATA

Unpacking the FitnessGram Black Box

SHANNON JETTE, KATELYN ESMONDE,
DAVID L. ANDREWS, AND CAROLYN PLUIM

We begin with a story:

Joe's mother saw she had an email from his teacher. The subject line read "FitnessGram Results." She clicked on the envelope icon, and when the email opened, she saw the familiar computer-generated report she had been receiving once a semester since Joe was in the fifth grade. Scanning the left-hand side of the report, she was pleased to see that Joe remained in the "Healthy Fitness Zone" (HFZ) for aerobic capacity and was no longer in the "Needs Improvement—Health Risk" (NI-HR) zone on the body fat test but rather just the "Needs Improvement" (NI) zone. Looking at the personalized messages on the right-hand side, generated by an algorithmic assessment of Joe's response to three survey questions about his activity level, she was annoyed by the suggestion that although he was in the HFZ, Joe "could be more physically active." But the message about Joe's body fat confirmed her suspicions: She would need to further limit his screen time. Joe's teacher, as well as information on the Fitness-Gram website, had informed her that FitnessGram was based on scientific research connecting the test results to Joe's risk of future disease. However, she also understood that the FitnessGram testing process was intended to teach Joe how to test himself and plan a personal program throughout his life; at the same time, it could be used by school administrators to determine students' levels of activity and fitness and guide the development of curriculum in the fight against childhood obesity.[1]

This is by no means the only story that could be told about a parent's understanding of and interaction with FitnessGram's testing and reporting procedure. It is, however, a narrative that elucidates how FitnessGram—a web-based fitness assessment tool being rolled out in tens of thousands of schools across the United States and reportedly reaching 10 million children[2]—is intended to operate: as a soft form of power functioning through the techniques of persuasion, enticement, and education to passively compel Joe, with the help and guidance of his mother and teachers, to engage in certain health practices (Jette, Bhaghat, & Andrews, 2016; Pluim & Gard, 2016, 2018). It is not an inconsequential story, since the scale and scope of FitnessGram's reach renders it a persistent element in the lives of millions of individual children and in shaping popular, programmatic, and policy approaches to children as a collective category.

According to marketing material published by the Cooper Institute for Aerobics Research (herein called the Cooper Institute)—the nonprofit research and education organization that designed and markets the FitnessGram testing system—FitnessGram is used in all 50 states and 14 different countries, with over 22 million students tested in 2012 alone (Corbin, Lambdin, Mahar, Roberts, & Pangrazi, 2013). Positioned by some public health advocates as an important tool in the fight against childhood obesity, in 2013 FitnessGram also became the assessment instrument of the newly revamped Presidential Youth Fitness Program (PYFP), an initiative of the United States' President's Council on Fitness, Sports and Nutrition (PCFSN), and part of former First Lady Michelle Obama's Let's Move! antiobesity initiative (Jette et al., 2016). Because of the FitnessGram-PYFP partnership, the Cooper Institute benefits from $10 million in funding from General Mills (the processed food manufacturer) to implement the PYFP program (and FitnessGram testing) in schools nationwide. The Cooper Institute also partners with NFL Charities, which, starting in 2009, has funded FitnessGram assessment in more than 1,120 schools nationwide (35 schools for each of the 32 teams) as part of a project to gather health-related data and improve the FitnessGram protocol (Plowman et al., 2013).

Given the pronounced sociocultural ubiquity and influence of FitnessGram, our aim within this chapter is to offer a preliminary explication and exploration of the complex network of actors (human and nonhuman) responsible for making FitnessGram the dominant physical education (PE) fitness testing protocol in the United States. While there are a multiplicity of theoretical positions and approaches within physical cultural studies (PCS), our analysis is informed by a particular form of radical contextualism (originating from British cultural studies) whereby the physical cultural practice of body quantification is understood in relation to broader social, political, economic, and technological forces, relations, and contingencies (Andrews, 2002, 2008). Central to PCS is the theory/method of "articulation," which refers to a way of understanding both the "contingent connection of different elements that, when connected in a particular way, form a specific unity" (Slack & Wise, 2015, p. 152) or articulation and how these elements do or do not come to be articulated at specific conjunctures or historical moments (Hall, 1986). However, while the (in)active human body remains central to our analysis, we are also concerned with tracing the role of nonhuman actants in the network of relations that constitute FitnessGram. In doing so, we expand the PCS project beyond its heretofore largely anthropocentric boundaries (see Esmonde & Jette, 2018; Millington & Wilson, 2017).[3] This necessitates drawing on insights from the field of science and technology studies (STS) and, in particular, questioning the very idea that the spheres of the human and nonhuman, the "social" and "natural," are separate and that agency is the domain of humans (Latour, 1999a, 1999b; Law, 1999). By collapsing dualisms and disentangling agency from its "human" association with intentionality, an STS perspective helps us examine the role that nonhuman entities (e.g., measuring devices, statistical calculations) play in the creation of FitnessGram—albeit in conjunction with human actors and their subjective decisions. Through our analysis, we render visible the often shrouded processes of data appropriation and aggregation through which individual bodies are measured, collective bodies become represented, and ab/normalizing bodies become constituted.

We begin with a brief overview of existing scholarship that critically examines the history of fitness testing and body quantification in PE in Western society, with a focus on scholars investigating the emergence of digital technology in PE, including FitnessGram. We explain how we build on this existing body of research by attending to the materiality of FitnessGram and further outline the theoretical insights guiding our investigation. This includes a discussion of how the material-semiotic approach of STS scholars and the PCS theory/ method of articulation might be brought into conversation, although we acknowledge that our attempt to do so is partial and exploratory. Turning to our empirical analysis, we first examine how, despite dissension among exercise scientists and physical education pedagogues regarding the best mode of fitness testing, FitnessGram eventually sedimented its position as the most widely used fitness testing tool in schools, albeit one contingent on a particular network of relations. We then seek to excavate the multiple steps undertaken to create FitnessGram in its current form: a series of field tests (e.g., skin-fold measure, 1-mile run, curl ups, push-ups) that quantify body composition, aerobic capacity, and musculoskeletal health such that performance on a particular field test (e.g., skin-fold measure) is used to evaluate whether a student is at risk of future disease. In doing so, we demonstrate how FitnessGram works to constitute the world in ways that often escape detection when being used, producing what Law (2004) calls "collateral realities" whereby particular ways of thinking about the world (and bodies) are often unknowingly privileged over others. In our conclusion, we further reflect on the complementarities and tensions that arise from bringing together PCS approaches with those from the field of STS.

SITUATING FITNESSGRAM

The measurement of bodies, health, and fitness has long been a part of the PE curriculum in the United States, tied to wider nation-building projects and their related political objectives. For example, as PE developed into an organized professional entity in the late 19th century, anthropometry (the scientific study of body measurement) became an important component of the curriculum, legitimizing the field of PE as it drew on the science of the day (and eugenic ideology) in establishing an ideal form to which students' bodies could be compared, abnormalities identified, and a program of corrective physical training implemented (Couturier, 2005; Park, 2006). Half a century later, Cold War anxieties over the fitness of American schoolchildren and, in particular, the "softness" of young (white) males led to the creation of the President's Council on Youth Fitness and the subsequent implementation of the first formal U.S. government system of physical fitness testing focused on strength, agility, and power—the characteristics needed to protect the United States from Communist invasion (Bowers & Hunt, 2011; Montez de Oca, 2009).

Over the past decade, yet another trend has emerged: the increased use of digital technology (e.g., pedometers, heart rate monitors) and software-based data management systems in PE settings (Gard, 2014). The goal of these digital PE initiatives has been to engage an increasingly sedentary—and digitally oriented—youth population while simultaneously allowing for the measurement, collection, and assessment of body and fitness-related data that can be used in programmatic and policy settings to fight childhood obesity. However, given the history of fitness testing as a cultural practice informed by—and simultaneously mobilizing—ideologies underpinning the operation of modern power and power relations, the incorporation of digital measurement technologies in PE has become the focus of a small but growing number of critical health and pedagogy researchers. Among the issues raised by these scholars are the manner in which tracking devices make surveillance an intrinsic part of people's lives (Lupton, 2015); how and by whom data generated by such technologies are used (i.e., issues of dataveillance; Gard, 2014; Lupton, 2015); and relatedly, the involvement of commercial and corporate interests that drive the inclusion of digital technologies in PE (see Gard, 2014; Powell & Gard, 2015; Vander Schee & Boyles, 2010). Adding to this are questions about the changing role of the PE teacher—as students are now able and oftentimes encouraged to "self-teach" using digital technology (Gard, 2014). Within this digital PE environment, teachers are replaced by computer programmers and statisticians who create the algorithms that constitute the digital "teachers" and, in doing so, substantiate a new kind of expert whose claim to authority rests on the technical capacity to shape education through data collection, aggregation, and analysis (Williamson, 2015).

Scholars also question how digital self-tracking devices might shape children's understandings of and relationships to their bodies and physical movement as well as their enjoyment of PE classes, especially if they are pressured to use such devices (Drew & Gore, 2016; Lupton, 2015, 2016). These concerns about digital technology add to longstanding questions about fitness testing in PE more generally (digital or not), including the reliability and validity of fitness testing (Armstrong, 1989; Rice & Howell, 2000; Safrit, 1990), the impact of fitness testing on students' enjoyment of and motivation for physical activity (Cale & Harris, 2009; Tinning, 2000), and the influence of measurement technologies such as BMI testing on students' body image and self-esteem (Evans & Colls, 2009).

Against the backdrop of the growing usage of digital technology in PE, FitnessGram is a particularly important site of study because, as noted by Pluim and Gard (2018), it is "physical education's first major experimentation with 'big data'" (p. 263; see also Gard, 2014). The term *big data* refers to the use of digital technology to generate, manipulate, and interpret large digital data sets (boyd & Crawford, 2012). While seen as offering exciting opportunities in the realm of education, big-data uses of educational information also raise important technological, policy, and ethical questions that include privacy of individual student information, anonymity of students in the aggregate, data accuracy, potential other uses of the data, and data quality and curation (Bertot, Gorham, Jaeger, Sarin, & Choi, 2014). Scholars have also called into question the authority given to "big data" as a generator of empirical fact, which obfuscates the subjective decision-making that informs the creation of the algorithms that drive data collection, aggregation, and analysis, and ultimately produce particular bodily "norms" to which individuals are compelled to conform (Williamson, 2015).

Although FitnessGram has received relatively little attention from critical health and pedagogy scholars (see Gard & Pluim, 2016), examinations to date (Jette et al., 2016; Pluim & Gard, 2018) have pointed to the potential of the FitnessGram testing system to operate as a form of biopower, a concept put forth by Michel Foucault (2003) to describe the working of power in modern society.[4] Discursively constituted yet materially enacted, biopower (power over life) developed around two complementary poles: the disciplining of the individual body through the governing regimes of various social institutions, including the prison, school, factory, and the gymnasium; and the regulation of the population through a range of techniques, such as the collection of demographic information, life expectancy, and birth and death rates that then informed population-level policies and interventions.

Pluim and Gard (2018) make the case that the mass collection of Fitness-Gram data produces new forms of knowledge pertaining to normative targets and modes of intervention at the societal level while also pressing young people (at the individual level) to understand their personalized FitnessGram data and seek to work on themselves in order to achieve optimal health. Elsewhere (Pluim & Gard, 2016), they outline how FitnessGram implicates parents in its biopolitical project by encouraging the intensification of parental surveillance in the lives of their children (as we tried to capture in our introductory story of Joe and his mother). Jette and colleagues (2016) situate the FitnessGram testing apparatus as one of the biotechniques mobilized within, and through, the Let's Move! antiobesity campaign that itself operates as a neoliberal biopolitical strategy in the fight against childhood obesity. The FitnessGram system can thus be understood as part of a series of technologies and related interventions developed in order to "know" and govern the population—in this case, children's physical activity and health-related practices in the context of the obesity epidemic.

MATERIALIZING FITNESSGRAM

While these critical analyses of digital technologies in the field of PE bring to the fore important questions and critiques, what has largely been omitted to date is a close investigation of the materiality of the FitnessGram testing system itself. To this end, we examine how FitnessGram is constituted via a network of human and nonhuman actors that, despite its fluid (contingent) nature, has achieved a certain durability as evidenced by its widespread implementation in schools across the United States. To aid such an investigation, we draw on the work of STS scholars who challenge modernist assumptions about the separation between nature and the social—or "out there" and "in here"—by mapping the work (both material and subjective) performed to produce scientific statements subsequently accepted as fact (Latour & Woolgar, 1986; Latour, 1987, 1993, 1999b). In their ethnography of a research laboratory, Latour and Woolgar (1986; see also Latour, 1987) demonstrate a process whereby an inscription device (a system that typically includes but is not reducible to a machine) converts a material substance (e.g., an extract taken from a rat) into a figure or diagram that is then taken to represent or "stand in" for the original material substance and eventually becomes accepted as scientific "fact." They make the crucial point that at this stage, the process of producing the figure or diagram (the inscription)—with the various human and nonhuman interactions involved in this process—disappears into the background such that the materiality (and subjectivity) of the process is rendered invisible. Thus the inscription device has constructed a particular reality described as an objective entity. Once an inscription device is standardized, or becomes a routinized "black box," it becomes a building block on which other scientific knowledge is based, making it more difficult (and expensive) to introduce other possibilities or realities.

The laboratory ethnography of Latour and Woolgar (1986; see also Latour, 1987), along with additional case studies by their STS colleagues (see Law, 2009), has developed into what has been termed *actor-network-theory* (ANT), or a material-semiotic approach[5] that treats everything in the social and natural worlds as a "continuously generated effect of the webs of relations within which they are located" (Law, 2009, p. 141). With this commitment to relationality, the distinction between human and nonhuman loses its analytic relevance, as do other dualisms (e.g., meaning and materiality, macro and micro, nature and culture). Following this logic, all actors of the network (both human and nonhuman) are able to "act," but only by virtue of their connection to other actors in the network of which they are constituents—a process of relational materialism (Abrahamsson, Bertoni, Mol, & Martín, 2015).[6] Moreover, the erosion of dualisms means that the (sociological) tendency to attribute phenomena to foundational factors such as the "social" is no longer valid; rather, social explanations (patriarchy, nation-state, class) are understood as the effects of relational materiality as opposed to explanatory foundations (Latour, 2005; Law, 2009). According to Law (2009), "This is not to say that they are not real—they may indeed be made real in practice—but they offer no framework for explanation" (p. 147).

Yet despite this antifoundational approach, the focus of material semiotics is to understand how it is that entities (such as FitnessGram) are performed into relations that are relatively stable and stay in place, especially since these relations are not necessarily fixed and could connect otherwise (Latour, 1999a; Slack & Wise, 2015). Here, the theory/method of articulation helpfully draws attention to the unequal distribution of agency and power in networks and how it is that some are more enduring than others—something that a material-semiotic approach does not tend to facilitate (Slack & Wise, 2015). As Slack and Wise (2015) remind us, "The work of articulation has effects: it empowers possibilities and disempowers others; legitimizes some identities and delegitimizes others; makes some things happen and other things not" (p. 155). Thus, in this chapter, we draw on the STS concepts of relational materiality and the black box as we attempt to unpack the network of actors (human and nonhuman) that constitute FitnessGram as a black-boxed technology. To do so, we explore the material (and subjective) work performed to transform the material body of a student such as Joe into a representation designated by FitnessGram health risk classifications. Recognizing that FitnessGram as a network is inherently unstable and fluid, we also use the theory/method of articulation to trace the context that constitutes the conditions of possibility for FitnessGram to form a durable network at this particular moment (Law, 2009; Slack & Wise, 2015).

According to Latour (1987), when studying the network of human and non-human actors that constitute scientific facts, it is much simpler to explore the object of interest before the black box closes. He explains that "the impossible task of opening the black box is made feasible (if not easy) by moving in time and space until one finds the controversial topic on which scientists . . . are busy at work" (p. 4). With this in mind, we begin with a controversy surrounding fitness testing in the 1980s (before the "black box" was closed) in which FitnessGram was a central player. We share how FitnessGram, along with a health-based definition of fitness and the criterion-referenced standard, emerged as the tentative victor of this controversy to eventually become, according to the Cooper Institute, "the most widely used youth physical fitness assessment, education and reporting tool in the world" (http://www.fitnessgram.net/teachers.asp#a01). We then peek into the black box of FitnessGram, illustrating how it is built on numerous other black boxes that, following Latour (1987), make it all the more difficult to call it into question—as long as it produces the desired effect.

It is important to note that the FitnessGram assessment tool has been refined over time based on scientific research concerning the reliability and validity of the tests (Welk, 2006; Welk, Going, Morrow, & Meredith, 2011) as well as advances in data visualization and analytics and the rise of big data (the Cooper Institute is currently marketing FitnessGram 10.0). When significant changes have been made, the black box has been "opened" to allow researchers and developers to tinker with the elements (i.e., increase the scientific validity and reliability of the test) before trying to close it again. In the process of such tinkering, they have left a trail for us to follow in the form of special issues in peer-reviewed journals (e.g., one in 2006 in the *Journal of Physical Activity and Health*, 3[S2], and one in 2011 in the *American Journal of Preventive Medicine*, 41[4, S2]) and various test administration manuals. Press releases and marketing materials also inform our analysis.

THE CONTROVERSY: FROM MEASURING ATHLETICISM TO ASSESSING HEALTH

In Chicago in October 1986, a public hearing was held that focused on youth fitness testing in the United States (Jackson, 2006; Plowman et al., 2013). The hearing was in response to a controversy that had been bubbling up since the 1970s, when a faction of exercise scientists challenged the testing protocols being used by leading physical educators at that time, particularly the President's Youth Fitness Test (PYFT) with its national fitness award program (Franks,

Morrow, & Plowman, 1988; Jackson, 2006). The PYFT was criticized because it "did not include a formal definition of physical fitness, had a limited physiological basis, and was not developed through sufficient measurement procedures" (Franks et al., 1988, p. 188; see also Jackson, 2006). Also at issue was its focus on athleticism as opposed to health and, relatedly, the evaluation of student performance based on normative referenced standards whereby a student's performance was ranked relative to the performance of a reference group of peers. Instead, the exercise scientists advocated the evaluation of youth fitness using health-based, criterion-referenced standards that would indicate if a student was at risk for future disease and that were viewed as more objective and meaningful than the percentile-based norms (Jackson, 2006; Plowman et al., 2013).

Key players at the hearing were members of the American Alliance for Health, Physical Education, Recreation and Dance (AAHPERD; now called the Society of Health and Physical Educators, or SHAPE), the President's Council on Physical Fitness and Sports (now called the President's Council on Fitness, Sports and Nutrition), and the Cooper Institute for Aerobics Research, commonly known as the Cooper Institute and creator of FitnessGram, which at that time was not its own test battery but rather a system for reporting results of state and national fitness testing to children and parents (the name FitnessGram played off the idea of a "telegram"; Plowman et al., 2006; Plowman et al., 2013). At the time of the hearing, AAHPERD had two tests—the Health Related Physical Fitness Test (HRPFT) and the Youth Fitness Test (YFT)—both of which were percentile referenced and the latter of which was affiliated with the PCPFS and its award system. The Cooper Institute provided a FitnessGram report card for both of them. The goal of the hearing was to find a test-and-award system that could be endorsed by all key players. Upon the close of the hearing, it appeared that a compromise had been reached: Software engineers would program two tests with norms that were criterion referenced for the AAHPERD awards *and* percentile referenced for the Presidential Award (Plowman et al., 2013).

Ultimately, however, the agreement was never ratified, and the compromise was unsuccessful. A spokesperson for the Campbell Soup Company, the main funder of the Cooper Institute (and FitnessGram) at that time, explained that "reprogramming more than one test is not an economic reality" (Sterling, personal communication [October 30, 1986], cited by Plowman et al., 2013, p. 1.6). A decision had been made by the Cooper Institute—under the advisement of a group of exercise scientists—to utilize a single test consisting of five items: the one-mile run, modified sit-ups, sit-and-reach, pull-up/flexed-arm hang, body composition, and an optional shuttle run (or pacer test). FitnessGram would no longer be just a reporting system but consist of its own test battery.

Utilizing the theory/method of articulation (King, 2005; Slack, 1996), it is possible to represent the aggregate and determining relations that account for this push toward a health-based, criterion-referenced standard testing system in the late 1980s that, now firmly embedded in the FitnessGram black box, appears commonsensical. While space limitations preclude a more expansive discussion, crucial considerations include a wider societal shift whereby the welfare state was being replaced by neoliberal ideology and its related policies that placed a premium on individual responsibility for one's health and well-being via lifestyle practices such as physical activity (Howell & Ingham, 2001; Ingham, 1985) as well as the inclusion of private and nonprofit entities such as the Campbell Soup Company and the Cooper Institute, respectively, in the realm of (physical) education policy and pedagogy (Ball, 2012). Also occurring during the 1980s was the increasing "scientization" of the field of PE such that objective knowledge generated in the research lab came to be privileged over practice/pedagogy (Andrews, Silk, Francombe, & Bush, 2013; Ingham, 1997). Finally, this was a time in which physical activity epidemiology as a field began to achieve recognition within the field of public health, bolstered by longitudinal health and physical activity studies such as that carried out at the Cooper Institute for Aerobics Research (Dishman, Heath, & Lee, 2012). These entities thus sutured together a network of relations that provided the context for a push to rearticulate the practice of fitness testing in the United States and the creation of what is now the FitnessGram black box.[7]

Yet despite these connections, the Cooper Institute's FitnessGram test was not an immediate success. While AAHPERD and the President's Council were both invited to adopt the new test and program, the President's Council continued to maintain its percentile-based awards system (the PYFT). It was not until 2012 (nearly 25 years later) that it would adopt FitnessGram and its criterion-referenced standards, shortly after President Obama renamed it the President's Council on Fitness, Sports and Nutrition and rebranded the PYFT as the Presidential Youth Fitness Program, which became part of Michelle Obama's Let's Move! antiobesity initiative (Jette et al., 2016). AAHPERD also declined the FitnessGram battery in favor of its existing percentile-based systems until 1993, when a strategic partnership between AAHPERD, the Cooper Institute, and new FitnessGram funder Prudential Insurance was formed (Plowman et al., 2013).

Regardless of the initial lack of interest by AAHPERD and the President's Council to adopt a new FitnessGram test battery, following the 1986 public hearing, the Cooper Institute began to move forward with the finalization of its test methodology, manual writing, and refinement of the awards program, contacting experts to form an advisory committee to contribute to the effort (Sterling, personal correspondence, cited in Plowman et al., 2013, chap. 1, p. 6). The FitnessGram Advisory Committee, later renamed the FitnessGram Scientific Advisory Board, was formed and consisted of prominent members of the exercise science community. The board, as one committee member recollected, was "committed to health-related fitness based on research evidence that would dictate the test items and program" (Plowman et al., 2013, chap. 1, p. 6), and FitnessGram's health-based, criterion-referenced standards were reportedly the first for youth fitness put into widespread national and international use (pp. 1–10). In order to create an evidence-based fitness test program, the researchers needed to come up with a system that would move from the material bodies of youth to a representation of these bodies and their health captured in the FitnessGram report card (see figure 5.1). Or, from an STS perspective, an inscription device had to be assembled to move from "matter" to "representation." This was a multistep process that we attempt to unpack in what follows, illustrating how this device can be considered a black box given that the process of producing the representation (i.e., the inscription) has mostly disappeared into the background such that the materiality (and subjectivity) of the process is invisible when presented to the recipients of this testing procedure, such as Joe and his mother.

UNPACKING THE FITNESSGRAM BLACK BOX

Within the remainder of this chapter, we offer a preliminary sketch of the FitnessGram black box, further complicated by our finding that FitnessGram is itself built on numerous black boxes, which also require further unpacking. In figure 5.2, we have mapped the translations—or the relations and associations—performed to move from "matter" (the child's body) to "representation" (the FitnessGram report; see Law, 1999).[8] When discussing the criterion-referenced standards, we focus on the fitness component of body composition in order to make our discussion of the hybrid network more manageable.

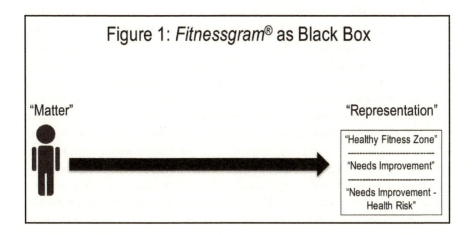

Figure 1: *Fitnessgram®* as Black Box

"Matter"   "Representation"

"Healthy Fitness Zone"

"Needs Improvement"

"Needs Improvement - Health Risk"

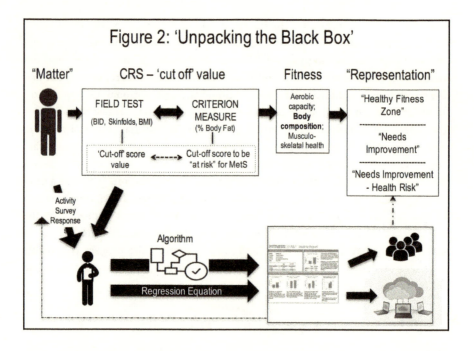

Figure 2: 'Unpacking the Black Box'

In order to create FitnessGram, a number of subjective decisions had to be made. First, the advisory board had to define the components of the construct "health-related physical fitness." They identified cardiovascular endurance, body composition, and musculoskeletal health as the key elements that would reduce the risk of diseases caused by a sedentary lifestyle. This combination of three factors was taken from the "classic definition" provided by exercise scientists in 1985 in the *Public Health Report* journal, although a range of other definitions were also options (Eisenmann, Welk, Morrow, & Corbin, 2013). Perhaps most significant, however, was the creation of the criterion-referenced standards or cutoff scores for each of the fitness components, which, as indicated in figure 5.2 (and captured by the box labeled CRS—"cutoff value"), entailed a number of steps. To begin, the researchers had to identify a criterion measure for each of the components (e.g., body fatness for body composition). Following this, the advisory board looked to epidemiological data to identify the minimal level of each criterion measure consistent with good health—in other words, the "single value that separates those at health risk from those who are at less risk" (Morrow, Zhu, & Mahar, 2013, chap. 4, p. 3). For body composition, this was the value of percentage body fat above which one would be "at risk" of metabolic syndrome. Significantly, chronic disease epidemiology (also known as risk-factor epidemiology) has been critiqued by epidemiologists as constituting a black box given the focus on inputs (i.e., the counts of diseased and nondiseased classified by exposure status) and outputs (i.e., the relative risk estimate) without examining what is happening inside the box—in other words, without understanding the mechanisms shaping associations that are observed (see Weed, 1998). Moreover, in using epidemiological methods, population-level risk statistics (calculated for adults) are then interpreted as individual-level risk factors and, in the case of FitnessGram, applied to the individual bodies of children. In other words, they do the work of translating some children's bodies into the category of normal and others into the category of abnormal. This direct translation of population-level data to individuals is problematic, as risk from an exposure is a quantitative concept that is calculable over a study population, whereas in the clinical context, risk from an exposure to an individual patient cannot be calculated (see Weir, 2006, p. 65).[9]

Next, a field test was identified that could be used as a proxy for each criterion measure. Again using body composition as an example, FitnessGram creators identified three field tests to translate a child's body measurements into a number that represents future risk of metabolic syndrome—namely, bioelectric impedance devices (BID), skin fold measures using calipers, and body mass index (BMI). All three of these nonhuman entities can be considered black boxes. The BID is a quintessential black box that sends a small electrical current through a segment of the body and reports back percent body fat. The skin-fold measures and the accompanying regression equations used to translate to percent body fat also constitute a black box that needs to be further unpacked. Finally, BMI is a black box whereby the input of one's height and weight produces a categorization of underweight, normal, overweight, or obese. While the black box of BMI is being pried open by critics (see Gard & Wright, 2005), it is used rather unproblematically in the FitnessGram test.[10]

In the final step of the creation of the CRS, the experts used a range of statistical procedures to identify the cutoff score for each field test that "validly separates" those at risk from those not at risk (Morrow et al., 2013, chap. 4, p. 4), thus connecting the field test back to the criterion measure cut-off, a point to which we return below. The field test measurement tools (e.g., BID, skin-fold calipers, BMI)—which come into direct contact with the student's body to translate it into a figure representing body fat—thus generate a number that, based on the cutoff determined by the researchers, becomes a proxy for the criterion measure cutoff and ultimately determines the health risk categorization the student receives. This was initially represented in the FitnessGram report by a single cutoff deemed "acceptable" for each component, although advances in statistical methodology eventually facilitated the identification of two separate thresholds and the current zones of HFZ, NI, and NI-HR for body composition and aerobic capacity and HFZ and NI on the musculoskeletal component (Meredith, Lambdin, Roberts, Welk, & Morrow, 2013, chap. 9, p. 2; Welk et al., 2011).

Thus all three of the field tests play an important role in the FitnessGram network, as these material entities are delegated the task of translating an aspect of a child's body into a number that quantifies future risk and prescribes tasks back to humans (i.e., the need to notify a parent of his or her child's risk; the need for the child to engage in physical activity and other health related behaviors).

The field tests, criterion measures, and cut-offs are not the only actors at play in the move from matter to representation, however. The FitnessGram system has utilized computer technology since its inception (see Plowman et al., 2013), but as FitnessGram has developed into an online digital platform, teachers can now input the field test data collected in PE classes into the FitnessGram system, delegating to the computer the work of translating the field test data into categorizations related to the three components. Children's responses about their activity levels can also be entered into the FitnessGram system that, via an algorithm, generates personalized comments on the report card. Algorithms, too, have been characterized as black boxes in that data are entered and a value or decision produced with only the creator of the algorithm knowing the choices, opinions, and assumptions embedded in the mathematical formula. As O'Neill (2016) explains, algorithms have the potential to become "Weapons of Math Destruction" (WMD) depending on the assumptions and ideologies embedded within. The PE class's aggregated information generated by the computer software can then delegate particular tasks to human recipients: Teachers and administrators can use it to shape PE curriculum, and it can also be fed back to the Cooper Institute to aid further software development. A feedback loop is thus created, also leading back to the child to prescribe certain health-related actions and (potentially) shape his or her subjectivity and physical activity practices.

The criterion-referenced standards, which are the purported strength of the FitnessGram testing system, therefore consist of a range of entities (human and nonhuman) that are connected in what seems to be a durable network. However, as we have attempted to illustrate, various entities in this network can themselves be considered black boxes. Latour (1987) suggests that such a stacking of black boxes serves the purpose of creating a scenario in which one who wishes to dissent a "fact" is "confronted with facts so old and unanimously accepted that in order to go on doubting he or she will be left alone . . . isolated from the community, or maybe, still more awful, sent to an asylum!" (p. 59). Such a scenario seems to be the case with FitnessGram: On the FitnessGram website, the testing system is presented as objective and scientific, created by a scientific advisory board that developed "criterion-based standards, carefully established for each age and gender" (http://www.fitnessgram.net/teachers.asp#ao1). In the end, we are left with a seemingly direct and obvious line between the student's body and his or her FitnessGram classification without a view of the fundamentally social and subjective decisions made to generate the labels attached to the child's aerobic capacity, body composition, and flexibility.

CONCLUSION

In this chapter, we have begun to trace the work done by scientists—in conjunction with a range of technical devices and calculations—to constitute Fitness-Gram as an inscription device that translates the "matter" of a child's body to a representation of that body (see Latour & Woolgar, 1986). We have attempted to peek into the inscription device that is FitnessGram in an attempt to make evident the numerous relations of human and nonhuman actors involved in this process that we argue have been lost from sight as FitnessGram has achieved a degree of stability (for now) that has made it the dominant fitness assessment tool in the United States. However, much work remains given the complex network of actors (human and nonhuman) at play in the FitnessGram case study. For instance, further examination of the nonhuman actors is required, including the technologies used in the field tests, such a skin-fold measurement (the calipers or BID would be nonhuman actors), and the work performed to translate body fat into a particular risk categorization when used by children in the classroom. Also needed is further research into how teachers and administrators understand and use the FitnessGram testing system as well as how children (and their parents) make sense of the testing process and the risk categorizations that they receive. Another pressing question relates to privacy issues and how the data generated through FitnessGram testing is (or is not) being protected. However, we have taken a step toward complicating the FitnessGram black box, and its constituent black boxes, and problematizing its intended goal of "encouraging" children (with the help of parents and teachers) to change their health related practices and objectives based on the data generated via the FitnessGram testing system.

We have also brought together PCS approaches to articulation with insights from the field of STS and, in particular, a material-semiotic approach. Arguably, both approaches are relational in nature. A material-semiotic approach seeks to trace the network of relations (human and nonhuman) that constitute a phenomenon or the "social" (i.e., the social is an effect of the relations as opposed to a causal explanation). The theory/method of articulation similarly seeks to map the context of a phenomenon or, in other words, attempts to arrive at the context as opposed to using it as an explanatory backdrop. However, they are also complementary in that each approach brings a perspective that the other lacks: Articulation helps us understand the durability of the network by drawing attention to power relations that allow for a particular articulation (or group of articulations) to "hold"; a material-semiotic approach allows us to move beyond the anthropocentric tendency of (physical) cultural studies by attending to the role of human *and* nonhuman actors in constituting FitnessGram and "facts" about children's bodies. However, their combination also presents a challenge in that it is easy to slip into a pattern of using articulation to attend to the macro power relations that constitute the context of the FitnessGram network (thus treating "power" as something of a black box) and a material-semiotic approach to explore the more micro elements of FitnessGram (i.e., the material minutiae that generate the social). Although in doing so we risk reproducing the ontological distinctions and their resultant dualisms that STS scholars have attempted to erode, we feel that the overall effect is a pushing of boundaries (of both PCS and STS) as well as a challenge to other scholars to further explore the tensions and complementarities of these theoretical approaches.

NOTES

1. This story is based on information from the *Fitnessgram/Activitygram: Reference Guide* (4th ed.) by Plowman and Meredith (2013) as well as "Parents as Pawns in Fitnessgram's War on Obesity" by Pluim and Gard (2016a).
2. These numbers are provided in the promotional material on the Cooper Institute's website: http://www.cooperinstitute.org/fitnessgram.
3. As Millington and Wilson (2016) argue, PCS scholarship has tended to privilege radically contextualized analyses of active or sporting bodies as opposed to "flattening" physical cultural contexts so as to account for the associations between active human bodies and active nonhuman entities and possibly expanding what else counts as "physical" in PCS-oriented research.
4. Although new materialist scholars approach with interest the developments in the natural and life sciences that are providing new ways of understanding matter, agency, and vitality, their analyses are typically sensitized toward contemporary modes of power within which this knowledge (especially biological data) is produced, disseminated, and oftentimes translated into biocapital (Coole & Frost, 2010). As such, Foucault's (2003) notion of biopower remains central to their analyses. Moreover, and as argued by Lemke (2015, p. 5), Foucault's (2007) notion of a "government of things" facilitates a consideration of "the interrelatedness and entanglements of men and things, the natural and the artificial, the physical and the moral," thus illustrating that Foucault's scholarship can move beyond anthropocentric analyses, although we have not sought to develop it in such a manner in this chapter.
5. Material semiotics takes the semiotic insight that words give each other meaning and extends it to all material entities (human and nonhuman) with the effect that entities in a network of relations give each other being or "enact" each other (Law, 1999).
6. In taking this relational view of materiality, agency is disentangled from its "human" association with intentionality; rather, an entity counts as an actor if it makes a perceptible difference, and within the relationally linked web, actors (human and nonhuman) do exactly this, bringing other entities into being while also themselves being constituted within the web (Knappett & Malafouris, 2008). Crucially, however, and as discussed here, the network of material relations can be forgotten as agency is ascribed to one particular actor in the network, often human.

7. While the articulation of these entities form what Law (2009) calls a "strategically durable configuration" (p. 148) or, in other words, the context that has allowed FitnessGram to become the dominant fitness testing tool in the United States, we recognize that the articulations outlined here can be considered black box(es) in that we have not unpacked the (human and nonhuman) networks from which they arise.

8. While a nonhuman entity can be the starting point or "matter" when studying an inscription device, because FitnessGram produces a numerical reading (representation) of children's bodies, the matter with which we begin is human (i.e., the child's body).

9. Notably, a major critique of the FitnessGram CRS, acknowledged by the creators, is that they apply population-level data for adults to individual youths (particularly in the early versions) given the lack of epidemiological data on youths.

10. It does, however, warrant a footnote on the FitnessGram report card warning about its potential inaccuracy.

REFERENCES

Abrahamsson, S., Bertoni, F., Mol, A., & Martín, R. I. (2015). Living with omega-3: New materialism and enduring concerns. *Environment and Planning D: Society and Space, 33*(1), 4–19.

Andrews, D. L. (2002). Coming to terms with cultural studies. *Journal of Sport & Social Issues, 26*(1), 110–117.

Andrews, D. L. (2008). Kinesiology's inconvenient truth and the physical cultural studies imperative. *Quest, 60*(1), 45–62.

Andrews, D. L., Silk, M., Francombe, J., & Bush, A. (2013). McKinesiology. *Review of Education, Pedagogy, and Cultural Studies, 35*(5), 335–356.

Armstrong, N. (1989). Is fitness testing either valid or useful? *British Journal of Physical Education, 20*, 66–67.

Ball, S. J. (2012). *Global education inc: New policy networks and the neo-liberal imaginary.* New York, NY: Routledge.

Bertot, J. C., Gorham, U., Jaeger, P. T., Sarin, L. C., & Choi, H. (2014). Big data, open government, and e-government: Issues, policies, and recommendations. *Information Polity, 19*(1–2), 5–16.

Bowers, M. T., & Hunt, T. M. (2011). The President's Council on Physical Fitness and the systematisation of children's play in America. *International Journal of the History of Sport, 28*(11), 1496–1511.

boyd, d., & Crawford, K. (2012). Critical questions for big data: Provocations for a cultural, technological, and scholarly phenomenon. *Information, Communication & Society, 15*(5), 662–679.

Cale, L., & Harris, J. (2009). Fitness testing in physical education: A misdirected effort in promoting healthy lifestyles and physical activity? *Physical Education and Sport Pedagogy, 14*(1), 89–108.

Coole, D., & Frost, S. (eds.). (2010). *New materialisms: Ontology, agency, and politics.* Durham, NC: Duke University Press.

Corbin, C. B., Lambdin, D. D., Mahar, M. T., Roberts, G., & Pangrazi, R. P. (2013). Why test? Effective use of fitness and activity assessments. In S. A. Plowman & M. D. Meredith (eds.), *FitnessGram/Activitygram reference guide* (4th ed.) (pp. Internet Resource). Dallas, TX: Cooper Institute, 2-1-2-21.

Couturier, L. E. (2005). The influence of the eugenics movement on physical education in the United States. *Sport History Review, 36,* 21–42.

Dishman, R., Heath, G., & Lee, I. (2012). *Physical activity epidemiology* (2nd ed.). Champaign, IL: Human Kinetics.

Drew, D. L., & Gore, J. M. (2016). Measuring up? The discursive construction of student sub-jectivities in the Global Children's Challenge. *Sport, Education and Society, 21*(3), 374–395.

Eisenmann, J., Welk, G. J., Morrow, J., & Corbin, C. (2013). Health benefits of physical activity and fitness in youth. In S. A. Plowman & M. D. Meredith (eds.), *FitnessGram/Activitygram reference guide* (4th ed.) (pp. Internet Resource). Dallas, TX: Cooper Institute, 3-2–3-13.

Esmonde, K., & Jette, S. (2018). Fatness, fitness, and feminism in the built environment: Bringing together physical cultural studies and sociomaterialisms, to study the "obeso-genic environment." *Sociology of Sport Journal, 35*(1), 39–48.

Evans, B., & Colls, R. (2009). Measuring fatness, governing bodies: The spatialities of the body mass index (BMI) in anti-obesity politics. *Antipode, 41*(5), 1051–1083.

Foucault, M. (2003). *"Society must be defended": Lectures at the Collège de France, 1975–1976.* M. Bertani & A. Fontana (eds.) (D. Macey, trans.). New York, NY: Picador.

Foucault, M. (2007). *Security, territory, population: Lectures at the Colle`ge de France, 1977–78.* New York, NY: Palgrave.

Franks, B. D., Morrow, J. R., & Plowman, S. A. (1988). Youth fitness testing: Validation, plan-ning, and politics. *Quest, 40*(3), 187–199.

Gard, M. (2014). eHPE: A history of the future. *Sport, Education and Society, 19*(6), 827–845.

Gard, M., & Pluim, C. (2016). Where is the critical health education scholarship in the United States? An examination of the FitnessGram. *Sport Education and Society.* Advance online publication. doi:10.1080/13573322.2016.121371.

Gard, M., & Wright, J. (2005). *The obesity epidemic.* New York, NY: Routledge.

Hall, S. (1986). The problem of ideology: Marxism without guarantees. *Journal of Communi-cation Inquiry, 10*(2), 28–44.

Howell, J., & Ingham, A. (2001). From social problem to personal issue: The language of life-style. *Cultural Studies, 15*(2), 326–351.

Ingham, A. G. (1985). From public issue to personal trouble: Well-being and the fiscal crisis of the state. *Sociology of Sport Journal, 2*(1), 43–55.

Ingham, A. G. (1997). Toward a department of physical cultural studies and an end to tribal warfare. In J. M. Fernandez-Balboa (ed.), *Critical postmodernism in human movement, physi-cal education, and sport* (pp. 157–182). Albany: State University of New York Press.

Jackson, A. S. (2006). The evolution and validity of health-related fitness. *Quest, 58*(1), 160–175.

Jette, S., Bhagat, K., & Andrews, D. L. (2016). Governing the child-citizen: "Let's Move!" as national biopedagogy. *Sport, Education and Society, 21*(8), 1109–1126.

King, S. J. (2005). Methodological contingencies in sports studies. In D. L. Andrews, M. L. Silk, & D. S. Mason (eds.), *Qualitative methods in sports studies* (pp. 21–38). New York, NY: Berg.

Knappett, C., & Malafouris, L. (2008). Material and nonhuman agency: An introduction. In C. Knappett & L. Malafouris (eds.), *Material agency: Towards a non-anthropocentric approach* (pp. ix–xix). Berlin, Germany: Springer.

Latour, B. (1987). *Science in action: How to follow scientists and engineers through society.* Cam-bridge, MA: Harvard University Press.

Latour, B. (1993). *We have never been modern.* Cambridge, MA: Harvard University Press.

Latour, B. (1999a). On recalling ANT. In J. Law & J. Hassard (eds.), *Actor network theory and after* (pp. 15–25). Oxford, England: Blackwell.

Latour, B. (1999b). *Pandora's hope: Essays on the reality of science studies.* Cambridge, MA: Harvard University Press.

Latour, B. (2005). *Reassembling the social: An introduction to actor-network theory.* New York, NY: Oxford University Press.

Latour, B., & Woolgar, S. (1986). *Laboratory life: The construction of scientific knowledge.* Princeton, NJ: Princeton University Press.

Law, J. (1999). After ANT: Complexity, naming and topology. In J. Law & J. Hassard (eds.), *Actor network theory and after* (pp. 1–14). Oxford, England: Blackwell.

Law, J. (2004). *After method: Mess in social science research.* New York, NY: Routledge.

Law, J. (2009). Actor network theory and material semiotics. In B. Turner (ed.), *The new Blackwell companion to social theory* (pp. 141–158). Malden, MA: Wiley-Blackwell.

Lemke, T. (2015). New materialisms: Foucault and the "government of things." *Theory, Culture & Society, 32*(4), 3–25.

Lupton, D. (2015). Data assemblages, sentient schools and digitised health and physical education (response to Gard). *Sport, Education and Society, 20*(1), 122–132.

Lupton, D. (2016). The diverse domains of quantified selves: Self-tracking modes and dataveillance. *Economy and Society, 45*(1), 101–122.

Meredith, M. D., Lambdin, D. D., Roberts, G., Welk, G. J., & Morrow, J. (2013). Interpreting FitnessGram and Activitygram reports. In S. A. Plowman & M. D. Meredith (eds.), *FitnessGram/Activitygram reference guide* (4th ed.) (pp. Internet Resource). Dallas, TX: Cooper Institute, 9-1–9-18.

Millington, B., & Wilson, B. (2017). Contested terrain and terrain that contests: Donald Trump, golf's environmental politics, and a challenge to anthropocentrism in physical cultural studies. *International Review for the Sociology of Sport, 52*(8), 910–923. doi:10.1177/1012690216631541.

Montez de Oca, J. (2009). "As our muscles get softer, our missile race becomes harder": Cultural citizenship and the "muscle gap." *Journal of Historical Sociology, 18*(3), 145–172.

Morrow, J. R., Zhu, W., and Mahar, M. T. (2013). Chapter 4—physical fitness standards for children. In S. A. Plowman & M. D. Meredith (eds.), *FitnessGram/Activitygram reference guide* (4th ed.) (pp. Internet Resource). Dallas, TX: Cooper Institute, 4-1–4-12.

O'Neill, C. (2016). *Weapons of math destruction: How big data increases inequality and threatens democracy.* New York, NY: Crown.

Park, R. J. (2006). "Taking their measure" in play, games, and physical training: The American scene, 1870s to World War I. *Journal of Sport History, 33*(2), 193–217.

Plowman, S. A., Sterling, C. L., Corbin, C. B., Meredith, M. D., Welk, G. J., & Morrow, J. R., Jr. (2006). The history of FitnessGram. *Journal of Physical Activity and Health, 3*(s2), S5–S20.

Plowman, S. A., Meredith, M. D., Sterling, C. L., Corbin, C. B., Welk, G. J., and Morrow, J. R. (2013). History of the FitnessGram. In S. A. Plowman & M. D. Meredith (eds.), *FitnessGram/Activitygram reference guide* (4th ed.) (pp. Internet Resource). Dallas, TX: Cooper Institute, 1-1–1-22.

Pluim, C., & Gard, M. (2016). Parents as pawns in FitnessGram's war on obesity. In L. Burrows & S. Dagkas (eds.), *Families, young people, physical activity and health: Critical perspectives.* New York, NY: Routledge.

Pluim, C., & Gard, M. (2018). Physical education's grand convergence: FitnessGram, big-data and the digital commerce of children's health. *Critical Studies in Education, 59*(3), 261–278.

Powell, D., & Gard, M. (2015). The governmentality of childhood obesity: Coca-Cola, public health and primary schools. *Discourse: Studies in the Cultural Politics of Education, 36*(6), 854–867.

Rice, M. H., & Howell, C. (2000). Measurement of physical activity, exercise and physical fitness in children: Issues and concerns. *Journal of Pediatric Nursing, 15*(3), 148–156.

Safrit, M. (1990). The validity and reliability of fitness tests for children: A review. *Pediatric Exercise Science, 2*, 9–28.

Slack, J. D. (1996). The theory and method of articulation in cultural studies. In D. Morley & K. H. Chen (eds.), *Stuart Hall: Critical dialogues in cultural studies* (pp. 112–127). London, England: Routledge.

Slack, J. D., & Wise, J. M. (2015). *Culture and technology: A primer* (2nd ed.). New York, NY: Peter Lang.

Tinning, R. (2000). Seeking a realistic contribution: Considering physical education within HPE in New Zealand and Australia. *New Zealand Physical Educator, 33*(3), 8–22.

Vander Schee, C., & Boyles, D. (2010). "Exergaming," corporate interests and the crisis discourse of childhood obesity. *Sport, Education and Society, 15*, 169–185.

Weed, D. L. (1998). Beyond black box epidemiology. *American Journal of Public Health, 88*(1), 12–14.

Weir, L. (2006). *Pregnancy, risk and biopolitics: On the threshold of the living subject.* New York, NY: Routledge.

Welk, G. J. (2006). Strengthening the scientific basis of the FitnessGram program. *Journal of Physical Activity and Health, 3*(Suppl2), S1–S4.

Welk, G. J., Going, S. B., Morrow, J. R., & Meredith, M. D. (2011). Development of new criterion-referenced fitness standards in the FitnessGram program. *American Journal of Preventive Medicine, 41*(4), S63–S67.

Williamson, B. (2015). Governing software: Networks, databases and algorithmic power in the digital governance of public education. *Learning, Media and Technology, 40*(1), 83–105.

6 · THE POLITICS OF THE GLOVES

Finding Meaning in Entangled Matter

KIRI BAXTER

Blood pours and sweat drips, hearts rise to the pounding, beating crescendo of feet flying across the floor. One, two, three, the counts drive bodies forward, onward to a goal of satisfying displeasure. Bodies dodge and duck, divulging their senses to the carnal impulse to stab once, twice, three times. A fist full of raging pleasure lashes uncontrollably to the side, hair entangled in the stench of sweat flicks painfully across the face, stabbing the willing participant unseeingly in the eye. Bodies hum as the whistle blows, gloves fall limp at the side, and the pleasure of pain is etched as if in a time-lapse across the faces of the participants.

The gloves come off. Training is over.   —field notes

This chapter is a theoretical move toward an entangled feminism. As a starting point, I evoke the boxing gloves as a methodological object that enables a discussion of the breaking of the binaries in subject-object-body relations. I draw specifically on vignettes from my research undertaken with females at boxing gyms in the South East of England. With a call to arms to centralize the material body for feminism in mind, I focus on the materiality of the boxing gloves alongside the women who box. The boxing gloves evoked throughout this piece must be thought of in terms of how particular bodies' matter and meaning work in and through one another, how the self generates its meaning in relation to the gloves, and how these meanings come to be materially embodied (Castaneda, 2001).

This chapter consists of four main parts. The first two sections of the chapter introduce my reading of new materialism and how this focus can forefront the female body and the objects of her everyday life as a political symbol of change. I foreground the female sporting body in order to understand the body's intra-action with the objects of our worlds within a specific cultural and historical formation. The aim here is to put the sporting body into context in order to fully engage in a theoretical discussion that calls for the specificity of the moment to be recognized. Following this, I centralize the boxing gloves in order to put the sporting body and the sporting object into conversation with one another. This conversation will form the third section of the chapter, where I will highlight my understandings of bodily boundaries as essential to the discussion of an entangled ontology. I do this through the empirical examination of the female body and her relationship with the boxing gloves. Finally, I summarize the chapter by drawing together the arguments for bringing materialism back into the research process. The primary questions addressed in this chapter are as follows: How is it that the gloves exert their existence in the world? How do we approach these problems without falling into the concept that it is the sovereign human who constructs objects or that objects have their own agency outside of human existence?

In order to address these questions, I draw on ethnographic work in the form of a case study conducted over a period of 4 months in the Brighton area in the South East of England. This case study is part of a larger ethnographic project on women's corporeal understanding of their bodies within sport. This study is based on a group of 15 women, 7 of whom formed the core of this case study; the women varied in terms of boxing ability, age, and social class. Boxing was chosen as a case study because it is an ideal sport for focusing on the apparent boundaries of the body; in every moment of training, fighting or mere physical exertion, the body is on display. Furthermore, within boxing, the body is consistently in close contact with objects and the environment, while at the same time it is a lone sporting body. Thus boxing offers interesting questions surrounding women's nuanced entanglements and intra-actions. Existing feminist research on boxing has increased over the last 20 years with a growing corpus of female writers (Allen-Collinson & Owton, 2014; Gammel, 2012; Mennesson, 2000; Van Ingen, 2011) who are providing insight to the significance of female participation in boxing. Most of this work focuses on the meaning and the symbolism of the body in boxing, however; Allen-Collinson and Owton (2014) focus on the corporeality and physical heat of the body within running and boxing. The approach I take within this chapter extends and builds on these existing works by questioning the ontological status of the female body through the intra-actions with a specific object (Barad, 2007). The work does not strive for universal applicability; instead, I use it here to ground a theoretical discussion in the corporeal everyday of female boxers.

FEMINIST NEW MATERIALISM AND THE FEMALE SPORTING BODY

There are many versions and interpretations of new materialism. For William E. Connolly (2013), new materialism is

> the most common name given to a series of movements in several fields that criticise anthropocentrism, rethink subjectivity by playing up the role of inhuman forces within the human, emphasize the self-organizing powers of several nonhuman processes, explore dissonant relations between those processes and cultural practice, rethink the sources of ethics, and commend the need to fold a planetary dimension more actively and regularly into studies of global, interstate and state politics. (p. 399)

Many proponents of new materialism seek to give primacy to matter and subjects simultaneously. Objects are no longer passive matter that has to be represented; thus "meaning-making takes place on a two-way track" (Dolphijn & Van der Tuin, 2012, p. 15). Working in such a way means focusing on the ways in which bodies and systems/objects materialize alongside each other as what Donna Haraway (1991) deems material-semiotic agents. Thus, as Coole and Frost (2010) note, "thinking anew about the fundamental structure of matter has far-reaching normative and existential implications" (p. 5). This ontological positioning, regardless of the name it is given, significantly changes the playing field of an embodied piece of feminist research, for as Coole and Frost (2010) consider,

> materiality is always something more than "mere" matter: an excess, force, vitality, rationality, or difference that renders matter active, self-creative, productive and unpredictable . . . to recognize that phenomena are caught in a multitude of interlocking systems and forces and to consider anew the location and nature of capacities for agency. (p. 9)

A nuanced and detailed understanding of the capacities of agency that represents new materialisms, most notably applied by Karen Barad, segues into the larger debate of how feminists undertake empirical research. Demonstrating that women are influenced by the objects of the everyday through a materialist analysis shows that new materialism can extend a traditional materialist analysis beyond the macro level of social phenomena (Van der Tuin & Dolphijn, 2012, p. 159).

As a point of entry, the methodological concern of a new materialist ontology considers research as an assemblage. The research assemblage (Fox & Alldred, 2014; Coleman & Ringrose, 2013) comprises of both bodies and things as well as the abstractions that get caught up in social inquiry. This empirical focus on interactions between a body and object aids an authentic understanding of female sporting bodies intra-action with their objects of the everyday (Deleuze & Guattari, 1984). This ontological positioning shifts the concept of bodies as objects and material things from distinct entities to relational ones.

An assemblage "owes its agentic capacity to the vitality of the materialities that constitute it" (Bennett, 2010, p. 34). Furthermore, assemblages are "living, throbbing confederations that are able to function despite the persistent presence of energies that confound them from within" (p. 24). Fox and Alldred (2014) found via their review of 30 empirical studies using a new materialist ontology that what these studies have in common is that they are not a return to the reductionist materialism. That is to say, they are not a turn toward an overtly simplistic concept of our material being or an oversimplifying of the relationship between object and subject but instead

> foreground an appreciation of just what it means to exist as a material individual with biological needs yet inhabiting a world of natural and artificial objects, well honed micro powers of governmentality but no less compelling effects of international economic structures. (Coole & Frost, 2010, p. 27)

Drawing on such work, I understand the assemblage to be the relationship between the body and the material objects that form the sporting experience. I use the example of women boxers and boxing gloves to demonstrate the dynamic nature of social science methodological practice and to underline the notion that neither body nor object is fully formed prior to their entanglement with one another. This entanglement is further based, in part, on the micropowers of governmentality and structure that form the everyday (Foucault, 1991). In order to comprehend the micropowers of governmentality that exist as part of a relationship between the body and objects, we must understand that it is the history that the object "contains" that gives it power.

This relationship between the body and object can extend our comprehension of the female body and, when attended to empirically, can help develop feminist methodological comprehension on which to base future material endeavors. Sporting bodies are inherently messy bodies: they leak sweat, ooze blood, cry in pain, and are highly sexualized by a wider, often media-hungry society. Sporting bodies break every conception of what is so often conceptualized as feminine, and as such, they are so often othered, sidelined, forgotten, and denigrated (Longhurst, 2001). If as feminists we fail to attend to the boundaries and fluidities of the female sporting body as a material, corporeal being, then we are set to deny the very materiality of the body and the social constructs within which all bodies are placed. Critiquing our own position—indeed, our membership to feminism—through a theoretical positioning can be a powerful research tool. What I propose is that marrying this theoretical critique of the material with a feminist notion of the body has the ability to reveal something new about female sporting empowerment. Central to these moments is an analysis of the female sporting body and the objects that create the female sporting experiences of the everyday.

For years, Western discourse has relied heavily on bodies living and interacting as sovereign selves—that is, bodies that live discreetly within their own skins and remain rigidly secure within the boundary of a physical flesh (Classen, 2012). Yet the objects of our sporting endeavors throw the body's physical materiality and corporeality into question. We sweat and leak on "another," and as we do, a sporting object that concurrently sweats and leaks complicates the entangled relationship between a body and the everyday object. Bodies and objects have an entangled, highly politicized relationship. This relationship is based on the idea—proposed by many feminists, cyborg theorists, anthropologists, phenomenologists, and new materialists, among others—that the boundary of the body is not necessarily at the end of one's epidermis.

In centralizing the gloves in the research alongside the women who participate in boxing, I am placing importance on the object of the everyday. What I propose is that women cannot be taken as separate from their relationship to the gloves. The focus on the gloves is crucial to the understanding of a female boxer's experience, as you cannot have a boxer without boxing gloves; it is the subject-object entanglement of woman and gloves that makes the assemblage "female boxer." Furthermore, the gloves have a history that is impossible to understand without the women using the gloves; if we tried to understand the women without the gloves, we would be missing a key component as to what it is that entangles in order to create a reality. Using the gloves as a central object of research means that I am finding "situated knowledges" (Haraway, 1988) while also considering that a focus on objects has the capacity to "highlight, exhibit, and make evident the entangled structure of the changing and contingent ontology of the world, including the ontology of knowing" (Barad, 2007, p. 73).

The gloves, female bodies, and the environment in which they participate are entangled fields of knowledge. By taking a diffractive view to researching women in boxing, I ensured that the effects of differences were mapped, using women's bodies and the objects of their sporting experience to do so. In other words, the use of a diffractive methodology allowed a mapping of interference, not of replication, reflection, or reproduction (Barad, 2003). Haraway (1997) suggests that a diffractive methodology is a critical practice for making a difference in the world. By taking a diffractive approach to methodology, feminist research will unravel the concept of a binding skin and divulge the notion of a porous entangled environment, allowing research findings to be enfleshed.

Quite simply, instead of reading different ontological positions as if they were set against one another or moving laterally alongside one another, I wish to examine the object of study using their interconnected rationalities. When read diffractively, theory can piece together a larger puzzle and investigate the "material-discursive boundary-making practices that produce objects and subjects" (Barad, 2007, p. 93). I approach the methodological research and analysis for this chapter diffractively and elaborate the notion of diffraction as a mutated critical tool of analysis. Hence while diffraction is a phenomenon deriving from classical physics, the significance in social sciences—and more specifically, feminist methodologies—is a powerful example of how research can recognize the intrinsic nature of an ethics that is predicated not on externality but rather on entanglement (Barad, 2007).

A feminist methodology diffracts the ontology of knowing through the treatment of social inquiry as an assemblage; the gloves, the body, and the environment are not autonomous beings with no effect on one another. Instead, a feminist methodology that centralizes the relationships and intra-action of both organic and nonorganic matter can problematize the power relations present within gendered ontological knowledge. For example, in researching the women in boxing, I had a heightened ontological orientation toward the gloves. In practice, this meant my research methods were based on using the gloves in order to elicit responses from the participants. There is a burgeoning literature concerning the interlinking of walking with others as a research methodology (see Lee & Ingold, 2006); the ability to interact with research participants so as to create a sense of empathy or sense of belonging derives from classic anthropological works of the 20th century (Geertz, 1973; Turnball, 1961). In literally "walking with others" or, for the sake of this study, "boxing with others," the research presents an alternative form of participant-researcher engagement. It is not that participant observation itself is in any way new; rather, tactile, sensory moments are used to (re)examine spatial environments and bodily experiences. To borrow from Lund (2005), it is thought that boxing can be seen as

> a bodily movement that not only connects the body to the ground but also includes different postures, speeds and rhythms . . . [that] . . . shape the tactile interactions between the moving body and the ground, and play a fundamental part in how the surroundings are sensually experienced. (p. 28)

While I produced reflexive sensory observations, these observations themselves cannot be said to describe the actual embodied experiences of other female boxers. However, as a feminist piece, this research set out to expose the rawness of emotions and sensory experiences of the everyday so often hidden within what is usually assumed or taken for granted.

The gloves were central to everything the women did within the gym. They were, in short, a way of relating. Suggesting that both space and objects had a different ontological status before the women entered the gym or put on the gloves (Longhurst, Ho, & Johnston, 2008) means that female boxers were able to transform an environment around a transgressive identity of a female boxer, rather than adhering to traditional and stereotypical notions of a weak and docile woman. Later in this chapter, I discuss an incident whereby women were keen to bring their own gloves to the gym. The women often chose to buy pink gloves rather than take the gym's communal black or white ones; this choice of using their own gloves was based on the dual desire to feminize the object by making it pink while also avoiding the sweat of another found within the communal gloves. Thus while a materialist ontology and a feminist methodology may uncover the relationship between the gloves and the female body, it doesn't necessarily result in the challenging of gendered ontologies by the female boxers.

FEMINISM, OBJECTS, AND HUMAN INTRA-ACTION

Object and human intra-action are unsurprisingly contradictory and complex. A recent turn in feminism to new materialisms has seen objects recentered as part of our understanding of the everyday. This centralization of objects has come in the first instance through feminists attending to the phenomena of the body in order to understand how bodies intra-act with one another. In doing so, the very boundaries of the body have been brought into question (R. Coleman, 2009; Hickey-Moody, 2009). Within the material divisions of body and object—in this case, the boxer and gloves—lay an opportunity for feminists to understand women's material configurations and discursive formations (Barad, 2007). It is pertinent to consider what implication, if anything, this understanding of the body as neither atomistic nor solipsistic has for a feminist turn that seeks to deprivilege human agency while not oversimplifying, nor indeed giving agency to, the objects of our experiences themselves.

Karen Barad's notion of intra-action is thus somewhat helpful here. Instead of defining "an object" as a discrete entity to "a body," we are able to understand the object as part of a relationship whereby through "mere" intra-action with a body, they become different. In other words, the focus on either the gloves or the female boxer becomes instead a focus on an entangled relationship with one another. The lived body is profoundly entangled within the project of feminism.

How we orientate ourselves, or indeed how we are orientated, has direct implications on how we identify as female or other. I am not giving rise to a piece of work that favors the ability to give an object an agency outside of the realm of the human; rather, I place the boxing gloves at the center of methodological inquiry in order to discuss the relationship between the material body and the material object.

Part of the appeal of new materialism is turning theoretical perspectives into useable sociological methods for research (Fox & Alldred, 2014, p. 153). Within traditional feminist investigations into how things work, we often "perform 'agential cuts' that effect a separation between objects and subjects existing in interdependence and that thereby constitute each as having distinct, determinate boundaries" (Barad, 2003, p. 815). During fieldwork, the boxing gloves were used as an object of the everyday in order to elicit responses from participants. Thus the research had a heightened ontological orientation toward the matter of the gloves:

> The gloves are hot, certainly, and sticky too. The humidity of the boxing gym ensures that it is hot enough for clothes to stick to your body and pool at every given crevice. Inside the gloves, it's slimy, clammy, and I can feel sweat dripping down my wrist trying to escape the tight confines of the Velcro strap. I'm taken almost instantaneously by how central the gloves are to the women in the gym. Conversations peak when it's time to put the gloves on; some women have their own and wear them proudly as though the gloves have a direct implication on how good a boxer you are. Others borrow them from the crate at the far end of the gym, but with the daily ritual of finding a pair that is small enough for women's hands, it is not surprising that some of these women have invested in their own pair.
>
> —field notes

Within social sciences, there is an increasing acknowledgment of the researcher being personally involved in the fieldwork. The researcher is "situated within" the social world under study and the data and knowledge produced (Coffey, 1999; England, 1994). The objects of the everyday, the matter of sporting equipment, has a direct causality on the relationship between bodies and their experience. In line with current new materialist thought, the figuration of matter as an agent is distinctive to the environment in which it exists (Frost, 2011). Matter has agency. Subsequently, the gloves have agency. The concern here is that in an effort to consider the peculiar agency of organic or inorganic matter, new materialists might, "wittingly or unwittingly, read linguistic, cultural, or political facts and meanings into the material—that they might misrepresent as biological, physiological or natural what is actually social and historical" (Frost, 2011,

p. 76). Contemporary social sciences, however, accept that "the notion of the isolated, autonomous fieldwork site has been something of a convenient functionalist fiction" (S. Coleman, 2004, p. 26). Hence while researchers cannot take themselves out of their fieldwork any longer, they can no longer take the objects of study out of the field as well. The inorganic matter of the gloves shapes and actively participates within the space of the research due to the entangled relationship with the women that wear them. Therefore the matter of the gloves should be centralized as part of the research process.

Understanding that both a body and an object have an agential reality and presence simplifies a much bigger discussion of feminist methodological considerations and analysis, one I tend to in detail elsewhere (Baxter, 2018). I am not suggesting that we ignore wider gender, racial, or structural issues but instead that we take a more nuanced approach to the capacity of both human and objects that make up the everyday. With this in mind, I will turn to a discussion of the history of boxing gloves. The gloves are tied inextricably to a patriarchal past. Boxing is, as Wacquant (2004) argues, a "definite form of masculinity: plebeian, heterosexual and heroic" (p. 14). It is consequently unsurprising that women's boxing has been characterized as a radical activity that blurs traditional male and female images, identities, and class identities (Hargreaves, 1997, p. 125). Women's boxing is entrenched in binary logic (Scott, 2008); it is underpinned by discourses of the "heroic" and has a powerful ability to transgress outdated and socially constructed views of femininity. At the same time, access for women is limited, and popular media portrayals of female boxers have been sexualized in an attempt to retain power over women's bodies. Indeed, to imagine the female body as strong and athletic often requires a radical reconstruction of the idealist and traditional Victorian "feminine" (Hargreaves, 1997; Woodward, 2007). Thus the capacities of the gloves—that is, their agential capacities—form an assemblage with the wearer despite the gloves' entangled patriarchal past.

THE POLITICS OF THE GLOVES

Boxing gloves in their traditional form are a pair of heavily padded leather mittens laced on the palm side of the wrist and hand. These days, boxing gloves vary in quality, but their look and makeup are much the same. Importantly, boxing gloves are made with one thing in mind: the force of an impact. When the gloves strike the target, the force is reduced or absorbed as the gloves compress. The gloves should be tight around the hand but not to the point where the hands will be deprived of blood. The fingertips should fit tightly against the tops of the gloves, and each should lace up or Velcro tight around the wrist. The wraps extend around the hand and create an intimate relationship that enlarges the active radius of the human touch. The gloves cease to be an object external to the boxer at this point. They are not an extension of the body, which would predicate the inherent nature of bodily boundaries that I am questioning. Instead, the gloves' seemingly self-evident boundary of human and object lead to an ontological question: How do we distinguish between ourselves and objects?

> For many of the women there, "their" gloves literally mean the gym's, and they borrow them at each session. Three huge crates sit at the far end of the gym by the changing rooms, just outside of the ring. From them, D will usually pick pairs to suit people. Some of the gloves are so large, it is hard to use them. You wait for her to hand you a pair, hoping they will be as clean and new as possible. Wearing gloves that others wear is not a nice experience; the "universal" glove contains hidden sweat that mingles with others' sweat. When your own hands are hot, your own sweat seeps into the fabric that holds itself as your second skin. It's clammy, at times tight, and more often than not, uncomfortable. The gloves act as an extension of the body, but it's an extension you have to get used to, and within this extension lies others' sweat, "their" presence felt so entwined with your own hands, while at the same time, the gloves are there to protect you from others' touch.
>
> —field notes

Our ontological production of bodily boundaries relies on our ability to distinguish ourselves from objects. It is therefore not a question of how bodies exist within the world, as phenomenologists' accounts lead us to consider, but rather how the gloves and the body are mutually constituted with one another. In order to engage with this ontological questioning, it should be noted that if we consider the body to end at the commonly held notion of a bodily boundary, then we fail to note the body's situatedness in the world (Haraway, 1994). As Donna Haraway (1994) notes, the "situation is never self-evident, never simply 'concrete,' [but] always critical" (p. 67). Indeed, the gloves arouse lived feelings of sweat as well as the connotations of being touched by another (Paterson, 2007). There are indeterminate boundaries between the skin and the gloves; thus the women participating with the gloves are intra-actively co-constituted through the material-discursive practices in which they engage.

This practice is informed, in part, by touch. For Gibson (1966), in evoking the senses, the "haptic" is not limited by the usual notion of touch as cutaneous contact but instead incorporates the "body's felt states" when touching occurs. Much like a new materialist analysis, the body's felt state in this sense is based on the engagement between subject and object. The feeling of touch when an object brushes our skin or we punch a bag is simultaneously an awareness of the object and of our lived body (Paterson, 2007, p. 2). Through haptics, the lived environment has the ability to go from a singular being to an example of the expanse. The expanse of the body then is created through the very matter of an experience. Thus touch does not possess an inherent outside boundary that limits the dynamic reconfigurings of the world through which these sporting women are intra-actively materialized. The lived everyday experience is intrinsically corporeal (Simonsen, 2007, p. 121). Perception, for its part within the process of the embodiment, is based on sensuous practices, such as the haptic and the olfactory, to name a couple. Unlike phenomenological critiques of touch, feeling is a quality that can surface from the body. For example, sweat does not come up against the body; rather, it leaks through the body, seeping out. While my focus in this chapter is primarily on the gloves, the analysis of sweat and touch alongside this highlights how research into physical fluids can provide insights into wider physical cultures (Atkinson, 2017).

Through the capacities of the gloves that create the female boxers' environment, there is an ongoing ebb and flow of agency (Barad, 2007). The skin is the primary site through which the touch of the gloves is actively experienced. In thinking through the skin and also attending to a proximity and sensuality of moisture that drips from one body to another, the body of the female boxers can be attended to as sensually grounded and brought to the surface by being with others. Attending methodologically to the gloves as integral to the female experience of boxing highlights the physical impossibility of inhabiting another's skin. Intraembodiment allows for an understanding of embodiment as part of a wider meshwork of relationships. This connection between women and objects is relational. Thus intraembodiment is the state of embodiment that takes into consideration the state of knowing and experience as a material practice. I employ the term here as a way of thinking through the nearness of others, but it is a nearness that requires distinction and difference.

Touch is a sense that allows us to actively experiences the world, thus it constitutes an experience of the world that cannot be detached from its embodiment but is also not reducible to the body itself. The women found that when they used the gloves available to "borrow" at the gym, their sense of disgust was heightened, primarily by the thought that they were sharing others' sweat. If we consider for a moment a diffractive understanding of the matter of sweat, the gloves, and the females participating, then we are closer to understanding that the phenomenon of feeling can be attributed to differential patterns of mattering (Barad, 2010). Many of the women noted the smell the borrowed gloves left on their hands long afterward, even after vigorous washing. Some of the women chose to go out and buy their own gloves to avoid coming into contact with others' bodily fluids:

The woman's gloves are baby pink; she tells me that she acquired them because she didn't want to continue using the gym's gloves. In her description of the gym's gloves, she uses the words "disgusting" and "smelly." They are 10oz gloves and she offers them to me when I am punching her pads. I turn her down. I find it odd to put my sweat directly onto someone else's property. Ironic, as that is exactly what I am doing with the gym's gloves. Our conversation goes on, and she tells me she got the pink because she wanted something a bit different and "girly."

—field notes

Stereotypical attributions of gender aside with regard to her boxing gloves' color, the skin itself is a site of possible embodied and gendered encounters that upon further diffractive reading also demonstrates an insight into the world of boxing. For many of the women, the thought of mixing their own sweat and using the gloves of another was too much to take. There is a sense of engagement with someone else in the sensation of reaching into the gloves where another's sweat lies. The hand is wrapped and encased in another's sweat, and one feels moisture and warmth akin to the literal action of reaching inside the body of another. In the case of the gloves from the gym, it's not just "one body"; it is an amalgamation of "multiple bodies." The sweat-covered gloves produce a visceral response in the women who put their hands into them. The gloves exert their own agentic capacity based on the sensuous effect on the women's skin. This relationship between organic and nonorganic life forms is a connection that relies on the ability of research and scholarship to identify the differences produced through specific encounters. Undoubtedly in the case of the boxing gloves, it is the skin that is central to the analysis of understanding this complex and contradictory relationship. Through his research using ethnoesthesia, Michael Atkinson (2017) renders the sensuality of sweat as central to the understanding of physical culture. While specific to Ashtanga yoga, there are striking reasons to recognize that Atkinson's (2017) conclusion—that "physical fluids and their contextual sensation help shape and articulate a broader yoga aesthetic" (p. 65)—can be applied in broader brushstrokes to other areas of physical culture. To this end, Atkinson theorizes that sweat as an object is worthy as a subject of study in and of itself.

The significance of the skin in the relationship between the gloves and the body is not simply organic but also psychic. Thus the self is not defined solely by what is on the inside, as Butler (1993) notes. Bodies are constructed by a set of discursive practices based on a hegemonic understanding of the body whereby "that which protrudes, bulges, sprouts or branches off . . . is eliminated, hidden or moderated. All orifices of the body are closed [. . .] [it is] an impenetrable façade" (Bakhtin, 1984, p. 320, as quoted in Shildrick, 2001, p. 163). The body is thus understood as that which does not leak or seep. Normative discourses create a statuesque body—a division that can be traced back to the dualism of the self and the other.

That said, the surfaces of our hands, our skin, are not a protective envelope that defines and unifies our limits (Shildrick, 2001). Within the boxing gloves, the skin is forced into a patriarchal-based history, a being with the gloves' past users. It is not an indifferent exchange between autonomous beings. Rather, it is within the openings of touch that "our bodies overstep their bounds; our flesh is in flux" (Cataldi, 1993, p. 126). The relationship between boxer-as-subject and gloves-as-object is the contiguity between subjects. That is to say, in the sharing and passing of communal gloves, the skin, the protection (being the gloves), and the fluidity of the body are not a singular, bounded body given over to a delimited subject; rather, the gym's gloves are "communal skins," where the insides and outsides of the bodies' functions meet, come together, and penetrate one another. To quote Irigaray (1992), "The internal and external horizon of my skin interpenetrating with yours weaves away their edges, their limits, their solidity. Creating another space outside my framework. An opening of openness" (p. 59).

Feminism has explored in depth how women understand and experience their bodies within patriarchal structures. Iris Young (1990) understood that the space of the body was able to transcend beyond the envelope of the skin. However, while somewhat helpful to the analysis of the gloves, Young conjures a female bodily image that projects an existential barrier enclosed around the body in order to keep *the other* at a distance. The existential barrier Young describes seeks to keep (an)other body within a safe bodily space. Boxing and the gloves throw the notion of keeping another at a safe distance into question, the premise of which derives from the female notion of fear related with another invading her body space. The most extreme form of spatial and bodily invasion, as set out by De Beauvoir (1972), is the threat of rape. While De Beauvoir's realistic upholding of the anxiety of rape has a defined position within the understanding of a body created through the unease of an oppressive, patriarchal society, there are circumstances, such as in sport, where bodies are beyond their own and others' boundaries—so much so that the bodies' boundaries themselves are brought into question. In the case of the gloves, the potential for female emancipation through an understanding of the organic and nonorganic relationship highlights the ability to move between states of being, with the gloves acting as both weapon and protection:

The first time I trained up in London, my coach said to me, "If I hit you, it's your fault, not mine. If you can't protect yourself from me, that's your fault, not mine, and if you go down, that's your fault, not mine. Use your brain," he said. "Use your gloves." I thought that was a fair enough point. I had the same weapons he had—my fists with my gloves—but when the punch came, I wasn't expecting it, and I did go down, and the bruise around my eye looked like something someone might have painted on, it was so bad. It was tough, the gloves were tough, and yeah, learning that someone wasn't going to hold back was tough. Some are idiots and really go for you, but then at the same time, I'm pleased they don't hold back. If they held back, I think I would be more annoyed. To start with, I was scared. I probably still am—I don't think you can ever be a boxer and never expect to be scared. You have to learn how to fight somehow, . . . to protect yourself. I'm still learning, but I just put the gloves on and get on with it.

—field notes

Intensifications of feelings do not create a distinction between inside and out-side. In emotions such as physical pain, the boxer becomes aware of skin as a bodily surface and of the importance of the gloves in directing a future passage of thought. In this recognition lies the potential to mediate a relationship between the internal pain or emotion of the boxer and the external reality of the situation; in a heightened sense of physical awareness through emotional (dis)comfort lies the ability to reconstitute the very bodily space and our understandings of relationships (Ahmed, 2004). In the expression of emotions, the bodies of the boxers create an environment; they "affect" the very distinction of outside and inside, external and internal, as within shared and voiced emotions, body sur-faces are not only pressed on but also projecting. Barad (2007) backs up Young's (1990) own early work on the body, this time in relation to the body's boundary. It is ever clearer, as Barad (2007) explains, that the seemingly infallible nature of the body's boundary is a result of cultural and historical repetition (p. 155). Drawing on science, Barad (2007) goes further:

> Physics tells us that edges or boundaries are not determinate either ontologically or visually. . . . It [is] a well-recognized fact of physics that if one looks closely at an "edge," what one sees is not a sharp boundary between light and dark but rather a series of light and dark bands. (p. 156)

Thus while the relationship between two boxers and their gloves is a focus on the production of bodies, it is also attentive to the production of nonhuman phe-nomena. In the boxing example above, it is clear that we do not simply make the world in our own image; "reality is an ongoing dynamic of intra-activity" (Barad, 2007, p. 206).

As Judith Butler (1988) writes, we must reread the texts of Western philoso-phy from the various viewpoints of those who have been excluded to establish philosophy itself as a cultural practice and criticize its tenets from marginal-ized cultural positions. Poststructural feminists of the body have spent time deconstructing the textual "feminine," and this preoccupation has led to many critiques. Butler, among others, has summed up the critique's apparent lack of materiality. Indeed, Butler's (1993) own reporting of a common critique of her work, which asks "but what of the materiality of the body, Judy?" (p. ix), is a furthering of the argument that what matters about an object is indeed its matter, but a matter of experience and history.

MATERIALIST WAVES

This turn toward a "new" materialism that is becoming ever more prevalent in scholarship does not indicate a feminist disengagement with power and its subsequent effects on the body. Indeed, quite the opposite. The work of Barad (1999, 2001, 2003, 2007) and Grosz (1994), arguably feminists who sit at the base of new materialist theory, are prime examples of how we should be alert to the relationships of power that govern the everyday (Coole & Frost, 2010). The challenge for feminist new materialism is to answer how an object, such as boxing gloves, can prove to be a worthwhile article of study for female emancipation and the feminist social scientist. Quite simply, on its own, it is not. Like everything social scientists study, we have to be congruent to the external forces that create the everyday. Female boxers are an assemblage of the enmeshed relationship that occurs through being a woman alongside the wearing of the boxing gloves.

The assemblage of the gloves, the history attached to the gloves, the body and the body's living environment is the key to untangling an entangled engagement. The gloves are an extension of the body and the skin as well as being entangled within an environment. The gloves as a boundary of indeterminate solidity can be diffractively read as a solid yet intrapenetrative object. The body, the gloves, and the lived environment are folded into one another, enmeshed in the same routine. To relinquish the relationship between the bounded body and the surrounding space is to open up the possibility of reconfiguring a relationship based on the understanding that space itself is an environment—an environment built on the essence of a group of bodies entangled and enmeshed in a relationship of becoming (Ingold, 2008). It is an environment that would not exist in the same context if it were not for the body (or bodies).

Female corporeality is inextricably tied to the environment such that one would not exist in the same way if it were not for the other. Touch is thus fluid and open, is capable of being receptive and expressive, and brings distant objects and people into proximity (Paterson, 2007). Furthermore, the skin is not limited to the fleshy substance surrounding the body's internal organs; the subject-object-space relation might thus be rewritten as an indecipherability between continually fluid subjects. The gloves are something that no one being can fully occupy; they are a "communal skin" with the potential to mobilize an equality based on the communion of our specificities, fluidity, and intricacies rather than the sex/gender differences between us.

Within intraembodiment lies a way of thinking through the nearness of others, but this nearness requires distinction and difference. The skin is the primary site through which touch is actively experienced. Ahmed and Stacey (2001) write that the skin should be thought of as a border, but unlike a mirror, it does not reflect the truth of the inner self. A diffractive reading of the gloves, and indeed a diffractive ontology, attends to the feminist need to research a specific social context ensuring that the proximity and corporeality of the body are attended to. The messy, leaky body of both sense and emotion in academia has long since been ignored or abjected (Longhurst, 2001). What the gloves are and what they mean to the women who box are formed through their relationship with participants. These meanings are materially embodied and have the potential to highlight and distinguish the capacity of women to empower themselves through sport. This can only be recognized, researched, and realized if we acknowledge the importance of an object's existence and impact on an environment. I have centered my analysis on the boxing gloves and a sporting situation because as sport becomes ever more important to our everyday lives and as women seep ever so casually into one of the last standing male bastions of leisure time, the importance of understanding and legitimizing women's place within sporting structures becomes ever more relevant. I posit here that a turn toward matter (Barad, 2007; Coole & Frost, 2010) and, in this case, boxing gloves allows feminists to reintroduce the specifications of female differences between one another that in turn allow for a far greater understanding of all bodies.

During this chapter, I have diffractively cut through various positionings in order to discuss a seemingly mundane object within the relational understanding of the female boxer. The need for gender equality and understanding of female participation in sport remains central to my analysis. The messiness of bodies and the bleeding, crying, sexual bodies of women break the boundaries of what is so often conceptualized as feminine, and as such, they are othered in academic writing (Longhurst, 2001). I attend to the corporeal, messy body in full as part of a much larger project. Using a new materialist framework to diffractively research the body allows for the corporeal to make a return to the forefront of social science research. In order to reach beneath the skin of how women's bodies are lived, we must transcend traditional materialist analysis and recognize the material-discursive formations that form our ontological everyday. Furthermore, a new materialist methodology grounded in the phenomena of matter allows for the deconstruction of patriarchal structures and androcentric flaws due to the attention given to the politics of difference. An approach that draws on new materialism builds on existing feminist research in sport and allows us to recognize that all parts of matter, including the gloves, are part of an agential experience. Thus the gloves are not inert or passive matter that is acted upon. The relationship formed between human and object is a specific one between a body and an object, both of which have an agential reality and presence. For feminism to address the patriarchal structures of sport, both the corporeal-lived-body and the environment of the research must be recognized as part of an intra-action so that researchers can actively and appropriately represent women in sport today.

REFERENCES

Ahmed, S. (2004). *The cultural politics of emotion*. Edinburgh, Scotland: Edinburgh University Press.

Ahmed, S., & Stacey, J. (2001). *Thinking through the skin*. London, England: Routledge.

Allen-Collinson, J., & Owton, H. (2014). Intense embodiment. *Body and Society*, 21(2), 245–268.

Atkinson, M. (2017). Astanga yoga and the sensuality of sweat. In A. Sparkes (ed.), *Seeking the senses in physical culture* (pp. 63–81). Oxford, England: Routledge.

Bakhtin, M. (1984). *Rabelais and his world* (H. Iswolsky, trans.). Bloomington: Indiana University Press.

Barad, K. (1999). Agential realism: Feminist interventions in understanding scientific practices (1998). In M. Biagioli (ed.), *The science studies reader* (pp. 1–11). New York, NY: Routledge.

Barad, K. (2001). Re(con)figuring space, time, and matter. In M. DeKoven (ed.), *Feminist locations: Global and local, theory and practice* (pp. 75–109). New Brunswick, NJ: Rutgers University Press.

Barad, K. (2003). Posthumanist performativity: Towards an understanding of how matter comes to matter. *Signs: Journal of Women in Culture and Society*, 28(3), 801–833.

Barad, K. (2007). *Meeting the universe halfway: Quantum physics and the entanglement of matter and meaning*. Durham, NC: Duke University Press.

Barad, K. (2010). Quantum entanglements and hauntological relations of inheritance: Dis/continuities, spacetime enfoldings, and justice-to-come. *Derrida Today*, 3(2), 240–268.

Baxter, K. (2018). Coming out of containment in transforming sport: Knowledge, structures, practices. In T. F. Carter, D. Burdsey, & M. Doidge (eds.), *Transforming sport: Knowledges, practices, structures* (pp. 50–65). London, England: Routledge.

Bennett, J. (2010). *Vibrant matter: A political ecology of things*. Raleigh, NC: Duke University Press.

Butler, J. (1988, December). Performative acts and gender constitution. *Phenomenology and Feminist Theory Theater Journal*, 40(4), 519–531.

Butler, J. (1990). *Gender trouble*. London, England: Routledge.

Butler, J. (1993). *Bodies that matter: On the discursive limits of "sex."* New York, NY: Routledge.

Castaneda, C. (2001). Robotic skin: The future of touch? In S. Ahmed & J. Stacey (eds.), *Thinking through the skin* (pp. 223–236). London, England: Routledge.

Cataldi, S. L. (1993). *Emotion, depth and the flesh: A study of sensitive space*. New York: State University of New York Press.

Classen, C. (2012). *The deepest sense: A cultural history of touch*. Urbana: University of Illinois Press.

Coffey, A. (1999). *The ethnographic self: Fieldwork and representation of identity*. London, England: Sage.

Coleman, R. (2009). *The becoming of bodies: Girls, images, experience*. Manchester, England: Manchester University Press.

Coleman, R., & Ringrose, J. (eds.). (2013). *Deleuze and research methodologies*. Edinburgh, Scotland: Edinburgh University Press.

Coleman, S. (2004). *ASA Annual Conference 2004, abstracts. Locating the field: The multi-sited ethnographer*. [Online] Retrieved from http://www.theasa.org/conferences/asa04/panels/panel03.shtml

Connolly, W. (2013). The "new materialism" and the fragility of things. *Millennium: Journal of International Studies, 41*(3), 399–412.

Coole, D., & Frost, S. (eds.). (2010). *New materialisms: Ontology, agency, and politics*. Durham, NC: Duke University Press.

De Beauvoir, S. (1972). *The second sex* (H. Parshley, trans.). Harmondsworth, England: Penguin Books.

Deleuze, G., & Guattari, F. (1984). *Anti-Oedipus: Capitalism and schizophrenia* (R. Hurley, M. Seem, & H. R. Lane, trans.). London, England: Athlone.

Dolphijn, R., & Van der Tuin, I. (2012). *New materialism: Interviews and cartographies*. Ann Arbor, MI: Open Humanities Press.

England, H. V. L. (1994). Getting personal: Reflexivity, positionality, and feminist research. *The Professional Geographer, 46*(1), 80–89.

Foucault, M. (1991). Governmentality. In G. Burchell, C. Gordon, & P. Miller (eds.), *The Foucault effect: Studies in governmentality* (pp. 87–104). Chicago, IL: University of Chicago Press.

Fox, N., & Alldred, P. (2014). New materialist social inquiry: Designs, methods and the research-assemblage. *International Journal of Social Research Methodology, 18*(4), 399–414. doi:10.1080/13645579.2014.921458.

Frost, S. (2011). The implications of the new materialisms for feminist epistemology. In H. E. Grasswick (ed.), *Feminist epistemology and philosophy of science: Power in knowledge* (pp. 69–83). Dordrecht, Netherlands: Springer.

Gammel, I. (2012). Lacing up the gloves. *Cultural and Social History, 9*(3), 369–390. doi:10.2752/147800412X13347542916620.

Geertz, C. (1973). *The interpretation of cultures: Selected essays*. New York, NY: Basic Books.

Gibson, J. (1966). *The senses considered as perceptual systems*. Boston, MA: Houghton Mifflin.

Grosz, E. (1994). *Volatile bodies: Toward a corporeal feminism*. Bloomington: Indiana University Press.

Haraway, D. (1988). Situated knowledges: The science question in feminism and the privilege of partial perspective. *Feminist Studies, 14*(3), 575–599.

Haraway, D. (1991). Situated knowledges: The science question in feminism and the privilege of partial perspective. In *Simians, cyborgs, and women: The reinvention of nature*. New York, NY: Routledge.

Haraway, D. (1994). A game of cat's cradle: Science studies, feminist theory, cultural studies. *Configurations, 1*, 59–71.

Haraway, D. (1997). *Modest_Witness@Second_Millenium, FemaleMan©_Meets_OncoMouse™: Feminism and technoscience*. London, England: Routledge.

Hargreaves, J. (1997). Women's boxing and related activities: Introducing images and meanings. *Body and Society, 3*(4), 33–49.

Hickey-Moody, A. (2009). *Unimaginable bodies*. Rotterdam, Netherlands: Sense.

Ingold, T. (2000). *The perception of the environment: Essays on livelihood, dwelling and skill.* London, England: Routledge.

Ingold, T., & Lee Vergunst, J. (eds.). (2008). *Ways of walking: Ethnography and practice on foot.* Aldershot, England: Ashgate.

Ingold, T. (2008). Bindings against boundaries: Entanglements of life in an open world. *Environment and Planning A, 40,* 1796–1781.

Irigaray, L. (1992). *Elemental passions* (J. Collie & J. Still, trans.). London, England: Athlone.

Lee, J., & Ingold, T. (2006). Fieldwork on foot: Perceiving, routing, socializing. In S. Coleman & P. Collins (eds.), *Locating the field: Space, place and context in anthropology* (pp. 67–86). Oxford, England: Berg.

Longhurst, R. (2001). *Bodies: Exploring fluid boundaries.* London, England: Routledge.

Longhurst, R., Ho, E., & Johnston, L. (2008). Using "the body" as an "instrument of research": Kim'chi and Pavlova. *Area, 402,* 208–217.

Lund, K. (2005). Seeing in motion and the touching eye: Walking over Scotland's mountains. In R. Bendix & D. Brenneis (eds.), *The Senses, a guest edited issue of Etnofoor: Anthropological Journal, 18*(1), 27–42.

Mennesson, C. (2000). "Hard" women and "soft" women. *International Review for the Sociology of Sport, 35*(1), 21–33. http://dx.doi.org/10.1177/101269000035001002

Paterson, M. (2007). *The senses of touch: Haptics, affects and technologies.* Oxford, England: Berg.

Scott, D. (2008). *The art and aesthetics of boxing.* Lincoln: University of Nebraska Press.

Shildrick, M. (2001). "You are there, like my skin": Reconfiguring relational economies. In S. Ahmed & J. Stacey (eds.), *Thinking through the skin* (pp. 160–173). London, England: Routledge.

Simonsen, K. (2007). Practice of spatiality and embodied emotions: An outline of a geography of practice. *Human Affairs, 17,* 168–181. https://doi.org/10.2478/v10023-007-0015-8

Turnball, C. (1961). *The forest people.* New York, NY: Simon & Schuster.

van Ingen, C. (2011). Spatialities of anger: Emotional geographies in a boxing program for survivors of violence. *Sociology of Sport Journal, 28*(2), 171–188. http://dx.doi.org/10.1123/ssj.28.2.171

Wacquant, L. (2004). *Body & soul.* Oxford, England: Oxford University Press.

Woodward, K. (2007). *Boxing, masculinity and identity: The "I" of the tiger.* Oxford, England: Routledge.

Young, I. M. (1990). *Throwing like a girl and other essays in feminist philosophy and social theory.* Bloomington: Indiana University Press.

7 · DIFFRACTING MIND-BODY RELATIONS

Feminist Materialism and the Entanglement of Physical Culture in Women's Recovery from Depression

SIMONE FULLAGAR

Finishing our hour-long interview about her insider knowledge of sport policy and the obesity agenda, I turn off my recorder as we leave the meeting room to enter the intimate space of the lift. As we descend slowly from the 17th floor, I casually mention other research that I am doing on women's recovery depression and everyday practices. After a long pause, she says quietly, "I wouldn't be here if it wasn't for sport. My depression became so bad at times. Playing sport kept me going somehow; it kept me alive."

The affective intensity of this research encounter has stayed with me for many years because that particular day, I did not think I was doing mental health research. It serves as a reminder of the ontological inseparability and entanglement of mind and body in all research that intends to focus on *either* mental health or physical practices. To grasp the connectedness of our thoughts, feelings, and senses as they are implicated in embodied movement, we need more nuanced ways of thinking of sport and physical culture as phenomena that materialize through complex biopsychosocial *relations* (as distinct from biopsy models of illness and treatment). In this chapter, I take up the question of how embodied movement matters in mental health with respect to the materiality of *all* knowledge practices (science and social science, qualitative and quantitative research). I draw on feminist and critical posthumanist scholarship to explore the material-discursive relations of depression and recovery (Blackman, 2012; Braidotti, 2013; Barad, 2007; Wilson, 2015). Such questions about *matter* require a thorough rethinking of the onto-epistemological assumptions that inform both naturalistic scientific and social constructivist research on physical activity.

Understanding this complexity means moving *with* the embodied trouble (Lather, 2015) to open up different ways of (un)knowing beyond dualistic categories of thought—mind/body, masculine/feminine, culture/nature, social/biological, objective/subjective. The value of new materialist feminisms lies in the *relational questions* that reorient our thinking about what gendered bodies can "do," how matter "acts," and what "effects" are produced through the production of difference. This approach stands in contrast to the pursuit of truth about what a body *is* or locating the source of meaning in unmediated experience or an essentialized notion of womanhood (Braidotti, 2013; Coole & Frost, 2010; Barad, 2007; Grosz, 1994). There are, of course, nuanced lines of thinking within new materialism, and critiques of supposed newness have importantly been raised by feminist and postcolonial, Indigenous scholars (see Barad, 2014; Ahmed, 2008).

At stake here is a biopolitical issue concerning how mental health problems materialize as culturally imagined, felt, and represented in terms of personal troubles rather than as "public feelings" (Cvetkovich, 2012). From a new materialist perspective, this complex bundle of distressing feelings is produced through an assemblage of forces that territorialize and deterritorialize subjectivity via diagnostic cultures of depression and therapeutic modalities of recovery. As Rose and Abi-Rached (2013) argue, contemporary configurations of depression are discursively positioned as biomedical and psy-problems of chemical and affective deficits located *within* the neurochemical self. Vidal and Ortega (2012) also identify how the sociotechnics of research practices (such as neuroimaging) are implicated in new diagnostic cultures where "depression is torn between biomedical and psychological accounts, between a search for neurological (including chemical and anatomical) causes, and the quest for contextualized explanations" (p. 351).

Yet the assemblage of expert, public, and personal knowledge about mental ill health is not fixed but rather continually reiterated through multiple and often competing practices of treatment, recovery, and prevention. These sites are constituted through the clinical encounter, digital self-help resources, familial narratives, complementary therapies, antipsychiatry and mental health service user movements, policies about well-being, hospital and community care models, and campaigns that challenge stigma. Physical activity and sport are positioned within this biopolitical context as nonpharmacological practices that can contribute to prevention and recovery via the growing expertise of the exercise sciences (Fullagar, 2017b). Exercise has become the new medicine for the soul in a very long genealogy of mind-body practices. This raises a number of questions about how embodied distress materializes through the knowledge practices of research as well as the complex social relations that are implicated in the everyday formation of (healthy or ill, normal or abnormal) subjectivities.

To work through these problematics, I draw together several trajectories of thought that emerge from Karen Barad's (2003, 2007) diffractive analysis as well as Deleuze and Guattari and feminist readings concerned with affective relations and assemblages (Fox & Alldred, 2016; Duff, 2014; Deleuze & Guattari, 1987; Coleman & Ringrose, 2013). Barad's background in quantum physics informs her approach to thinking of *matter* as more than an inert substance. She says, "All bodies, not merely 'human' bodies, come to matter through the world's iterative intra-activity—its performativity" (Barad, 2007, p. 141). In this sense, embodied movement practices are phenomena that are coimplicated in the performance of recovery from depression. As I shall argue through a diffractive analysis, recovery is far more than a process of transforming thoughts, chemicals, *or* interactions "with" the world. Recovery *is* an intra-active, entangled process through which agency is produced and performed in its embodied multiplicity. New materialist thinking troubles common tropes about sport and physical activity as inherently beneficial for mental health and empowering for women (e.g., the This Girl Can, UK campaign; Fullagar & Pavlidis, 2017c). I take a different turn in the latter half of the chapter to explore how embodied movement is coimplicated in the affective, entangled relations that shape women's recovery from depression in an Australian study.

DIFFRACTING KNOWLEDGE: ONTOLOGICAL POLITICS OF GENDERED MINDS AND BODIES

Barad's (2007) diffractive methodology offers a critical mode of reading science and social science together to surface "how differences get made, what gets excluded and how these exclusions matter" (p. 32). Hence a new materialist feminist politics seeks to unsettle, or diffract, the ontological assumptions informing biological, psychological, and sociological knowledge about recovery as it is entangled with depression as a gendered phenomenon. I consider how the gendering of knowledge is produced through three different but related knowledge practices (policy, exercise science, and social scientific research).

Public Health Discourses
Within global public health discourse about depression, gender has been positioned as a psychological or social *variable*. This tends to obscure the ways in

which research produces the phenomena it seeks to know by excluding critical questions about gender that matter. Depression has been identified by the World Health Organization (2017) as the leading cause of disability worldwide, with more women than men being diagnosed with different kinds of depressive "disorders." In these accounts, depression is produced rather unproblematically as a definitive illness category (legitimized via the *Diagnostic and Statistical Manual of Mental Disorders*, Fifth Edition [DSM-5]; see Fullagar, 2017c). Depression is gendered through the identification of individual and social factors that act upon the population of (cisgendered) women as biopsychosocial determinants of mental health. Women's minds and bodies are acted upon by many identified factors, and yet they are also positioned as active agents who require *empowering* programs to improve their own individual well-being. The identification of women as a homogenous population enacts a particular "cut" that aggregates gendered bodies into universal categories that erase understandings of how differences get made "between" bodies (race, sexuality, class, disability, religion, age, etc.; Barad, 2003).

Feminist and critical research have significantly contributed to a critique of how women's emotional lives have become subject to the medicalizing and colonizing imperatives of Western diagnostic cultures. Despair and sadness become articulated through psy-discourses that privilege a dysfunctional brain, problematic thought patterns, or emotional responses over critical analysis of the social forces that shape mental (ill) health in gendered, classed, and racialized ways (Blackman, 2012; Cvetkovich, 2012; Fullagar & O'Brien, 2012; McDermott & Roen, 2016; Metzl, 2010; Mills, 2014; Moncrieff, 2010; Stoppard, 2000; Ussher, 2011)—for example, poverty, childrearing, childhood and family violence, and sexist microaggression on and offline, along with normalized ideals of bodily perfection and success in all life domains. Importantly, scholars have also identified how gender relations are bound up with the mental health of men identifying with different masculinities (Caddick, Smith, & Phoenix, 2015; Cover, 2016) as well as the embodied distress of gender nonconforming and transgender persons (Roen, 2016).

The Scientific Benefits of Exercise

Exercise is now included in a range of government guidelines and mental health promotion campaigns as the new medicine for mental ill health. Elsewhere I have identified how these discourses mobilize a new *corporeal therapeutics* that has become institutionalized with the rise of sport and exercise sciences that promise nonpharmacological solutions to medicalized problems (Fullagar, 2017b). Exercise research (largely informed by various physiological and psychological approaches) promises to offer an important "evidence" base for the value of active embodiment in prevention, treatment, and recovery from depression.

For women, nonpharmacological recovery practices *can* be beneficial. However, exercise science offers a therapeutic mode of knowing that is recouped within a biopolitical context that brackets out the gendered conditions through which distress materializes (as mental illness; Fullagar, 2009).

The discursive formation of the psy-benefits of exercise is understood from a materialist perspective to be more than a social construction of meaning. Rather, knowledge is produced through the material-discursive practices of research across a range of natural science disciplines and techniques that include epidemiological, behavioral, cognitive, and physiological sciences, as well as and neuroscience (on mental health research, see Pickersgill, 2010; on exercise science, see Gibson, 2015). As Barad (2007) would say, the particular "cuts" made in research render knowable the relationship between varying measures (as apparatus) of depression-recovery symptomology and exercise benefits—for example, the aggregation of individual bodies into trial populations (exercise often achieves similar symptom reduction results as antidepressants, but so does the placebo), the measurement of psychobiological changes (such as noradrenergic and opioid effects, increased serotonin synthesis, and reduced levels of proinflammatory cytokines), brain visualization via neuroimaging, and psychological slicing of experience into cognition, mood, or self-efficacy perceptions, to name just a few (Hearing et al., 2016; Harvey, Hotopf, Øverland, & Mykletun, 2010; Wegner et al., 2014). Revealing how such cuts are produced, Barad (2003) demonstrates how diffraction identifies the "shadows in the light" in terms of how the social figures *within* the scientific as an unrecognized process shaping matter as we know it: "This is not a static relationality but a doing—the enactment of boundaries—that always entails constitutive exclusions and therefore requisite questions of accountability" (p. 803).

In this context, the psychologically empowering or biochemically enhancing processes of exercise identified in numerous studies (but certainly not all studies; see Chalder et al., 2013, on negative findings) position the sociocultural as a background against which *natural* bio-psych mechanisms are foregrounded. Questions concerning the coimplicated relations of embodied matter and social context cannot be asked within a naturalistic science paradigm, as they evoke the excluded. Even when social support is acknowledged in shaping exercise behavior, it is often positioned as an aspect of the individual—one's social psychology. Searching for the socioculturally stripped "facts" of matter reiterates mind-body, nature-culture oppositions and instrumental relations to the body, where the reduction of depression symptomology is positioned in a linear, causal relation to exercise. Hence debates often focus on identifying the ideal dose/response for prescribed types of exercise, or at the more humanistic end of the continuum, findings emphasize the role of self-efficacy in determining preferred exercise intensity (Khalil, Callaghan, Carter, & Morres, 2012; Parker et al.,

2016). Despite broader policies that reiterate the need for complex biopsychosocial models of mental (ill) health and recovery, the social becomes a ghostly presence in scientific knowledge that articulates a naturalistic ontology of the mind-body relation with little "accountability" (Barad, 2007) for sociomaterial relations (and the ensuing "evidence" taken up in policy and practice contexts).

From a feminist perspective, what is frequently excluded from naturalistic science is an ontological politics that recognizes the affective relations and social forces that produce and alleviate embodied distress as gendered phenomena. Where gender is acknowledged in exercise science, it is all too often evoked as a static variable and essentialist category of identity (the sameness of all women, sex/gender differences) that ignores differences between women and the normative power relations that constitute feminine subjectivity. For example, one of the first trials to compare the effects of exercise with antidepressant medication and a placebo identified similar rates of "remission" from depression (at 16 weeks) across several interventions: "supervised exercise = 45%; home-based exercise = 40%; medication = 47% compared with the placebo = 31% (p = .057)" (Blumenthal et al., 2007, p. 587). Despite its clearly gendered sample (153 women; 49 men) with a typical patient profile of "age 52 years, female, Caucasian, one prior major depressive episode," the study ignores gender (as well as class, race, sexuality, etc.) in all its knowledge claims (Blumenthal et al., 2007, p. 590). Reading these kinds of studies diffractively, we see how research practices (methodological apparatus, theories, representational forms) work to materialize knowledge of differences (gendered minds and bodies) through cuts that erase, reduce, and simplify complexity.

Social Constructivist Accounts of Active Embodiment

Against scientific reductionism, the qualitative literature on sport and leisure has contributed critical, interpretative, and feminist insights to the complex phenomenological, narrative, psychosocial, and discursive processes that shape mind-body experiences of distress, suffering, and pleasure (Allen-Collinson & Hockey, 2001; Atkinson, 2016; Caddick, Smith, & Phoenix, 2015; Carless & Douglas, 2012; Caudwell, 2014; Fullagar, 2008; Jette & Vertinsky, 2011; Lafrance, 2011; Lee, Frisby, & Ponic, 2014). In relation to exercise interventions, many studies have focused on the positive benefits of sport for mental health—in particular, men's experience of football programs. Yet feminist scholars have importantly identified the need for more critical engagement with the effects of *normalized* gender practices and masculine histories of sport cultures and medicalized categorizations of mental ill health (Spandler & McKeown, 2012). Despite differences in disciplinary perspectives, often common ontological assumptions are at play about the recovering subject as a voluntaristic, intentional self who is coping with neurochemical, cognitive, or interpersonal deficits. Within a constructivist approach, the subject may be more clearly articulated within the social world, but individual agency or social structures are positioned as the *source or determinants* of meaning about depression and recovery. The embodiment of this recovering subject is evoked in a variety of ways—an instrumental site of activation through exercise prescription (Medina, Jacquart, & Smits, 2015), a fleshy background against which cognitive change occurs (Wegner et al., 2014), or a site of therapeutic change where the effects of psychosocial trauma are narrated (Carless & Douglas, 2012). What various disciplines often share is the positioning of recovery from depression as an individualized experience that locates meaning within the interpretative self or the biological body. In this context, the insights of feminist science scholars usefully point to neglected ontological questions by asking how "foundational inscriptions of the nature/culture dualism foreclose the understanding of how 'nature' and 'culture' are formed" (Barad, 2003, p. 828).

Moving beyond simply offering critique, we can draw new insights from reading work across sciences and social sciences through a focus on onto-epistemological concerns about embodied experience as entangled *biopsychosocial* phenomena. The turn toward new materialist questions in sport and physical culture has also begun to open up new spaces of possibility. Arguments have been put forward for transdisciplinary approaches, theory-method formulations, and boundary crossings to examine the materialities of movement (Barker-Ruchti, Grahn, & Lindgren, 2016; Evans, Davies, & Rich, 2014; Fullagar, 2017a; Giardina, 2017; Giardina & Newman, 2011; Larsson & Quennerstedt, 2012; Markula, 2014; Thorpe, 2014, 2016). The matter of embodiment is shifting, and these debates have much to contribute to dialogue about movement and physicality that can reorient understanding beyond the limitations of binary thinking. Inventive theory-methods are also materializing new concepts, such as *naturecultures* (Barad, 2007) and *bodyminds* (Merrell, 2003). In their diffractive analysis of children's own bodymind accounts of eating disorders, Levy, Halse, and Wright (2016) push beyond privileging the discursive *over* matter. Such concerns have informed the diffractive analysis of feminist materialism for some time (on mental health, see Nikoleyczik, 2012; Pitts-Taylor, 2014). In light of these compelling critiques of "representationalist" assumptions (Barad, 2003), how do we think through theory and method together to move beyond a naive epistemology that assumes the world can be known apart from the material-discursive practices of research? Next I explore this question through new materialist thinking on recovery and depression as a jump-off point for the postqualitative research that I then present/perform to empirically orientate such concerns.

THEORIZING THE MATERIALITY OF GENDERED RECOVERY

Within the broader mental health literature, recovery is most commonly articulated in terms of the *personal meaning* that individuals develop to support their sense of emotional well-being beyond the limited, medicalized focus on the remission of symptoms and diagnostic classification (Slade, 2012). The recovering self is frequently positioned as ontologically separate (having personal meanings and embodied responses) but "interacting" with the social world that contributes to the conditions of both distress and change. In contrast to thinking about individual experience as a biological, psychological, or sociological phenomenon, a posthumanist perspective does not privilege human intentionality and instead situates the subject as constituted (normal, ill, well, pathological, successful, etc.) within assemblages of practices, objects, and relations of affect that connect depression and recovery in particular ways (Duff, 2016a, 2016b; Fox, 2015; McCloud, 2014). In this way, we can understand how the embodied specificity of distress (personal thoughts, feelings, actions) becomes *culturally intelligible* through the material, affective, and discursive relations of the contemporary moment (typically as a medical-psychological disorder one recovers from with expert treatment; Fullagar, 2017b).

Feminist debates have pushed the notion of assemblages further by emphasizing how matter is thoroughly *coimplicated* in the entangled relations that produce phenomena via the human and nonhuman, affects, objects, and practices (Barad, 2003). Barad (2007) moves beyond notions of interaction by referring to "intra-action" as a process of coimplication where embodied experiences (such as depression and recovery) are phenomena constituted *through* the relation rather than being positioned as external to it. Recovery practices matter in relation to how depression is materialized and discursively constituted as a gendered phenomenon—usually as a biochemical or psychological disruption to life centered on regaining control, normative responsibility for self (and others), and agentic choice. While the intra-actions of biopsychosocial processes are significant to the matter of recovery, they are not often thought of together in medical models (Blackman, 2012). Hence the problematic location of agency *within* the individual creates further dilemmas in contemporary biomedical health care systems that require the active performance of self-care. The rise of exercise as a social prescription assumes that individuals can invoke the inner "motivation" to overcome the embodied weight and stasis of depression to "do" things (run, ride a bike, go to the gym).

New materialist scholars have contributed to a nuanced understanding of these complex *relations of affect* that act upon the subject to produce and alleviate experiences of distress. Deleuze and Guattari (1987) write against the reduction of emotion to subjective feeling by conceptualizing affect as a relational force, "an ability to affect and be affected. It is a prepersonal intensity corresponding to the passage from one experiential state of the body to another and implying an augmentation or diminution in that body's capacity to act" (p. xvi). Feminist debates have underlined the complexity of affect in understanding injustice by arguing that "the distinction between sensation and emotion can only be analytic" (Ahmed, 2004, p. 6). Emotional states, moods, and thoughts all figure in the experience of embodied distress (Roen, 2016) as complex forces past and present enact gendered subjects who do not always know what is *wrong* with them. For individuals, the attractiveness of diagnostic cultures lies within the promise of certainty and cure, of having these troubling affects translated through expert sense making (e.g., DSM-5, clinical guidelines) to legitimize personal troubles as illnesses to be rationally managed (required by workplace sick leave policies and access to mental health services). In this biopolitical assemblage, the agentic capacities of the self are thus understood to be produced through social relations of different *scales* that connect personal feelings, health service systems, digital mood-tracking tools, and prescribed medication within the (Big Pharma) flows of global capitalism (Duff, 2016a, 2016b; McCloud, 2014). Fox (2015) articulates how this materialist orientation situates agency differently in order to focus

> upon assemblages of human and non-human relations rather than upon individual "emotional" bodies; on flows of affect within assemblages rather than notions of agency attached to humans; and on territorializations and de-territorializations of capacities to do and feel rather than deterministic social structures and fields. (p. 308)

These multiple social forces are negotiated simultaneously and enfolded in the process of territorializing women's subjectivities in often essentializing and contradictory ways. For example, antidepressants are a common medicalized response to women's emotional distress (whether the essentializing logic is biological or psychosocial deficit), and they act on the embodied self through complex, material-discursive forces that are also bound up with placebo effects as well as iatrogenic "side effects" of treatment itself (see Fullagar, 2009; McCloud, 2014; Wilson, 2015). Women's capacities for action, thought, and feeling are entangled somatically with the intense investment of hope, biologically in ways that produce unanticipated effects (numbness, agitation, reduced libido), and socially in terms of accessing support in a diagnostic culture that produces illness identities with little regard for broader inequalities. Within a gendered assemblage of recovery, women are constituted as "expert or irresponsible" biomedical consumers who respond to the moral and gendered imperatives to manage one's health and, significantly, the health of others (the emotional labor of familial care and paid work; Fullagar & O'Brien, 2012). Therapeutic practices of self-management are commonly articulated as important complements to expert treatments for depression to alleviate demand on public mental health care systems within advanced liberal societies. Paradoxically, the increased prescription of antidepressant medication for self-management is bound up with ever-expanding capitalist markets, and hence such a population health strategy increases costs to the state. Yet as Price-Roberston, Manderson, and Duff (2017) argue, such individualized treatment practices contribute to a limited imagining of recovery that ignores the complex assemblages of family life and distributed relations of care. In this sense agency is also refigured to emphasize

> an emergent capacity to manipulate the affects, spaces and events of a body's "becoming well." The always-unfinished event of recovery links human and non-human spaces, bodies, objects and forces in the joint expression of an enhanced capacity to affect (and be affected by) other bodies and spaces. (p. 1)

Taking their argument a step further, we can consider how gendered relations of recovery are intimately entangled with women's capacities, opportunities, and affective modes of becoming ill/well. The agency of the recovering self in this assemblage is performative and thus entangled with heteronormative relations and affects (shame, despair, failure, disconnection) that work to territorialize and reterritorialize certain categories of being (a woman). The movement of affect through Deleuze and Guattari's (1987) machine-like assemblage opens up the possibility of other lines of flight and possibilities of desiring subjectivity beyond the normative. Considering these dynamics, how might the embodied practices in women's lives (sport and physical culture) contribute to, or disrupt, the affective relations of depression through particular "affective and embodied rhythms of recovery" (Duff, 2016a, p. 176)?

Women's sporting bodies have been positioned as vulnerable to exertion, interiorized and racialized, desirable heteronormative objects, or undesirably less than heterosexy feminine, as well as full of fleshy possibility, queer strength, and unrealized capacities (Fullagar & Pavlidis, 2017c; Pavlidis & Fullagar, 2014; Thorpe, 2016). Hence women's sport and physical practices can be understood as produced within particular assemblages of movement and relations of affect—sensory, feeling, moving, fleshy, visceral matter is coimplicated in depression and recovery as embodied phenomena. Exploring this line of thought through empirical research, I ask how embodied movement is also implicated in affective relations that *territorialize* the depressed subject with respect to gender normativity (Deleuze & Guattari, 1987)? Also, what lines of flight are opened up through the materialities of movement to counter the stasis of depression (on metaphors of being stuck or trapped, see Fullagar & O'Brien, 2012)?

RESEARCH AS MATERIAL PRACTICE

In researching the materiality of movement in women's accounts of recovery from depression, I pursued the analytic possibilities informed by postqualitative inquiry (Fullagar, 2017a; St. Pierre, 2011). This theory-method approach values the visceral sense of materiality produced through the apparatus of qualitative research practices (Jackson & Mazzei, 2013). These practices include the affective relation of interviewing 80 Australian women (aged 18–75) about their stories; debriefing in the car, cafes, or airports with my colleague (Wendy O'Brien) after our respective interviews on field trips; engaging with theoretical and empirical literature; and reading-writing with the affective intensities, or glow spots, evoked in participant transcripts as an ongoing analytic process (MacLure, 2013b).

Against homogenizing data into aggregated themes, Jackson and Mazzei (2013) advocate a theoretical "reading of data that is both within and against interpretivism" (p. vii). As scholars working through new materialist approaches argue (Berbary & Boles, 2014; Fox & Alldred, 2016; MacLure, 2013a), the research process is itself an assemblage of forces, relations, and practices (talking, listening, questioning, recording, listening, reading, writing differently). This postpresentational approach attends to the forces of affect, as they are entangled with recognized emotions, to identify what they "do" through particular intra-active relations of embodied movement (MacLure, 2013a). As a critique of the rational, unified subject, analytic attention is turned toward the *relational* as constitutive of a more distributed sense of agency bound up with things, places, objects, senses, and the affective working of power.

As a feminist researcher, my ethical *claim to know* the experiences of other women is only ever partial, precarious, and produced relationally through multiple embodied connections and differentiations (whiteness, sexuality, age, education, embodied histories of trauma, dislocation, and pleasurable movement) and contextual mediations. Hence this feminist orientation is less concerned with creating knowledge *about* women (acknowledging the limits of essentialism and representationalism) and more concerned with how embodied matter contributes to ways of knowing that diffract and trouble normative, gendered practices of othering (Fullagar, 2017a; Barad, 2007). Central to this approach is the orientation of listening through the body as a multiplicity rather than privileging a notion of individuated agency, meaning, or corporeality that emerges through binary thinking. In this sense, postqualitative research is a material-discursive formation through which my own and the participant's bodies and stories are coimplicated in the production of meaning. Manning (2010) troubles the body as a singular entity by emphasizing the relational, "more assemblage than form, more associated milieu than being" (p. 118). To resist slipping into a normalized view of physical culture as either empowering or disempowering, thinking through multiplicity enables the exploration of how embodied movement figures in women's accounts of *doing* recovery practices over time. In this sense, we can trace the processes that constitute embodied recovery as transformation, a becoming or unfolding of spacetimemattering (Barad, 2007) where biopsychosocial forces materialize in gendered ways.

In the analytic example that follows, I worked with the multiplicity of material processes and affective relations that evoked different intensities (suffering, shame, anger, and also pleasure) as part of research. In this way, my analysis involved tracing the *affective intensities* that arose in the methodological-theoretical relation with empirical contexts (embodied listening, writing, reading; Ringrose & Renold, 2014). Responding to materialist critiques of representation, MacLure (2013a) also calls for approaches that involve creative "research practices capable of engaging the materiality of language itself" (p. 658; see also Chadwick, 2016, on embodied poetics). In this case, I write through the fragments of "data" from one interview with Anna—a 25-year-old, middle-class, Anglo-Australian woman—to pursue a method of analyzing the hot spots of depression and recovery that "glow" (MacLure, 2013b). These affective moments reveal how Anna articulated embodied practices (dancing, attending school, riding a bike, cooking) entangled through relations of depression *and* recovery that moved. Different practices, logics, and relations of embodied movement were coimplicated in Anna's account of becoming depressed and anxious as well as the process of transformation or recovery.

Working with this critical-creative analytic, I became attuned to how Anna's struggle with depression and transformation was rendered culturally intelligible, transforming her subjectivity in ways that involved conscious and nonconscious negotiations of biopsychosocial forces. The glow spots ("pressure" and "trying" in the achievement of normative ideals in education and dance) evoked the flows of affect that territorialized her life as a young woman ("worthless"). This account also reveals moments when gendered relations shifted through the process of recovery via embodied practices that generated new kinds of entanglements, desires, and hence agentic capacities. Two accounts are presented as a means of breaking with representational assumptions and attuning readers to multiple readings. The first is composed of fragments spoken by Anna that have been drawn from the interview transcript with very light editing, while the second is rewritten to reposition the subject ("I") within a multiplicity of (un)spoken relations that bring into view the nonhuman and human entanglements of agency. The affective traces of other women's accounts from the broader project are also (un)knowingly entangled with my embodied reading-writing as a researcher.

MATERIALIZING DEPRESSION AND RECOVERY

At 17 I was doing a lot of competitions with ballet.
Umm . . . there's all the pressure from school,
when you're trying
to get a good result.
Then all the pressure with dancing and
trying to pass dance exams.
I was trying to juggle both.
In the middle of grade 12,
I just dropped the ball big time.
I ran out of adrenalin.
Just crying, crying all the time,
crying to be free from pressure, expectation.
If I didn't get all these good marks and
I wasn't the best dancer at my ballet school,
then I actually wasn't worth anything.

My parents took me to the doctor.
He told me I was depressed,
I needed to see a psychiatrist, take the meds.
My bouts of quite severe depression came from
a long time of untreated anxiety.
If I don't get enough sleep,
I really start to unravel.
Now I manage ups and downs better; I self-manage.
Counseling helped for a while.
My psychiatrist knows my history
and my parents' histories; we have all struggled.
If anything goes awry,
she is there with a variety theories.
I've been taking medication for seven years.
It helped my anxiety; I had too many chemicals.
But the medication for depression became part of the numbness;
it was like a bad secret I couldn't share.
Anxiety felt like life was not worth living,
but I needed anxiety to achieve all those things.
I don't know how much the meds helped;
were they just a Band-Aid?
I feel more empowered; I'm coming off it now.
Recovery changes all kinds of body chemicals.

I was doing the female thing, helping out others too.
Now I put energy into myself,
turn it outward, to change things.
I wish I could say I did yoga, but I didn't!
Exercise, sleep, and eating well are pretty important.
I use exercise when I'm really feeling down;
it changes the body chemistry.
But if I'm anxious and I go for a jog or a run,
it kind of exacerbates all the symptoms.
Exercising can be a symptom of anxiety.
So I let my body recover and slowly built myself up for Uni,
but I would always compare myself to others.
I was working 12-hour days, couldn't sleep.
Ending up in hospital, exhausted was a turning point.
Frightened by my physiology,
I had weird body seizures from sleeping pills and herbal remedies.
Doctors accused me of drug taking.
Mum always said I should reduce stress.
I hated the idea that I wasn't like everyone else.

I just went through a period of not being a high achiever;
it was really emancipating.
It's that whole thing of slowing down—
cooking, music, bike riding
(yeah, bike riding is a good one),
riding to work and riding to Uni.
It is the whole point of doing something,
that takes time, rather than not.
Learning how to relax,
doing things that are not "important."

Anna, interviewed aged 25

In contrast, the following account works through a different analytic device diffracted through a new materialist emphasis to displace the speaking "I" as a way of opening up questions about distributed agency and embodied transformations. It makes visible the human and nonhuman relations, the visceral responses and normative modes of feminine subjectivity that are entangled with different affects (intense desires for perfection, the weight of despair and shame, the lightness of pleasure and moments of joy). In writing the following account, the practice of listening for multiplicity was deployed, listening for contradictions, the unspoken gestures and silences between the lines (Chadwick, 2016).

(DIS)ASSEMBLING DEPRESSION

An increasing whirl of ballet shoes, study notes, memorizing dance routines, texting,
  fast talking, reading with sticky notes, stark mirrors   looking.
      Trying harder, the tempo of pressure quickens,
        pen gripped more tightly, pulse racing,
      head swimming in class.
  Fighting fatigue that ripples under the skin.
      Back at the desk, eating quickly, sleep eludes, dance demands, study calls.
        Schedule organized, movement routinized,
            winning/losing, desiring/fearing, succeeding/failing.
        Affective logic pulsates through veins and arteries,
      trying harder to do perfection
for fear of being nothing, worthless woman.
      Movement falters, stumbling with the weight of depression,
        energy drained, vitality emptied.
          The blur of worried faces of friends and family,
        receding under the blanket,
            withdrawing from the world of impossible demands.
              The hands and voices of others pull me through.
                They feed, nourish and care.
          Doctors, therapists, friends, meds.
        The demands of perfection begin to fade,
      crowded out by learning different rhythms.
  Tentative movement, slow steps,
pleasurable, without trying or recriminating self-blame.
Riding to Uni with a gentle breeze, sensual warmth,
  machine-like propulsion, embodied capabilities.
      Pedals, muscles, wheels, streets, breathing harder,
        looking forward, outward, green vistas.
          Returning, immersing oneself in being,
            creatively, beyond the discipline of the desk.
          Cooking experimentally,
            exploring new tastes, textures, and
              the possibility of messing up,
              expectations.

Listening to the tension in Anna's story about the desire to try harder—to be
worth something as a woman—I hear a resonance with McRobbie's (2015)
"top girls" and how becoming agentic is bound up with affective relations,
class aspirations, and competitive success. Desire becomes a territorializing

force—perfecting education and work practices to achieve self-empowerment and embodied success where privilege and failure are closely entangled. Such desires reiterate a logic of empowerment through sport—or in this case, the feminized realm of dance—that is not necessarily "good" for mental health in postfeminist times. In my reading-writing of Anna's account, multiple forces and relations intensify and disrupt the normative desire to perform successful womanhood through education and dance (and be valued in terms of white, middle-class privilege). Pressure materializes through embodied distress, exhaustion of mind and body, as she becomes anxious and depressed—a subject who has failed to achieve an autonomous, singular self-identity. The normative gendered expectations of perfecting womanhood puts Anna in a bind, for "to be different-from means to be worthless-than" (Braidotti, 2013, p. 20). Anna's movement of self through the entanglement of anxiety and depression was shaped by multiple affects (shame in failure, desire to live differently, pleasure in different self-world relations) and material-discursive relations that produced different ways of doing feminine subjectivity. Biochemical discourses permeated Anna's articulation of the embodied intra-actions of energy levels, moods, medication, study, work, and exercise. Familial histories of mental ill health came into play along with negotiations and experimentation with desires for and meanings of study, work, and leisure.

Anna's practices were not articulated in terms of self-conscious "resistance" to gender expectations; rather, they involved a more complex entanglement of desires to perform and escape the exhausting pressures of entrepreneurial feminine self-hood—the subject "of" continual improvement. Hence the alternative logic of slowing down, riding her bike to university, or cooking afforded embodied experiences beyond calculative acquisition (education and dance marks). These generative practices reoriented her relation to self and the human/nonhuman through a different rhythm, immersion, and sense of space-time. Cycling produced an outward orientation, a means of moving away from preoccupying negative thoughts bound up with feelings of self-blame and comparison—the materiality of what psychologists refer to as the cognitive stasis of "rumination." Against the exhaustion of depression, Anna questions whether her recovery was produced by medication, self-awareness through counseling and the psychiatrist's theories, social support, or her capacity to self-manage through different embodied practices (eating, exercising, sleeping, not working). Exercise was entangled in a set of complex relations with depression and anxiety that both exacerbated and alleviated her embodied distress. Contra to policy guidelines that prescribe exercise as an instrumental relation to treat depression, it was not the expenditure of energy or "activity" that mattered. Rather, there were affective possibilities in new and forgotten embodied pleasures—relaxation, immersion in movement, doing "unimportant" things. As Ahmed (2004) says, "Pleasures open bodies to worlds through an opening up of the body to others. . . . Spaces are claimed through enjoyment" (p. 164).

Different modes of physicality are implicated in both depression (dance) and recovery (cycling, cooking), as they are produced through different logics, moral imperatives, and affective power relations, and hence assemblages configure the gendered conditions of agency in particular ways. While conventional medical categories of dis-ease seek to contain and *know* complex affects as depression and recovery, such practices paradoxically position the agentic self as both the source of the problem and the solution (exercising individual willpower, motivation to recover). Yet these feeling states are bound up in affective economies and complex gender relations (sexist and interiorizing institutionalized norms, disciplining affects of success—fear and desire, self-surveillance and othering practices) that produce conditions of uncertainty and locate agency *within* the individual. In their critique of neoliberal notions of empowerment, Harris and Dobson (2015) argue that women who have been victims of violence, harassment, or sexism are culturally positioned as having failed to perform the popular script of empowered, entrepreneurial selves in control of their lives.

Feminist materialism offers a means of extending the argument against privileging an agentic unified self to create spaces for examining how different women are subject to contemporary power relations that materialize and territorialize emotional life. Such relations are produced not simply through the abstract *mediations* of neoliberal culture but through the very material-discursive entanglement of gendered bodies, technologies, and institutional spaces that intensify the individualization of "failed" subjectivities (see also Taguchi & Palmer, 2013). These arguments have implications for how we think about sport and physical culture, as they are commonly articulated as empowering experiences for women's mental health in both community and clinical contexts. Yet as Anna's account suggests, the gendered embodiment of recovery is produced through multiple biopsychosocial intra-actions. In this way, material feminisms contribute to rethinking subjectivity beyond liberal assumptions of a voluntaristic, (ir)rational self by positioning agency as distributed and profoundly coimplicated in relational ontologies, multiple forces, and complex affects.

This question of agency is central to Barad's (2003) argument, and she importantly contributes to a more useful account of the entangled relations that can also shift, diffract, and open up gendered ways of moving through depression. In this sense, the presumption of agentic control, individualized responsibility, or blame for success/failure comes undone:

> Agency is a matter of intra-acting; it is an enactment, not something that someone or something has. . . . This is true not only of the surface or contours of the

body but also of the body in the fullness of its physicality, including the very "atoms" of its being. Bodies are not objects with inherent boundaries and properties; they are material-discursive phenomena. (p. 827)

Understanding how *bodyminds* are coimplicated in movement cultures can turn our attention to the *conditions of possibility* for facilitating different relations and practices of recovery that question, refuse, and transform the micropolitics of gender normativity (Merrell, 2003).

CONCLUDING REMARKS

Throughout this chapter, I have argued that sport and physical culture cannot be understood as neutral activities or instrumental relations to the body that are inherently good or bad for health and emotional well-being. Feminist materialisms offer a different way of thinking through the logics, practices, and affective relations bound up in creating and transforming gendered assemblages of depression and recovery. Taking inspiration from Barad's (2003, 2007) diffractive methodology, we can consider how all research performs agentic *cuts* in the way active embodiment and mental (ill) health are rendered intelligible. Despite the desire to accurately measure, carefully observe, or sensitively interpret personal accounts of depression within the world or the lab, the complex (political) interplay of biopsychosocial forces shaping embodied movement are often not acknowledged within the epistemic cultures of exercise science *or* social science. Feminist materialisms reveal the limitations of thinking embodied movement as an instrumental "exercise" of the body, a source of biochemical effect, or activated by inner motivation as an individualized form of empowerment. Much social science research in sport and physical activity has also largely assumed that matter is "mute," and hence there remains a challenge to think through the complex interplay between the *bodyminds* and *naturecultures* that has been missing in accounts of cultural "inscriptions" that shape "the body" (Barad, 2007; Merrell, 2003).

Rethinking embodiment opens up policy and practice questions about how everyday social relations within and beyond the domains of sport and physical culture can support a different kind of agentic capacity. How can we move beyond the rhetoric of individualized choice that pervades health and sport policies in an era of austerity that is profoundly changing the nature and extent of community leisure services, public mental health provision, and gender-based services? Individualized, humanist notions of agency only perpetuate a paradox where recovery from depression requires an inner motivation to move, which in turn is undermined by powerful affective states of despair, sadness, and disconnection. Understanding how recovery is produced through an assemblage of meanings draws attention to the multiplicity of experiences that need to be considered with questions of gender justice and inclusion: embodied movement becoming a habit over time; connecting confidence, skill and strength, know-how and diverse pleasures that involve humans and nonhumans; and negotiating discomfort, bodily change, and shifting moods (Duff, 2016a). Research on sport and physical culture can significantly contribute to new forms of praxis that acknowledge how "recovery is not solely about changing aspects of individual people but is about changing the places in which we all live and recover" (Yates, Holmes, & Priest, 2012, pp. 111–112). Refiguring depression and recovery as embodied matters can serve to make visible gendered experiences and complicate reductionist pharmacological "solutions" to open up lively modes of becoming well.

REFERENCES

Ahmed, S. (2008). Open forum imaginary prohibitions: Some preliminary remarks on the founding gestures of the "new materialism." *European Journal of Women's Studies, 15*(1), 23–39.

Ahmed, S. (2004). *The cultural politics of emotion.* New York, NY: Routledge.

Allen-Collinson, J., & Hockey, J. (2001). "Runners" tales: Autoethnography, injury and narrative. *Auto/Biography, 6* (1 & 2), 95–106.

Atkinson, M. (2016). The suffering and loneliness of the fell runner. In G. Molnar & L. Purdy (eds.), *Ethnographies in sport and exercise research* (pp. 96–110). New York, NY: Routledge.

Barad, K. (2003). Posthumanist performativity: Toward an understanding of how matter comes to matter. *Signs: Journal of Women in Culture and Society, 28*(3), 801–831.

Barad, K. (2007). *Meeting the universe halfway: Quantum physics and the entanglement of matter and meaning.* Durham, NC: Duke University Press.

Barad, K. (2014). Diffracting diffraction: Cutting together-apart. *Parallax, 20*(3), 168–187.

Barker-Ruchti, N., Grahn, K., & Lindgren, E. C. (2016). Shifting, crossing and transforming gender boundaries in physical cultures. *Sport in Society, 19*(5), 615–625.

Berbary, L., & Boles, J. (2014). Eight points for reflection: Revisiting scaffolding for improvisational humanist qualitative inquiry. *Leisure Sciences, 36*(5), 401–419.

Blackman, L. (2012). *Immaterial bodies: Affect, embodiment, mediation.* London, England: Sage.

Blumenthal, J., Babyak, M., Doraiswamy, M., Watkins, L., Hoffman, B., Barbour, K., . . . Herman, S. (2007). Exercise and pharmacotherapy in the treatment of major depressive disorder. *Psychosomatic Medicine, 69*(7), 587–596.

Braidotti, R. (2013). *The posthuman.* Cambridge, MA: Polity Press.

Caddick, N., Smith, B., & Phoenix, C. (2015). Male combat veterans' narratives of PTSD, masculinity, and health. *Sociology of Health & Illness, 37*(1), 97–111.

Carless, D., & Douglas, K. (2012). The ethos of physical activity delivery in mental health: A narrative study of service user experiences. *Issues in Mental Health Nursing, 33*(3), 165–171.

Caudwell, J. (2014). "Feeling blue": The ordinary pleasures of mundane motion. *Qualitative Research in Sport, Exercise and Health, 7*(3), 309–320.

Chalder, M., Wiles, N. J., Campbell, J., Hollinghurst, S. P., Haase, A. M., Taylor, A. H., . . . Winder, R. (2013). Republished research: Facilitated physical activity as a treatment for depressed adults: Randomised controlled trial. *British Journal of Sports Medicine, 47*(10), 629–629.

Chadwick, R. (2016). Embodied methodologies: Challenges, reflections and strategies. *Qualitative Research*. doi:1468794116656035.

Coleman, R., & Ringrose, J. (eds.). (2013). *Deleuze and research methodologies.* Edinburgh, Scotland: Edinburgh University Press.

Coole, D., & Frost, S. (eds.). (2010). *New materialisms: Ontology, agency, and politics.* Durham, NC: Duke University Press.

Cover, R. (2016). *Queer youth suicide, culture and identity: Unliveable lives?* London, England: Routledge.

Cvetkovich, A. (2012). *Depression: A public feeling.* Durham, NC: Duke University Press.

Deleuze, G., & Guattari, F. (1987). *A thousand plateaus: Capitalism and schizophrenia.* Minneapolis: University of Minnesota Press.

Duff, C. (2016a). G major to D major to a minor 7 (a progression to recovery). In J. Coffey, S. Budgeon & H. Cahill. (eds.), *Learning bodies: The body in youth and childhood studies* (Vol. 2, pp. 173–189). Singapore: Springer.

Duff, C. (2016b). Atmospheres of recovery: Assemblages of health. *Environment and Planning A, 48*(1), 58–74.

Duff, C. (2014). *Assemblages of health: Deleuze's empiricism and the ethology of life.* Rotterdam, Netherlands: Springer.

Evans, J., Davies, B., & Rich, E. (2014). We/you can tell talk from matter: A conversation with Håkan Larsson and Mikael Quennerstedt. *Sport, Education and Society, 19*(5), 652–665.

Fox, N., & Alldred, P. (2016). *Sociology and the new materialism: Theory, research, action.* London, England: Sage.

Fox, N. (2015). Emotions, affects and the production of social life. *British Journal of Sociology, 66*(2), 301–318.

Fullagar, S. (2017a). Post-qualitative inquiry and the new materialist turn: Implications for sport, health and physical culture research. *Qualitative Research in Sport, Exercise and Health, 9*(2), 1–11.

Fullagar, S. (2017b). Mind-body relations. In M. Silk, A. Andrews, & H. Thorpe. (eds.), *The Routledge international handbook on physical cultural studies* (pp. 401–441). London, England: Routledge.

Fullagar, S. (2017c). Foucauldian theory. In B. Cohen (ed.), *Routledge International Handbook of Critical Mental Health* (pp. 63–69). London, England: Routledge.

Fullagar, S., & Pavlidis, A. (2017c). Feminist theories of emotion and affect in sport. In L. Mansfield, B. Wheaton, J. Cauldwell, & B. Watson. (eds.), *Handbook of feminism in sport, leisure and physical education* (pp. 447–462). Houndsmills, England: Palgrave.

Fullagar, S., & O'Brien, W. (2012). Immobility, battles, and the journey of feeling alive: Women's metaphors of self-transformation through depression and recovery. *Qualitative Health Research, 22*(8), 1063–1072.

Fullagar, S. (2009). Negotiating the neurochemical self: Anti-depressant consumption in women's recovery from depression. *Health, 13*(4), 389–406.

Fullagar, S. (2008). Leisure practices as counter-depressants. *Leisure Sciences, 30*(1), 1–18.

Gibson, K. (2015). *A witness to fitness: Contextualising processes in exercise physiology* (Unpublished doctoral dissertation). University of Toronto, Canada.

Giardina, M. (2017). (Post?)qualitative inquiry in sport, exercise, and health: Notes on a methodologically contested present. *Qualitative Research in Sport, Exercise and Health,* 9(2), 258–270.

Giardina, M., & Newman, J. I. (2011). What is this "physical" in physical cultural studies? *Sociology of Sport Journal,* 28(1), 36–63.

Grosz, E. (1994). *Volatile bodies: Toward a corporeal feminism.* Sydney, Australia: Allen & Unwin.

Harris, A., & Dobson, A. S. (2015). Theorizing agency in post-girlpower times. *Continuum,* 29(2), 145–156.

Harvey, S. B., Hotopf, M., Øverland, S., & Mykletun, A. (2010). Physical activity and common mental disorders. *British Journal of Psychiatry,* 197(5), 357–364.

Hearing, C. M., Chang, W. C., Szuhany, K. L., Deckersbach, T., Nierenberg, A. A., & Sylvia, L. G. (2016). Physical exercise for treatment of mood disorders: A critical review. *Current Behavioral Neuroscience Reports,* 3(4), 350–359.

Khalil, E., Callaghan, P., Carter, T., & Morres, I. (2012). Pragmatic randomised controlled trial of an exercise programme to improve wellbeing outcomes in women with depression: Findings from the qualitative component. *Psychology,* 3(11), 979–988.

Jackson, A. Y., & Mazzei, L. A. (2013). Plugging one text into another: Thinking with theory in qualitative research. *Qualitative Inquiry,* 19(4), 261–271.

Jette, S., & Vertinsky, P. (2011). "Exercise is medicine": Understanding the exercise beliefs and practices of older Chinese women immigrants in British Columbia, Canada. *Journal of Aging Studies,* 25 (3), 272–284.

Lather, P. (2015). The work of thought and the politics of research. In N. K. Denzin and M. D. Giardina (eds.), *Qualitative inquiry and the politics of research* (pp. 97–107). Walnut Creek, CA: Left Coast Press.

Lafrance, M. N. (2011). Reproducing, resisting and transcending discourses of femininity: A discourse analysis of women's accounts of leisure. *Qualitative Research in Sport, Exercise and Health,* 3(1), 80–98.

Larsson, H., & Quennerstedt, M. (2012). Understanding movement: A sociocultural approach to exploring moving humans. *Quest,* 64(4), 283–298.

Lee, D. S., Frisby, W., & Ponic, P. (2014). Promoting the mental health of immigrant women by transforming community physical activity. In L. Greaves, A. Pederson, & N. Poole (eds.), *Making it better: Gender transformative health promotion* (pp. 111–122). Toronto, Canada: Canadian Scholars' Press.

Levy, G., Halse, C., & Wright, J. (2016). Down the methodological rabbit hole: Thinking diffractively with resistant data. *Qualitative Research,* 16(2), 183–197.

MacLure, M. (2013a). Researching without representation? Language and materiality in post-qualitative methodology. *International Journal of Qualitative Studies in Education,* 26(6), 658–667.

MacLure, M. (2013b). Classification or wonder? Coding as an analytic practice in qualitative research. In B. Coleman & J. Ringrose (eds.), *Deleuze and research methodologies* (pp. 164–183). Edinburgh, Scotland: Edinburgh University Press.

Manning, E. (2010). Always more than one: The collectivity of a life. *Body and Society,* 16(1), 117–127.

Markula, P. (2014). The moving body and social change. *Cultural Studies ↔ Critical Methodologies,* 14(5), 483–495.

McDermott, E., & Roen, K. (2016). *Queer youth, suicide and self-harm.* London, England: Palgrave Macmillan.

McLeod, K. (2014). The missing work of collaboration: Using assemblages to rethink antide-pressant action. *Contemporary Drug Problems, 41*(1), 109–142.

McRobbie, A. (2015). Notes on the perfect: Competitive femininity in neoliberal times. *Australian Feminist Studies, 30*(83), 3–20.

Medina, J. L., Jacquart, J., & Smits, J. A. (2015). Optimizing the exercise prescription for depression: The search for biomarkers of response. *Current Opinion in Psychology, 4,* 43–47.

Merrell, F. (2003). *Sensing corporeally: Toward a posthuman understanding.* Toronto, Canada: University of Toronto Press.

Metzl, J. M. (2010). *The protest psychosis: How schizophrenia became a black disease.* Boston, MA: Beacon Press.

Mills, C. (2014). *Decolonizing global mental health: The psychiatrization of the majority world.* New York, NY: Routledge.

Moncrieff, J. (2010). Psychiatric diagnosis as a political device. *Social Theory and Health, 8*(4), 370–382.

Nikoleyczik, K. (2012). Towards diffractive transdisciplinarity: Integrating gender knowl-edge into the practice of neuroscientific research. *Neuroethics, 5*(3), 231–245.

Parker, A., Hetrick, S., Jorm, A., Mackinnon, A., McGorry, P., Yung, A., . . . Purcell, R. (2016). The effectiveness of simple psychological and physical activity interventions for high prev-alence mental health problems in young people: A factorial randomised controlled trial. *Journal of Affective Disorders, 196,* 200–209.

Pavlidis, A., & Fullagar, S. (2014). *Sport, gender and power: The rise of roller derby.* Farnham, England: Ashgate.

Pickersgill, M. D. (2010). Psyche, soma, and science studies: New directions in the sociology of mental health and illness. *Journal of Mental Health, 19*(4), 382–392.

Pitts-Taylor, V. (2014). Cautionary notes on navigating the neurocognitive turn. *Sociological Forum, 29* (4), 995–1000.

Price-Robertson, R., Manderson. L., & Duff, C. (2017). Mental ill health, recovery and the family assemblage. *Culture, Medicine, and Psychiatry, 1–24.* doi:10.1007/s11013-017-9522-2.

Ringrose, J., & Renold, E. (2014). F** k rape! Exploring affective intensities in a feminist research assemblage. *Qualitative Inquiry, 20*(6), 772–780.

Roen, K. (2016). The body as a site of gender-related distress: Ethical considerations for gen-der variant youth in clinical settings. *Journal of homosexuality, 63*(3), 306–322.

Rose, N. S., & Abi-Rached, J. M. (2013). *Neuro: The new brain sciences and the management of the mind.* Princeton, NJ: Princeton University Press.

Slade, M. (2012). Everyday solutions for everyday problems: How mental health systems can support recovery. *Psychiatric Services, 63*(7), 702–704.

Spandler, H., & McKeown, M. (2012). A critical exploration of using football in health and welfare programs gender, masculinities, and social relations. *Journal of Sport & Social Issues, 36*(4), 387–409.

Stoppard, J. (2000). *Understanding depression: Feminist social constructionist approaches.* Lon-don, England: Routledge.

St. Pierre, E. (2011). Post qualitative research: The critique and the coming after. In N. K. Denzin & Y. S. Lincoln. (eds.), *Sage handbook of qualitative inquiry* (4th ed., pp. 611–635). Los Angeles, CA: Sage.

Taguchi, H., & Palmer, A. (2013). A more "livable" school? A diffractive analysis of the per-formative enactments of girls' ill-/well-being with (in) school environments. *Gender and Education, 25*(6), 671–687.

Thorpe, H. (2014). Moving bodies beyond the social/biological divide: Toward theoretical and transdisciplinary adventures. *Sport, Education and Society, 19*(5), 666–686.

Thorpe, H. (2016). Athletic women's experiences of amenorrhea: Biomedical technologies, somatic ethics and embodied subjectivities. *Sociology of Sport Journal, 33*(1), 1–13.

Ussher, J. (2011). *The madness of women: Myth and experience*. London, England: Routledge.

Vidal, F., & Ortega, F. (2012). Are there neural correlates of depression? In S. Choudhury & J. Slaby (eds.), *Critical neuroscience: A handbook of the social and cultural contexts of neuroscience* (pp. 345–366). Chichester, England: John Wiley & Sons.

Wegner, M., Helmich, I., Machado, S., Nardi, A., Arias-Carrion, O., & Henning Budde, A. (2014). Effects of exercise on anxiety and depression disorders: Review of meta-analyses and neurobiological mechanisms. *CNS & Neurological Disorders-Drug Targets, 13* (6), 1002–1014.

Wilson, E. A. (2015). *Gut feminism*. Durham, NC: Duke University Press.

World Health Organization. (2017). Depression and other common mental disorders: Global health estimates. Geneva, Switzerland: WHO. License: CC BY-NC-SA3.0 IGO.

Yates, I., Holmes, G., & Priest, H. (2012). Recovery, place and community mental health services. *Journal of Mental Health, 21*(2), 104–113.

8 · TOWARD A MULTISPECIES SPORT STUDIES

SAMANTHA KING

Over the past 20 years, the consumption of protein powder has moved from the niche realm of serious bodybuilding to the cultural mainstream. Alongside established claims about the role of protein in the production of muscle mass and recovery from exercise, the rapid expansion of this $7.9 billion market has been built on a growing list of new promises. Regular ingestion of protein powder is now recommended as a boost to a variety of bodily becomings ranging from weight loss to increased bone density, glossy hair to flawless skin, energy to satiety, and healthy pregnancy to active aging. This calorie-dense foodstuff is also promoted as a salve for the frantically busy and the culinarily challenged. As protein powder has become an everyday staple of North American diets, as both an advertised and hidden ingredient, punishing weight routines or endurance training are no longer customary precursors to protein supplementation—nor, for that matter, is physical activity of any kind.[1]

The consumer love affair with protein powder has developed within the context of a much broader cultural infatuation with protein in all forms. For the moment at least, sugar and milk are out and fat is back in favor, while protein continues its reign as the transcendent nutrient of our time, despite the environmental disaster and appropriation of Indigenous, forested, and subsistence farming land generated by its overconsumption in the form of meat, dairy, fish, and soy. Protein deficiency in the industrialized West is extremely rare, except among a small number of elderly adults, but the almost magical association of amino acids with lean limbs, boundless energy, and physical and intellectual performance is daily drawn tighter as manufacturers inject protein into substances ranging from bread to chocolate bars, and consumers keep coming back for more.

I began to think about protein powder as an object of study because of persistent encouragement from my fitness coaches to incorporate pre- and post-workout shakes into my daily routine. Initially, I was simply skeptical in the way one tends to be of nutritional fads and preoccupied with the incongruity of tall, strong, well-fed gym patrons lining up at the grocery store checkout to buy huge tubs of highly processed, chalky residue to supplement their "clean eating" regimens. To make sense of this, I could see the potential of a biopolitical-economic-discursive analysis of the emergence of protein powder as a culturally meaningful and marketable consumer product—the kind of analysis I have been wont to do over the course of my career. At a certain point, however, I began to think about the matter differently. I remained interested in demystifying the allure of this fine, granular substance by unveiling the human-powered social structures that have shaped its emergence, but I also became intrigued by the work performed by other life forms—from dairy cows to protein molecules—to produce nutritional shakes.

I was a vegetarian of more than 30 years but had only recently started to think about the disconnect between my dietary practices and my scholarship. How this came to be is a long and complex story that changes each time I try to pin it down. But suffice it to say that it emerges from a conjuncture of affective, intellectual, and social experiences that include falling in love with a dog; reading Jonathan Safran Foer's best-selling book *Eating Animals*; grading an intriguing paper by a graduate student, Victoria Millious, on the "vegan killjoy"; experimenting with increasingly elaborate plant-based cooking techniques; and being sucked into the giant vortex that is contemporary bourgeois ethical eating talk. During this time, I found myself drawn to the burgeoning literature in critical animal studies, food studies, and political ecology, increasingly aware of the anthropocentrism of my work and more cognizant of the intellectual and political limits of centering human bodies, actors, and experiences in my analyses. What this conjuncture has amounted to—gradually, haltingly, and circuitously—is a commitment to try to not look away from questions of the animal, or at least to notice when I do. And so it was the dairy cows who coproduce protein powder (or the whey version of it that is the most popular kind) that ultimately prompted me to move beyond my methodological and theoretical comfort zones.

Notwithstanding the existence of a small corpus of critical literature on animals in sport, and a much larger poststructuralist literature that takes humanistic approaches to sport as its focus of critique, scholars of physical culture trained in Eurocentric traditions, myself included, have tended to assume the essential humanness of the moving body that constitutes our object of study. This is true even in the present volume, where the ontological status of corporeal matter is open to question yet often divorced from its own animality. When nonhuman animals are centered analytically in sport scholarship (Gillett & Gilbert, 2014;

Young, 2014), it is rare that their vulnerability and suffering, their passions and pleasures, or their embodied affectivity are understood to cross the species line. That the notion of species remains largely unhistoricized and untouched by analyses of power despite prevailing scientific and philosophical views that reject hard-and-fast lines between life forms is one of the challenges we face as a field. What would it mean to set aside our species essentialism as part of our ongoing effort to grapple with a range of critical theoretical perspectives that challenge universalist, Eurocentric narratives about the essence and quality of the human?

Neel Ahuja (2016) is not concerned with sport per se, but he usefully categorizes these perspectives into two camps: the biopolitical and the posthuman. Biopolitical analysis of the sort spearheaded by Michel Foucault tends to foreground the contested category of the human and the configurations of race, sex, and class that differentially discipline bodies and populations. Foucault did acknowledge in a series of lectures on the "government of things" that rationalities and technologies of control cross—and indeed help define—boundaries between human and more-than-human worlds (Foucault, 2007; Lemke, 2015), but he and those he continues to inspire have only occasionally noted the work of nonhuman forces in the operation of power. In posthumanist analysis, on the other hand, the more-than-human world is taken seriously, and demarcations among species (and other organic and inorganic things) are understood as permeable, fluid, and contingent, with the contested category of the human understood as just one life form among many. While posthumanist perspectives draw attention to the liveliness and agency of nonhuman life and matter, they have a tendency, Ahuja (2016) argues, "to project upon an outside, the nonhuman, the possibility of resistance to anthropocentrism" (p. viii). He and other scholars (Belcourt, 2015; Kim, 2015; Weheliye, 2014; Weisberg, 2009) have noted how posthumanist work of this sort can work as a license to brush aside longstanding feminist, queer, black, Indigenous, and left humanist critiques of subjectivity, power, and economic and social injustice as if these forces are somehow exterior to the "worlds of bodies, physical matter, and interspecies exchange" (Ahuja, 2016, p. ix). Rather than abandon biopolitical and posthumanist traditions, however, Ahuja sees potential in their cross-fertilization. He argues that expanding and reinvigorating the proper zone of the political to include interspecies relations is vital to understanding the durability of a system of empire that severs the natural from power and history while concomitantly "managing bodies in unequal planetary conjunctions of life and death" (p. x). Furthermore, a postcolonial biopolitical analysis infused with posthumanist sensibilities allows for an understanding of these bodies and environments not as "empty containers of human political subjects" but as "lively, transitional assemblages of political matter" (p. x).

Inspired by the epistemological and political potential of theorizing across dynamic life kinds but wary of the traps that Ahuja outlines, this chapter addresses the following questions: What happens to conceptualizations of multispecies sport if unequivocal definitions of the "animal" and the "human" are problematized? What might a critical, political, multispecies approach offer to the study of physical culture? What difference does a shift from "the (human) body" to "(animal and other) bodies" make? I attempt to answer these questions in three stages: First, I establish a conceptual framework that deemphasizes human-animal differences while concomitantly recognizing how resources and the capacity to flourish are unevenly distributed among living things; second, I use this framework as a lens through which to review the extant literature on animals and sport; and third, I offer a preliminary, multispecies analysis of the more-than-human foodstuff that is whey protein powder.

Located at the intersection of four fields—sport studies, animal studies, environmental studies, and food studies—my argument draws most explicitly on the work of Neel Ahuja, Val Plumwood, and Matthew Calarco to imagine a scholarly approach in which the human/animal binary no longer operates as a fundamental organizing principal of thought and practice; in which the human is decentered, ecologized, and understood in relation to other beings; in which nonhuman agency, culture, and creativity are taken seriously as objects of study; and in which the colonial, racial, gender, and sexual capitalism that constrains both human and nonhuman capacities is critiqued and resisted.

This work is motivated by a defining challenge of the contemporary world: the environmental crisis and the urgent need to challenge the aggressive harnessing of broad swathes of planetary life to the accumulation of capital and territory. It is a primary contradiction of our time that those steeped in Enlightenment notions of a fundamental divide between nature and culture exhibit a growing acceptance of shared sentience, emotion, and culture among humans and animals, a shift perhaps most obviously exemplified in the mass conversion of companion animals into fully fledged members of the bourgeois nuclear family. And yet this reorientation has emerged in tandem with an unprecedented level of suffering among nonhuman animals, including those incarcerated and tortured in the factory farming system and those displaced and destroyed by the decimation of their natural habitats (the latter often in the service of the former). Inspired by decolonial thinkers such as Billy-Ray Belcourt (2015), who names settler-colonial expansion and white supremacy as the conditions of possibility for animal domestication, speciesism, and other contemporary forms of human-animal relations, I am curious about ways of thinking, models of research, and modes of practice that enable movement beyond the assimilation or disavowal of animal subjectivity and thus the settler state. This is a tentative analysis. It is proffered not as a way to highlight the failings of existing approaches or as a call to abandon a focus on human physical culture but as a modest inquiry into a number of ideas that may have the potential to enrich our understandings of embodiment, relationality, and matter in a manner that probes rather than denies the incorporation of multiple bodies into assemblages of inequality and violence.

ECOLOGICAL EMBODIMENT AND INDISTINCTION

My review of the literature on animals in sport is inspired by feminist ecophilosopher Val Plumwood (1993, 2002, 2012), whose life's work focused on challenging the anthropocentrism that shapes human relationships with animals and the environment. Plumwood's writing largely preceded the emergence of the term *posthumanism* within critical humanities and social science discourse, but her insistence on acknowledging the animality and embodiment of humans alongside the culture and minds of animals (Mathews, Rigby, & Rose, 2012) has been crucial to the development of contemporary research in this tradition. She was not the first author to challenge the Eurocentric theoretical assumption that humanity is fundamentally distinct from and superior to nature by highlighting the intelligence and agency of the latter, but her work was unique in two related ways: first, she recognized that human domination and exploitation of nature works in tandem with colonial capitalism and heteropatriarchy; and second, she argued that ecological struggles must reflect this complexity if they are to close the conceptual gulf between nature and culture and otherwise succeed in provoking social change. Across a variety of sites, and drawing on Indigenous ontologies of relational personhood, Plumwood (2012) sought to envisage humans as "ecologically embodied beings" akin to other animals (p. 16) even as systems of economic rationality and colonial rule work ruthlessly against such imaginings.

Plumwood's work is particularly useful for my purposes because of her interest in food as a primary site through which human embodiment is denied. She turned her attention in this direction following a traumatic, near-death attack by a large crocodile while she was kayaking alone in Australia's Kakadu National Park. Plumwood (2000) writes of her struggle with the crocodile as a "shocking reduction" in which she was transformed from a "complex human being to a piece of meat" (n.p.). She continues, "Reflection has persuaded me that not just humans but any creature can make the same claim to be more than just food. We are edible, but we are also much more than edible. Respectful, ecological eating must recognize both of these things." Plumwood's hope was that an acknowledgment that humans can be used as food would present an opening to "affirm principles emphasizing human-animal mutuality, equality, and reciprocity" (p. 18).

Matthew Calarco (2015) draws on Plumwood's work and that of scholars such as Donna Haraway, Giorgio Agamben, and Gilles Deleuze to develop what he calls an "indistinction" approach. Like Ahuja (though with much less emphasis on colonial systems), he sees virtue in drawing out the biopolitical dimensions of multispecies relationships. For Calarco, this means interrogating the effects of the foundational split between animal life and "properly political" human life that Agamben (2004) understands to determine who is admissible into the sphere of governance (as problematic as that might be) and who is not. Drawing attention to the way that biopolitics works to deem disposable those judged to be inhuman or insufficiently human, Calarco argues that the human must be displaced as the privileged subject position and the center of ethical reflection. Calarco (2015) elaborates the concept of indistinction by emphasizing humans' "shared zone of exposed embodiment" with animals (p. 115). He offers this approach in contrast to "identity-based" theories that have dominated traditional animal rights philosophy and that make arguments for animal justice based on animals' "morally relevant" similarity to humans in terms of such capacities as sentience, subjectivity, and intentionality (Singer, 1975; Regan, 1983; Cavalieri, 2001). In their attachment to the human as the measuring stick against which ethical consideration is extended to other living beings, identity-based approaches are limited in their ability to address the exclusion and precarity of humans and other life forms who are not sufficiently like "us" to warrant consideration. In contrast, an indistinction approach encourages scholars to consider the ways humans are like their animal kin and to "appreciate the richly differentiated modes of existence found among animals" (Calarco, 2015, p. 100).

Sport studies boasts a longstanding critique of how mind/body, culture/nature, civilized/primitive dualisms have shaped how human physical activity is understood—and often denigrated—as an object of study and a social practice. But this critique has operated on a parallel track to work on animals in sport. The result is that connections of the kind that Plumwood and Calarco draw—which proceed from the assumption that anthropocentrism as a value system shapes embodied practices and relationships across all spheres of life and that we have much to learn from thinking of animal and human bodies indistinctly—have for the most part not been made. Nor has the literature probed in any sustained manner how multispecies relationships are bound up in projects of empire and the government of race, gender, and sexuality, as Ahuja's work inspires scholars to do. As I will explain in what follows, while radically relational, antianthropocentric, anticolonial approaches have not yet permeated the sport studies literature, some provocative resonances are beginning to emerge.

MORE-THAN-HUMAN SPORT: FROM
THE MARGINS TO THE CENTER?

James Gillett and Michelle Gilbert (2014) write in their introduction to *Sport, Animals, and Society*, the only edited collection of its kind, "More so than in many areas of study, there is a history of scholarship in sport that considers animals" (p. 3). The literature to which they refer was until recently largely historical and philosophical, a state of affairs Kevin Young (2014) bemoans in the introduction to a special issue of the *Sociology of Sport Journal* titled "The Sociology of Animals in Sport." Calling sociologists of sport "blinkered" about the "animal-sport complex," Young describes the issue as "one small corrective gesture aimed at encouraging a less speciesist Sociology of Sport and a commitment to a steadier flow of human-animal research" (p. 388). While I sympathize with the politics of Young's project, his strictly disciplinary view of what counts as proper sociology results in a rather narrow rendering of extant work and overlooks the considerable overlap among sociological, historical, anthropological, and philosophical approaches that has defined this area of study from the very start and that I seek to capture here.

It is possible to parse extant literature on animals and sport into two broad camps: one that explores animal questions as a route to a better understanding of human experiences and another that approaches animals in relation to humans but also as interesting and important in and of themselves. The first, more anthropocentric set of scholarship tends to use pursuits such as hunting (Carr, 1986; Egan, 2014; Elias & Dunning, 1986; Franklin, 1998; Griffin, 2007; Holt, 1989; Howe, 1981; Marvin, 2001; May, 2013; Ridley, 1990; Stokvis, 1992), horseracing (Riess, 2011; Vamplew, 1976), or dog- and cockfighting (Geertz, 1973; Hawley, 1989; Holt, 1989; Worden & Darden, 1992;) as vehicles through which to explore questions pertaining to nationalism, class history and politics, and the role of the state in managing, justifying, or resisting human violence toward nonhuman animals. Despite the key role played by the human treatment of animals in motivating this work, animal experiences and emotions are usually overlooked. Instead, the focus is on understanding the ritualistic, social, or political elements of these activities for their human participants—an approach that has the effect of rendering animal subjectivities, suffering, and pleasure largely invisible.

In contrast, a broader and more diverse literature centers animal subjectivities or at least acknowledges the co-constituted character of multispecies sport. This literature includes ethnographic research on such topics as the relations between human and nonhuman animals that comprise activities such as foxhunting (Marvin, 2001, 2002; Hurn, 2007); the less discernable forms of abuse that occur in the breeding, training, competing, and discarding of racing greyhounds (Atkinson & Young, 2005); and intersectional analyses of narratives about pit bulls who are trained to fight, their owners, and their rescuers (Nair, 2016; Tarver, 2014; Weaver, 2015). It also includes more abstract philosophical inquiries about animal rights (Morris, 2014; Wade, 1990, 1996), the virtue of compassion (Tymowski, 2014), and ecofeminism (Kheel, 1996) and representational critiques of animal celebrity, animal athletes, and animal mascots (Carey, Millington & Prouse, 2014; Gilbert & Gillett, 2011; Nance, 2014). A final strand within this literature explores interspecies communication and cooperation in the mutual ethics of caring between women adventurers and their horses (Gilbert & Gillett, 2011); efforts to transform or theorize equestrianism toward less hierarchical, more reciprocal human-horse relationships (Dashper, 2016; Gillett & Gilbert, 2014; Hansen, 2014; Smith, 2014; Thompson & Birke, 2014); and the promise of dog agility as a model of playful, mutually attuned cross-species companionship (Haraway, 2003; Lund, 2014). Although some work within this broad category highlights the abuse, exploitation, and objectification of animals, there is also a notable emphasis on the emotional, relational, and agentic character of animal interactions with humans within sporting contexts. To my knowledge, none of these analyses has been explicitly articulated as a challenge to the anthropocentric emphases of earlier work in the field. But the influence of Donna Haraway's (2003, 2008) writings and her rejection of the "foolishness of human exceptionalism" are readily apparent as scholars of human-animal sport work with the assumption that "becoming is always becoming *with*" and share Haraway's (2008) insistence on the pleasurable intimacy, shared vulnerability, and co-composition of multispecies subjectivities, embodiments, and relationships (p. 244).

Even sport scholarship that gives greater consideration to animal experiences, multispecies relationships, and forms of community that humans and animals develop and share tends to participate in what Calarco (2011) calls a "kind of anthropocentric extensionism" insofar as it grants priority to those animals who mostly closely resemble human beings (p. 45). That is to say, the disproportionate emphasis on domesticated mammals, specifically horses and dogs, reflects not simply the fact that these are among the animals most frequently engaged in human-driven sporting endeavors but that these animals display what are taken to be superior communication skills and intelligence, a heightened sense of subjectivity and awareness, and the ability to form and maintain familial bonds (Calarco, 2011). These are the animals that are most like humans, in other words, and thus deemed worthy of scholarly attention and moral and legal standing.

Michael Atkinson and Kass Gibson's (2014) schema, which is designed to encourage scholars to see beyond the obvious and offer a fuller account of the numerous ways that animals are enfolded into interspecies sport practices, is useful in making visible the inevitable and consistent co-presence of humans and animals in all aspects of physical culture. They write of animals as combatants (e.g., in pitfighting or agility), as prey (e.g., in sport hunting or fishing), as food (e.g., at tailgate parties), as paraphernalia (e.g., as the providers of skin and other tissue to make sports equipment), as transport (e.g., in equestrian sports or dog sledding), as mascots, as the dispossessed (e.g., when displaced by the development of recreational spaces such as golf courses), and as passive receptors (a broad category that includes the use of animals in sport science research). Such conceptual structures help build a sense of the multitude of life forms with which humans interact (consciously or not) when undertaking physical activity and begin to point to the implications of sporting humans in complex ecosystems.

Reflecting a broader "ecological turn" within the critical social sciences and humanities, Atkinson and Gibson's (2014) essay forms part of a small body of research that has started to explore human physical culture as ineluctably distributed through and dependent on so-called natural life, animal and otherwise (Atkinson, 2014; Millington & Wilson, 2014, 2016; Schaffner, 2009; Weedon, 2015). In this work, existing frameworks are expanded to consider plant species and the more mundane, less perceptible, "beyond the field" (Atkinson, 2014, p. 427) participation of nonhumans in sporting cultures and spaces such as hunting, golfing, birding, and mud running. As such, this scholarship acts as a bridge between the environment and sport literature, in which animal subjectivities and culture tend to fade into the background, and the animals and sport literature, in which the environment tends to do the same. It also represents, as Brad Millington and Brian Wilson (2016) suggest in their analysis of sand dunes as environmental actors in disputes about golf courses, a challenge to the "anthropocentric pathway" that "belies the dynamism of physical cultural contexts" (p. 12).

That my own tentative retreat back down that pathway was prompted by an experience with protein powder—a substance that appears as the most processed, most human-driven of foods, the deadest of previously living stuff—may seem ironic. But as I hope to show in the discussion that follows, an analysis of the bodies and lives that are enlisted, often at great cost, in the constitution and metabolism of desiccated whey reveals much about the theoretical and political potential of taking seriously the more-than-human matter of sport.

TOWARD A POLITICAL ECOLOGY OF PROTEIN POWDER

From my initial foray into the history of protein powder, I learned that the production of desiccated whey emerged from an attempt to squeeze surplus value from the toxic effects of dairy industry by-products. A key source of the industry's waste management problems is declining milk consumption. The popularity of milk reached its peak in the United States during World War II and has declined quite sharply in both the United States and Canada over the past 30 years. This was also the period in which artificial insemination, hormone treatment, and other new technologies resulted in each cow producing four times the quantity of milk they generated under nonindustrialized conditions (Wiley, 2014). Moreover, a decades-long public health campaign to persuade people to eat less saturated fat has resulted in a shift toward skimmed milk consumption and a surplus of whole milk and milk fat (Moss, 2010). Due in no small part to the marketing strategies of Dairy Management, an arm of the U.S. Department of Agriculture whose mission is to "build demand for dairy products," this surplus has slipped back into North American diets in the form of cheese (https://www .dairy.org/; Moss, 2010). The necessary campaign to encourage people to eat more cheese has been remarkably successful: In 2010, Americans ate triple the quantity they consumed in 1970 (Colvington, 2014). While this shift in eating habits has helped dairy farmers stay in business, because only 10% of the milk used in cheese production can be turned into curds, the move from whole milk to skim milk and from milk to cheese produced another undesirable side effect: a massive surplus of whey. Cheese manufacturers thus began dumping untreated liquid whey into waterways and municipal sewers or spreading it on agricultural land with little regard for the effect on groundwater. With its high protein and acid content and thus heavy biochemical oxygen demand, dairy wastewater is about 10 times the strength of municipal wastewater and 175 times more potent than untreated sewage (Smithers, 2008). Whey waste disrupts water ecology by spurring aquatic plant growth and denying fish and other forms of aquatic life the oxygen they need to thrive. Its dumping has thus been environmentally devastating, creating ecological dead zones across the planet. As public complaints and hearings led to the establishment of water protection agencies and increasingly strict environmental legislation, researchers began trying to refine filtration, concentration, and drying techniques to divert waste whey into animal feed, food additives, and powder for nutritional supplements.

With Gavin Weedon, who has now joined me as a collaborator on this project, I have begun to conceptualize the history of protein powder as part of a broader set of relationships that we are calling "protein ecologies." Describing our project in this way signals our interest in exploring not only what humans do with animal protein or what meanings we assign to it but what animal protein as a dynamic and lively material-discursive subject does to humans and other life forms as it circulates in and through diffuse assemblages of bodies and environments. This may sound like a physiological inquiry, and in one sense, it is: We must educate ourselves about how natural scientists understand the substance and work of protein if our attempt to take its agentic capacities seriously is to be credible. The scope of our analysis exceeds the biological, however, and our purpose is not harnessed to biomedical therapy or enhancement. Instead, we seek to weave a metabolism of amino acid motion and mutation with a political, ecological rendering of the animal labor, biopolitical human-body projects, and technoscientific knowledges and practices that produce and process whey powder in a commodity form. This means following, through ethnographic and textual methods, the chains of amino acids that travel and mutate through and with bodies and machines in a complex dance of dispersion and assemblage; the embodied labor of human farmworkers and cows whose exploitation are mutually imbricated; the unruly and environmentally destructive proteins constitutive of whey effluent and disciplined through elaborate processes of food engineering; and the ongoing appropriation and contamination of Indigenous land and water through industrial agriculture in the name of satiating settler appetites (Belcourt, 2015).

As protein circulates through multiple bodies, as it is composed, ingested, broken down, secreted, excreted, and regenerated, it creates surpluses and deficits of value, whey, and waste, strengthening some bodies and compromising others, including those of nonhuman animals. Consider how the reproductive capacities of cows and their bodily organs and physiological processes, together with the labor of the human dairy workers (largely racialized migrants in the United States) who toil in this most dangerous of industries, are put to work in order to supply other bodies with protein. Consumer bodies, human and animal, are in turn instrumentalized as living garbage disposal units charged with consuming recycled toxic waste. For those privileged humans for whom such protein-waste is surplus to their nutritional needs, it can be converted into other forms of capital, such as improved health, antiaging aesthetics, and musculature. Yet the cultivation and satiation of strong, lean, efficient bourgeois bodies are indelibly connected to the profound alienation of the human and animal workers who help provide such nourishment—alienated, that is, from both social relations and their own bodies and bodily functions. Moreover, the regeneration of whey powder does not stop once it mutates into muscle, or indeed into fat if the powder is not taken as prescribed and instead incorporated into the body without appropriate amounts of energy expenditure.

To follow protein around as it moves and mutates through multiple bodies, to consider what happens to the energy human bodies cannot absorb because they consume, on average, more than they need, to trace where protein surplus lands when it "comes out the other end" is to open the analysis up to a whole new layer of questions. Dr. Stu Phillips, a prominent figure in research on the impact of nutrition and exercise on human muscle protein turnover, consistently emphasizes that there is a limit to how much protein the body can use. Protein is the only thing that humans eat that has nitrogen in it, but once the body has done what it can with the protein it ingests, the excess nitrogen is, according to

Phillips, "fundamentally toxic" (Ledin, 2014). Here he is referring to the excess nitrogen that forms into ammonium ions as amino acids are broken down in the liver. Some of these ions are used in ongoing processes of biosynthesis and some—"a large percentage," according to environmental scientist Arnold Mulder (2003)—are converted to urea and excreted. Nitrogen, like the whey protein that begets it, is a compound that is "recalcitrant" in nature (Westgate & Park, 2010, p. 5352), and its removal from wastewater is costly and energy intensive. This new layer of questions might thus take up the inability of current treatment processes to respond with ecological sensitivity to increased human consumption of whey and other forms of animal protein. Regardless of where such inquiry ultimately leads, it is evident that whey protein is a resilient and undisciplined biocultural subject (Frost, 2016; Myers, 2015). As such, it compels an analytic approach that acknowledges the implication of sport and exercise projects in ecologies that exceed and complicate individual, human-centric, or discursive renderings of embodiment.

CONCLUSION

Trailing protein's movements and metamorphoses with a critical sensitivity of the sort that Ahuja displays results in a rather dystopic rendering, one that might seem rather removed from Plumwood's or Carlarco's imaginings of an ethics of shared ecological embodiment with its attendant pleasures and vulnerabilities. But all three authors' focus on challenging human exceptionalism, tracking relationality, exposing indistinctions, and attending to the uneven distribution of precarity, suffering, and death is what makes an analysis of this type—and the imaginings of alternative futures it implies—possible. Their work encourages scholars to conceptualize animal subjectivities and agentic matter not beyond but *as* power and history and to be attentive to the often subtle, nonandrocentric pressures and limits that make "life itself an ongoing site of social and political contestation" (Ahuja, 2016, p. 26). Such an approach entails not simply adding animals to existing anthropocentric analyses of sport and physical culture but reorienting our theory and practice so that we might account for the complex and dispersed forms and trajectories of more-than-human physical culture.

This being said, the present chapter is not offered as a wholesale challenge to existing approaches to the study of (animals and) sport. It is perhaps best understood as drawing attention to a potential opening—a way for scholars to expand and complicate their conceptualizations of relational embodiment in the context of the virulent exploitation and degradation of more-than-human life that defines the contemporary world. Put differently, while ecological indistinction, like other modes of thought, will not be appropriate for grappling with all times and all places, the transcorporeal circulation of protein lends itself to this analytic; it even perhaps demands it.

NOTE

I am grateful to Adam Ali for research assistance and to Gavin Weedon for ongoing conversations about protein and its afterlives.

1. For examples of this discourse in both popular and academic contexts, as well as critiques of the preoccupation with protein powder, see Chai, 2017; Mathews, 2016; Phillips, Chevalier, and Leidy, 2016; and Robinson, 2017.

REFERENCES

Agamben, G. (2004). *The open: Man and animal.* Stanford, CA: Stanford University Press.

Ahuja, N. (2016). *Bioinsecurities: Disease interventions, empire, and the government of species.* Durham, NC: Duke University Press.

Atkinson, M. (2014). The terrier [men]. *Sociology of Sport Journal, 31*(4), 420–437.

Atkinson, M., & Gibson, K. (2014). Communion without collision: Animals, sport, and interspecies co-presence. In J. Gillett & M. Gilbert (eds.), *Sport, animals, and society* (pp. 268–290). New York, NY: Routledge.

Atkinson, M., & Young, K. (2005). Reservoir dogs greyhound racing, mimesis and sports-related violence. *International Review for the Sociology of Sport, 40*(3), 335–356.

Belcourt, B. R. (2015). Animal bodies, colonial subjects: (Re)locating animality in decolonial thought. *Societies, 5*(1), 1–11.

Calarco, M. (2011). Identity, difference, indistinction. *CR: The New Centennial Review, 11*(2), 41–60.

Calarco, M. (2015). *Thinking through animals: Identity, difference, indistinction.* Stanford, CA: Stanford University Press.

Carey, R. S., Millington, R., & Prouse, C. (2014). Branding boundaries: Colonial sporting identities and the racialized body. In J. Gillett & M. Gilbert (eds.), *Sport, animals, and society* (pp. 209–232). New York, NY: Routledge.

Carr, R. (1986). *English fox hunting: A history.* London, England: Weidenfeld and Nicolson.

Cavalieri, P. (2001). *The animal question: Why nonhuman animals deserve human rights.* Oxford, England: Oxford University Press.

Colvington, C. (2014, March 11). Cheese consumption's phenomenal increase: Will it continue? *Progressive Dairyman.* Retrieved from http://www.progressivedairy.com/blogs/guest-blog/cheese-consumptions-phenomenal-increase-will-it-continue

Dairy Management Inc. (2017). *About DMI.* Retrieved from http://www.dairy.org/

Dashper, K. (2016). *Human–animal relationships in equestrian sport and leisure.* New York, NY: Routledge.

Egan, S. (2014). National identity and cultural resonance in English foxhunting movements. *National Identities, 16*(1), 71–90.

Elias, N., & Dunning, E. (1986). *The quest for excitement: Sport and leisure in the civilizing process.* Oxford, England: Blackwell.

Foucault, M. (2007). *Security, territory, population: Lectures at the Collège de France, 1977–78.* New York, NY: Palgrave.

Franklin, A. (1998). Naturalizing sports: Hunting and angling in modern environments. *International Review for the Sociology of Sport, 33*(4), 355–366.

Frost, S. (2016). *Biocultural creatures: Towards a new theory of the human.* Durham, NC: Duke University Press.

Geertz, C. (1973). *The interpretation of cultures: Selected essays.* New York, NY: Basic Books.

Gilbert, M., & Department of Sociology, McMaster University, Hamilton, Ontario, Canada. (2014). Trust in interspecies sport. *Sociology of Sport Journal, 31*(4), 475–491. doi:10.1123/ssj.2013-0084.

Gilbert, M., & Gillett, J. (2011). Equine athletes and interspecies sport. *International Review for the Sociology of Sport, 47*(5), 632–643.

Gillett, J., & Gilbert, M. (2014). *Sport, animals, and society.* New York, NY: Routledge.

Griffin, E. (2007). *Blood sport: Hunting in Britain since 1066.* New Haven, CT: Yale University Press.

Hansen, N. C. (2014). Young equestrians: The horse stable as a cultural place. In J. Gillett & M. Gilbert (eds.), *Sport, animals, and society* (pp. 233–250). New York, NY: Routledge.

Haraway, D. J. (2003). *The companion species manifesto: Dogs, people, and significant otherness* (Vol. 1). Chicago, IL: Prickly Paradigm Press.

Haraway, D. J. (2008). *When species meet.* Minneapolis: University of Minnesota Press.

Hawley, F. F. (1989). Cockfight in the cotton: A moral crusade in microcosm. *Crime, Law and Social Change, 13*(2), 129–144.

Holt, R. (1989). *Sport and the British: A modern history.* Oxford, England: Oxford University Press.

Howe, J. (1981). Fox hunting as ritual. *American Ethnologist, 8*(2), 278–300.

Hurn, S. (2007). *Clan of the fox? "Hunting" subculture in a rural Welsh farming community* (Unpublished doctoral dissertation). University of London, University College London (United Kingdom).

Kheel, M. (1996). The killing game: An ecofeminist critique of hunting. *Journal of the Philosophy of Sport, 23*(1), 30–44.

Kim, C. J. (2015). *Dangerous crossings: Race, species, and nature in a multicultural age.* Cambridge, England: Cambridge University Press.

Ledin, E. (2014, April 17). The protein interview: An interview with Dr. Stuart Phillips. Retrieved from http://www.leanbodiesconsulting.com/articles/the-protein-interview-an-interview-with-dr-stuart-phillips/

Lemke, T. (2015). New materialisms: Foucault and the "government of things." *Theory, Culture & Society, 32*(4), 3–25.

Lund, G. (2014). Taking teamwork seriously: The sport of dog agility as an ethical model of cross-species companionship. In J. Gillett & M. Gilbert (eds.), *Sport, animals, and society* (pp. 101–126). New York, NY: Routledge.

Marvin, G. (2001). Cultured killers: Creating and representing foxhounds. *Society and Animals, 9*(3), 273–292. doi:10.1163/156853001753644417.

Marvin, G. (2002). Unspeakability, inedibility, and the structures of pursuit in the English foxhunt. In N. Rothfels (ed.), *Representing animals* (pp. 139–158). Bloomington: Indiana University Press.

Mathews, F., Rigby, K., & Rose, D. (2012). Introduction. In V. Plumwood (ed.), *The eye of the crocodile* (pp. 1–6). Canberra, Australia: ANU E Press.

May, A. N. (2013). *The fox-hunting controversy, 1781–2004: Class and cruelty.* Farnham, England: Ashgate.

Millington, B., & Wilson, B. (2014). The masters of nature: Golf, nonhumans, and consumer culture. In J. Gillett & M. Gilbert (eds.), *Sport, animals, and society* (pp. 52–68). New York, NY: Routledge.

Millington, B., & Wilson, B. (2016). An unexceptional exception: Golf, pesticides, and environmental regulation in Canada. *International Review for the Sociology of Sport, 51*(4), 446–467.

Morris, S. P. (2014). The ethics of interspecies sports. In J. Gillett & M. Gilbert (eds.), *Sport, animals, and society* (pp. 127–139). New York, NY: Routledge.

Moss, M. (2010, November 6). While warning about fat, US pushes cheese sales. *New York Times*. Retrieved from http://www.nytimes.com/2010/11/07/us/07fat.html

Mulder, A. (2003). The quest for sustainable nitrogen removal technologies. *Water Science and Technology, 48*(1), 67–75.

Myers, N. (2015). *Rendering life molecular: Models, modelers and excitable matter*. Durham, NC: Duke University Press.

Nair, Y. (2016, August 11). Racism and the American pit bull. Retrieved from https://www.currentaffairs.org/2016/08/racism-and-the-american-pit-bull

Nance, S. (2014). A star is born to buck: Animal celebrity and the marketing of professional rodeo. In J. Gillett & M. Gilbert (eds.), *Sport, animals, and society* (pp. 173–191). New York, NY: Routledge.

Plumwood, V. (1993). *Feminism and the mastery of nature*. New York, NY: Routledge.

Plumwood, V. (2000, July–August). Surviving a crocodile attack. *Utne Reader*. Retrieved from https://www.utne.com/arts/being-prey

Plumwood, V. (2002). *Environmental culture: The ecological crisis of reason*. Abingdon, England: Psychology Press.

Plumwood, V. (2012). *The eye of the crocodile*. Canberra, Australia: ANU E Press.

Regan, T. (1983). Animal rights, human wrongs. In H. B. Miller & W. H. Williams (eds.), *Ethics and animals* (pp. 19–43). New York, NY: Humana Press.

Ridley, J. (1990). *Fox hunting*. London, England: Collins.

Riess, S. A. (2011). *Major problems in American sport history: documents and essays*. Boston, MA: Houghton Mifflin.

Schaffner, S. (2009). Environmental sporting: Birding at superfund sites, landfills, and sewage ponds. *Journal of Sport & Social Issues, 33*(3), 206–229.

Singer, P. (1975). *Animal liberation*. New York, NY: Avon Books.

Smith, S. J. (2014). Human-horse partnerships: The discipline of dressage. In J. Gillett & M. Gilbert (eds.), *Sport, animals, and society* (pp. 35–51). New York, NY: Routledge.

Smithers, G. W. (2008). Whey and whey proteins—from "gutter-to-gold." *International Dairy Journal, 18*(7), 695–704.

Stokvis, R. (1992). Sports and civilization: Is violence the central problem? In E. Dunning & C. Rojek (eds.), *Sport and leisure in the civilizing process* (pp. 121–136). Basingstoke, England: Palgrave Macmillan.

Tarver, E. C. (2014). The dangerous individual('s) dog: Race, criminality and the "pit bull." *Culture, Theory and Critique, 55*(3), 273–285.

Thompson, K., & Birke, L. (2014). "The horse has got to want to help": Human-animal habituses and networks of reliability in amateur show jumping. In J. Gillett & M. Gilbert (eds.), *Sport, animals, and society* (pp. 69–84). New York, NY: Routledge.

Tymowski, G. (2014). The virtue of compassion: Animals in sport, hunting as sport, and entertainment. In J. Gillett & M. Gilbert (eds.), *Sport, animals, and society* (pp. 140–154). New York, NY: Routledge.

Vamplew, W. (1976). *The turf: A social and economic history of horse racing.* New York, NY: Viking.

Wade, M. L. (1990). Animal liberationism, ecocentrism, and the morality of sport hunting. *Journal of the Philosophy of Sport, 17,* 15.

Wade, M. L. (1996). Sports and speciesism. *Journal of the Philosophy of Sport, 23,* 10.

Weaver, H. (2015). Pit bull promises inhuman intimacies and queer kinships in an animal shelter. *GLQ: A Journal of Lesbian and Gay Studies, 21*(2–3), 343–363.

Weedon, G. (2015). Camaraderie reincorporated: Tough Mudder and the extended distribution of the social. *Journal of Sport & Social Issues, 39*(6), 431–454.

Weheliye, A. G. (2014). *Habeus viscus: Racializing assemblages, biopolitics, and black feminist theories of the human.* Durham, NC: Duke University Press.

Weisberg, Z. (2009). The broken promises of monsters: Haraway, animals and the humanist legacy. *Journal for Critical Animal Studies, 7,* 22–62.

Westgate, P. L., & Park, C. (2010). Evaluations of proteins and organic nitrogens in wastewater treatment effluents. *Environmental Science and Technology, 44*(14), 5352–5357.

Wiley, A. S. (2014). *Cultures of milk.* Cambridge, MA: Harvard University Press.

Worden, S., & Darden, D. (1992). Knives and gaffs: Definitions in the deviant world of cockfighting. *Deviant Behavior, 13*(3), 271–289.

Young, K. (2014). Toward a less speciesist sociology of sport. *Sociology of Sport Journal, 31*(4), 387–401.

9 · REIMAGINING THE DANCING BODY WITH AND THROUGH BARAD

MARIANNE CLARK

We take our places at the barre, fingertips resting gently on the smooth, rounded surface. The pianist starts to play, and the swell of melody fills the room. Breath mingles with music. Muscles contract and release. Knees soften to bend. Through a series of productive contradictions, our bodies perform the plié. There is no identifiable starting point for movement; rather, movement materializes through the multiple flows of energy and connections created among bodies, music, and space. As we move, these flows of energy intensify, and boundaries among individual bodies, the barre, the floor, and music seemingly dissolve.

This description of performing a plié is reconjured from extensive field notes recorded throughout my dissertation inquiry examining how adolescent recreational ballet dancers experience and make sense of their dancing bodies (Clark & Markula, 2017; Clark, 2017). Guided by Foucault's (1972, 1979) theoretical concepts of anatomo-political (disciplinary) power and discourse, I sought to expand on existing feminist analyses that focus primarily on the representative ballet body and that seek to determine its meaning. These scholarly critiques, deeply haunted by aesthetics, generally conclude that the ideologically and discursively constructed ballet body is highly feminized, disciplined, and therefore oppressive for girls and women. However, as a feminist scholar interested in the moving, physically active body and as a dancer with an extensive history of studying and performing ballet, I was left wondering what else we might know about the ballet body beyond the discursive meanings ascribed to it.

What about the ballet body in motion and its sweat and its sinews? What about its ability to exert, express, and exhilarate? At the time, I was immersed in Foucault's writings, and his interest in the body and its practices resonated with my curiosities. Therefore, drawing on Foucault, I sought to exceed critiques of the representational ballet body by examining its everyday practices, the contexts in which these practices occur, and the body's entanglements with power relations.

Throughout the study, which involved prolonged participant observations and interviews with adolescent girl dancers in a recreational dance setting, I was struck by the vitality of the ballet body and the relationships it created with the other bodies, things, and spaces around it. A few weeks into data collection, I began to participate in ballet class with participants as part of a methodological improvisation of sorts. Dancing alongside my participants prompted complex and powerful corporeal and emotional responses and a messy but vitalizing flood of rememberings, yearnings, and new relationships with my body, the participants' bodies, and the spaces and things around me. These responses prompted me to notice ballet bodies as somehow evocative, as other than passive entities that can be somehow observed, known, and represented. Yet when it came to analyzing data (field notes, interview transcripts) and writing up my "results," I found myself unable to adequately articulate and theorize what had so clearly shimmered when conducting the inquiry. I distrusted my corporeal and emotional responses, concerned they might lead me down the slippery slope of essentialism or to an exercise in navel-gazing. And I could not think of where these evocative pieces of data might otherwise "fit" within my theoretical framework. As a result, the liveliness of the ballet body and its doings, its capacity to evoke responses and alter other bodies around it (including my own), continued to shimmer—but just beyond my periphery, beyond my ability to articulate and theorize.

Grappling with this tension prompted a theoretical and methodological rethinking, and as a result, I believe the project yielded insights that elaborate on existing research (I outline these insights later in this chapter). However, there remained (and remains) a persistent sense that there is more to think and know about the ballet body and its doings, its capacity to exert and evoke. Yet the question of how to go about this knowing remains. How do we engage in inquiry that privileges the life of bodies and things and acknowledges and explores materiality? How does the material come to matter?

As the editors of this book suggest, the new materialist turn provides a fruitful and invigorating space in which to explore these very questions. This growing corpus of work (see Alaimo & Hekman, 2008; Barad, 2003, 2007; Bennett, 2010; Coole & Frost, 2010; Dolphijn & Van der Tuin, 2012) shares a commitment to theorizing the dynamism of matter. It seeks to achieve this by refusing dualist understandings of matter and discourse as conceptually distinct from each other and by recognizing matter (bodies, nature, things) as agentic, as always engaged in mutually co-constitutive relations with the discursive. In so doing, new materialisms disrupt the privileging of discourse that accompanied the linguistic turn without opposing or refuting its importance, a move that would see us resort back to dualisms (Alaimo & Hekman, 2008). Rather, new materialisms urge us to attend to the ways in which matter and discourse become entangled and to what emerges through these entanglements (Barad, 2007, 2012).

This acknowledgment of the vitality and agency of matter resonates deeply with my lingering curiosities about how we might think and know the ballet body in previously unthought ways. In this chapter, I revisit selections of data from my dissertation project through a new materialist lens in order to reimagine the ballet body and its doings. Specifically, I reread these data "chunks" by "plugging into" (Jackson & Mazzei, 2012, p. 1) the theoretical writings of feminist scholar and physicist Karen Barad (2003, 2007) in an effort to elaborate on and theorize the energetic life of the body. Barad (2003, 2007, 2012) insists on accommodating the material, pointing out that without acknowledging the material, our understanding of the discursive itself is limited. This is because material and discursive forces are in fact inseparable—something she compellingly demonstrates through quantum physics—and always co-constitutive of each other.

This ontological shift provides new, if conceptually challenging, ways to acknowledge the dynamism of the ballet body without essentializing the body as a starting point for meaning (Hekman, 2010). The goal is not to produce knowledge that stands against or trumps previous insights or theorizations. Rather, plugging in allows for movement of thought and the putting to work of multiple ways of thinking. The material I plug into Barad is based on three encounters with ballet bodies (in the change room, in the studio space, and in performance) that I returned to over and over again when writing up my dissertation, that filled pages of field notes, and that troubled and excited me but with which I did not know what to do (Jackson & Mazzei, 2012). In this chapter, I engage plugging in as a process that allows me to rethink these encounters and reimagine the vitality of ballet bodies and their agentic capacities. To contextualize this exercise in reimagining, I briefly outline how the ballet body has been understood within current dance literature. I then outline the utility of new materialism to meaningfully extend these understandings.

CONTEMPORARY UNDERSTANDINGS OF BALLET BODIES

As dance creeps stealthily from the realm of high art into the mainstream, dancing bodies increasingly inhabit our media landscape. Visible on the silver screen, on prime-time television, and splashed across the pages of women's fitness magazines, dancing bodies have shimmied into the spheres of health and fitness and entertainment and popular cultures. Dance is also one of girls' and women's favorite recreational physical activities (Clark, Spence, & Holt, 2011; Dowda, Dishman, Porter, Saunders, & Pate, 2007; Greiser et al., 2006), and research suggests that in Canada, where my dissertation project was set, close to one million girls are enrolled in dance classes each year (Solutions Research Group, 2014). Given its popularity and the growing presence of dance within physical cultures, it presents a fascinating and important area of feminist physical cultural study. Ballet is particularly intriguing, as it is one of the most recognizable forms of dance and considered "foundational" for young dancers starting out.

Despite the popularity of recreational ballet, much ballet scholarship has focused on the representations and experiences of professional and elite ballet dancers (Aalten, 2004; Alexias & Dimitropoulou, 2011; Dryburgh & Fortin, 2010; Green, 2001, 2003; Kleiner, 2009; McEwen & Young, 2011; Pickard, 2012, 2013; Ritenburg, 2010; Wainwright & Turner, 2004; Wainwright, Williams, & Turner, 2005). This research has yielded myriad critiques of the discursive and ideological constructions of ballet bodies as well as in-depth examinations of the social environments in which these bodies are produced. The extraordinary aesthetic of the ballet body—easily conjured as exceedingly thin, flexible, and graceful—emerges as one of the most salient concepts through which the ballet body is imagined (Aalten, 2004; Alexias & Dimitropoulou, 2011; Green, 2001, 2003; Kleiner, 2009; Ritenburg, 2010). Through this imagining, the ballet body is generally posited as a passive substance that is shaped, disciplined, and oppressed by culture and discourse (Markula & Clark, 2017).

The critical and discursive approaches of the above reviewed research largely emphasize the problematic aesthetics of the discursively constructed ballet body and leave the biological ballet body unexamined. In contrast, researchers from health disciplines have examined the biological body closely to determine the incidence, prognosis, and treatment of injury and eating disturbances among (mostly elite) ballet dancers (Bowerman, Whatman, Harris, & Bradshaw, 2015; Caine, Goodwin, Caine, & Bergeron, 2015; Campoy et al., 2011; Gamboa, Roberts, Maring, & Fergus, 2008; Ringham et al., 2006). Pickard (2015) notes that much of the research focusing on ballet injuries objectifies the body through processes of measurement of diagnosis, resulting in an understanding of the biological body as inert, a machine to be acted upon and understood only through scientific processes conducted by rational human agents. These studies rarely consider the social or material contexts in which the ballet body experiences injury or the social, material, and emotional implications of injury. From this research we know that ballet injuries tend to occur primarily in the ankle, foot, and hips and that female dancers experience injury at a higher rate than male dancers. Considerable effort is spent identifying risk factors, prevention and treatment strategies, and pathologies of these body parts. Consequently, the ballet body is further objectified as it becomes fragmented and disassembled. The body as an interrelated system remains unacknowledged, as does the relationship between the body and its social and physical environment.

Although the human science perspective acknowledges the fleshy, biological body, as in the critical research reviewed above, the body is conceptualized as inert and passive. Interpretive dance scholars have taken yet a different "meat and bones" (Foster, 1997, p. 235) approach to studying the lived experiences and contexts of (elite) ballet dancers through ethnographic (Aalten, 2004; Wulff, 2008) and interview studies (Alexias & Dimitropoulou, 2011; Pickard, 2015). While the ballet body aesthetic is also emphasized in these studies, other experiences of ballet bodies emerge. For example, ballet is often described as a means through which dancers can express emotions and an authentic self (Aalten, 2004; Pickard, 2015). Ballet training and performance, even when grueling, can also give way to intense and pleasurable physical experiences. Although less research has focused on recreational ballet dancers, the limited studies that exist report similar experiences of pleasure (O'Flynn, Pryor, & Gray, 2013; Pickard, 2013, 2015; Wellard, Pickard, & Bailey, 2007). Young ballet dancers (ages 11–13) in a study by Wellard and colleagues (2007) expressed feeling "free" when they danced, and Pickard (2015) notes that ballet can provide a means of escape from the mundane demands of everyday life.

Although this overview is brief, these accounts of enjoyment and creative self-expression provide a reprieve from the overarching critiques that position ballet as an oppressive and feminizing practice. However, much of this research imagines the ballet body as an unmediated surface through which an authentic inner self can be located and expressed. The actual fleshy body and its doings remain underexplored, and it is unclear how the materiality of the body is actively implicated (if at all) in the production of expression and pleasure. Nevertheless, there is something about the ballet body and the complex ways it interacts with the cultural and material world that seems to prompt meaningful experiences for girls and women. There is something about the ballet body that matters.

THINKING BALLET BODIES BEYOND DOCILITY: WHAT ELSE CAN WE KNOW?

In an early essay, cultural scholar Angela McRobbie (1991) called for dance to be recognized as an important feminist project, as it provides a means through which the private culture of femininity, particularly adolescent femininity, can be extended into the public sphere. She argues that dance provides an embodied opportunity for social fantasy and the active negotiation of the material and social constraints of conventional femininity. For McRobbie (1991), dance is "both a dramatic display of self and the body, with an equally dramatic negation of the self and the body" (p. 144). I argue that McRobbie's call resonates even more deeply today given the increasingly complex sociopolitical and technological landscapes in which adolescent girls' bodies and subjectivities are produced. In my dissertation work, dance practice emerged as a productive means through which girls could navigate these complex social worlds. Participants described the physical practice of ballet as one way they actively negotiated and navigated social expectations to look, dress, and act according to particular discursive norms (Clark, 2017; Clark & Markula, 2017). They further expressed that the physical performance of ballet allowed them to use and move their bodies forcefully, expressively, and in ways not accommodated or encouraged in other social spaces.

Therefore, the totalizing understanding of ballet as a feminized practice that produces docile bodies was challenged in small but important ways. And it was through bodily *practices* that these negotiations were enabled, prompted, and often experienced. Therefore ballet bodies *interacted* with and *responded* to power in unexpected and unpredictable ways. This animated Foucault's relational understanding of power in ways not emphasized in previous Foucauldian studies of ballet bodies (Green, 1999, 2001; Ritenburg, 2010). Consequently, I argue it is these responses and their effects that merit further attention.

However, Foucault's framework does not explicitly acknowledge the *active* materiality of the body; nor does it explore the relations *between* the discursive and the material (Barad, 2007). Therefore, further thinking about ballet bodies—specifically their materiality, multiplicities, and capacities—is needed to elaborate on current understandings of the ballet body in meaningful ways. Barad, whose work is deeply informed by Foucault's, provides concepts that allow me to elaborate on my Foucauldian analysis. I now situate Barad's thinking in the greater context of new materialisms and outline the utility of her ontological framework (agential realism) for reimagining ballet bodies.

MATERIALITY AND MOVEMENT: THINKING WITH BARAD

The emerging work in new materialism provides rich conceptual tools for those interested in the study of moving, physically active bodies (see Fullagar, 2017; Fullagar, chapter 7 in this volume; Thorpe, 2016). While feminist scholars such as Elizabeth Grosz and Rosi Braidotti have long insisted on the importance of the corporeal in the production of subjectivity, Hekman (2010) notes how the linguistic turn, while yielding vital critiques, resulted in a privileging of the discursive at the expense of the material. She laments, "A whole generation of feminist scholars has been taught to put 'matter' in square quotes. Removing those square quotes . . . will be no mean feat" (p. 72). However, a growing number of feminists are insisting, with increasing urgency, that emphasizing the discursive at the expense of the material constrains the ways we are able to think about and respond to the increasingly complex biological, social, political, and technologi-cal concerns facing contemporary society (Barad, 2007; Grosz, 2005; Haraway, 1991, 2003; Mol, 2002; Tuana, 2008). Barad in particular is explicitly critical of the assumption that language and discourse are responsible for producing mean-ing, while matter and nature tend to be viewed more suspiciously. She pointedly asks, "How did language come to be more trustworthy than matter?" (Barad, 2003, p. 801). This concern underpins her ontological project, which theorizes the inseparable entanglement of discourse and matter.

Barad (2007) presents the neologism "intra-action" to refer to these entangle-ments and the ways "meaning and matter are mutually constituted" in the pro-duction of knowledge (p. 152). Importantly, intra-action refuses the assumption that discrete boundaries exist between bodies, agents, and forces. She argues, "The relationship between the material and the discursive is one of mutual entailment. Neither is articulated/articulable in the absence of the other; mat-ter and meaning are mutually articulated" (p. 151). Instead, intra-action insists on the generative co-constitutive capacity of the material and the discursive and understands material-discursive forces as constantly relating, transforming, and acting inseparably. It is through these inseparable relations, or intra-actions, that knowledge is produced.

In its refusal of inseparability and dualisms, intra-action has important impli-cations for thinking itself. Barad (2003, 2007) suggests an onto-epistemological shift is necessary to disrupt the categorical understandings of existence that mark modern thought (e.g., discourse/matter, culture/nature, male/female). This shift involves rethinking our own relationship to knowledge and acknowl-edging that we are never separate from processes of thinking and knowing because knowing and being are inextricably entangled with each other. Barad (2007) elaborates, "Practices of knowing and being are not isolable; they are mutually implicated. We don't obtain knowledge by standing outside the world; we know because we are of the world" (p. 185). Therefore, in Barad's ontology, humans and things constitute each other, and knowing is an active, performa-tive practice, not a cognitive process that occurs within the inner workings of the human mind.

Thus taking up Barad to think about ballet bodies means understanding one's own subjectivity, or "bodymind" (Merrell, 2003), as always already in the middle, or what Jackson and Mazzei (2012) refer to as on the "threshold" of pro-cesses of knowing. This differs from reflexivity, a common practice in qualitative inquiry through which researchers acknowledge their own subjectivity and the way it may shape the inquiry process. To practice reflexivity means to engage in a critical self-positioning. But as Barad (2007) points out in her critique, "reflexivity, like reflection, still holds the world at a distance" (p. 87). It assumes the researcher is or can be separate from what it is he or she is observing and reflecting on. Barad's ontology instead suggests that "knowledge is never done in isolation, but is always effected by different forces coming together" (Hultman & Lenz Taguchi, 2010, p. 536).

Putting this ontology to work requires decentering the researcher (and the participants) as the primary sources and arbiters of meaning and instead attending to "events of activities and encounters [that] evoke transformation and change in the performative agents involved" (Hultman & Lenz Taguchi, 2010, p. 235). Therefore, plugging into Barad accommodates the "different forces" at work at the site of research without resorting to reflexivity. I now turn to Barad to reimagine ballet bodies by attending to the multiple forces encountered in this inquiry. I first revisit an encounter in the change room with a participant named Hannah to reimagine ballet bodies through Barad's concepts of knowing in being and intra-action. I then revisit an encounter with pointe shoes that explores the material-discursive intra-actions of ballet bodies and outline how these encounters help us think through the previously unthought.

STEPPING INTO THE THRESHOLD: ANIMATING INTRA-ACTION AND SENSING INSEPARABILITY

The distinct scent of well-worn leather ballet slippers mingles with particles of hairspray, catching sharply in the back of my throat. A steady hum of girlish chatter fills the air, vibrant and effusive. Here in the change room, ballet bodies fidget and stretch, adorned in black bodysuits, pink tights, and various layers of woolen warmers. The satin ribbons of ballet shoes are tied and adjusted, and long locks of hair are coaxed into tidy buns. Bodies are draped across the sparse wooden benches, arranged on the floor in nonchalant but extraordinary displays of flexibility, in an easy intimacy, limbs overlapping. As the girls chat to one another, their conversations are punctuated by physical exuberances and exclamations—physical and verbal expression become one and the same. The vitality of bodies in this space is palpable.

This excerpt from my field notes captures the dynamics of a change room of a recreational dance studio. Through the continuous relations created with other bodies (human and nonhuman) around them, these ballet bodies exceed descriptions of feminized bodies wrapped in pink tulle and satin and disrupt fixed images of ethereal slenderness and flexibility. These are ballet bodies that move, connect, relate, animate, emote, stretch, and sweat.

I plug this vital entanglement of bodies, things, and affects of the change room into Barad's (2003, 2007) ontological position of knowing in being to further elaborate on what this might look/feel like when conducting qualitative inquiry. Barad suggests we should understand "knowing in being" as mutually articulated. In other words, we become what we know "in a steadily ongoing flow of material-discursive events" (Lenz Taguchi, 2011, p. 39). Rereading these data through Barad therefore requires a shift from thinking about and seeking to interpret what the bodies, practices, and relationships in the change room might mean to noticing of what these relationships produce. In addition, particular attention must be paid to the material forces so as to resist the seductive tendency to resort to the descriptive and the discursive.

The material is everywhere in this anecdote. Hairspray, used to fix hair in place in a bun and discursively understood as a feminine beauty product, intra-acts with the cells and saliva of my tongue and throat to produce an acerbic tingle, an unpleasant taste that slightly burns the back of my throat. This "tingle" is not produced solely by the chemical makeup of the hairspray, nor by the taste buds and nerves of the mouth and throat, but rather through the intra-action of the two. In turn, this blurs the boundaries among my body, the bodies I encounter, and what it is I seek to know. With this blurring, it becomes impossible to know ballet bodies (or my researching body) as discrete entities that are separate from other material and discursive forces. Rather, the multiple forces at work in the studio (i.e., participants' bodies, my researching/dancing body, nonhuman bodies, and discursive understandings that shape ballet and research practices) emerge as "mutually active agents" engaged in these performative practices of knowing (Lenz Taguchi, 2013, p. 708). I further explore the performativity of knowing in a Baradian-inspired analysis of the above anecdote:

> In the midst of this cacophony of vitality and chatter, I struggle to fix my hair in the requisite ballet bun, strands of hair defying my fingers, the rubberized ends of bobby pins digging bruisingly into my scalp. Beside me you giggle and offer your help, my hair yields to your deft fingers, and the assaulting bobby pins shift out of the tender grooves formed in my flesh. I shift my sit bones on the hard wood of the bench and watch in the mirror as you facilitate my transformation. I take in our reflections in the mirror: black bodysuits, pale pink tights, ballerina buns. It is now time for class.

Through these intra-actions with Hannah, one of the dancers in my study, the boundaries of my researching body dissolve and the discursive categories of "researcher" and "participant" blur as we are both produced and transformed through material-discursive intra-actions. As Hannah helped me wrestle my hair into a bun, we lamented and laughed about the age-old requisite ballerina hairstyle. A dancer's greatest fear is for her hair to become loosened during class or performance, as it disrupts the ballet aesthetic and the lines of the body and is distracting for dancer and viewer alike. Therefore, much time is spent disciplining the hair with bobby pins, hairnets, and hairspray. Yet Hannah notes,

> When my hair is back in a tight bun, it sometimes hurts, like your hair is tight and bobby pins can dig in, but I feel like a real dancer, like I'm ready to go out there and do amazing things. It means my hair is not in my face, so there is nothing to distract me; nothing is in my way.

In this data fragment, matter and discourse become "active participants in . . . ongoing intra-activity" (Barad, 2007, p. 136). The material (Hannah's hair) is constrained by other material entities (bobby pins and hair spray) in order to comply with discursive understandings of ballet aesthetics. Yet these material-discursive intra-actions produce more than Hannah's bun; they create a body with an increased capacity to dance, create, and participate in new intra-actions. Hannah says there is "nothing" in her way, nothing constraining her ability to dance. Therefore, material-discursive intra-actions are transformative, co-constitutive, and generative. The way Hannah feels with her hair in a bun shapes her physical way of being in the studio and prompts a specific manner in which she then continues to move, dance, and create relationships with other bodies in the studio space.

SATIN RIBBONS, SWEAT, AND SINEWS: BALLET BODIES AS MATERIAL-DISCURSIVE FORCES

If the ballet bun is an identifiable aspect of the ballerina aesthetic, so too are the pink satin pointe shoes and the pink ribbons that adorn dancers' feet. Designed to enable dancers to perform *en pointe*, such shoes are often coveted by young ballerinas. As Hannah explains, "Every little girl can't wait to get her first pair of pretty pointe shoes. They're the sign that you're really a ballerina now . . . but then they realize how much they hurt and how hard [pointe work] is."

Hanna's quote illustrates the salience of pointe shoes in shaping the discursive practices of young female dancers, noting how they inform understandings of what "real" ballerinas look like and do. However, she also alludes to the generative capacities of pointe shoes and their ability to provoke excitement and corporeal discomfort. Taguchi (2010) suggests that to think the material through Barad requires installing "ourselves in the event that emerges in our reading" (p. 16). In the excerpt that follows, I install (Hultman & Lenz Taguchi, 2010; Jackson & Mazzei, 2012) my researching/dancing self in order to explore ballet shoes as material-discursive forces (Barad, 2003, 2007). This is not a reflective exercise meant to account for one's own subjectivity but rather an "attempt to make sense of the blurring and viscous interactions" that emerge through processes of knowing (Jackson & Mazzei, 2012, p. 131):

> Putting my leather slippers on takes just a minute, the pink elastic sliding easily over the arch of my foot. They are soft and comfy and emanate a familiar, leathery smell. I perform a few simple *relevés*, rising up onto *demi-pointe* and lowering again, the leather supple and unyielding, moving with my foot like a second skin. Around me, the other dancers tie up their pointe shoes, wrapping and crisscrossing the pink ribbons around their ankles, securing them neatly in a small knot just above the inside of the ankle bone. Watching them, I recall the feel of the smooth satin ribbon between my fingers, the reassuring tension they create around the ankle, the ritual of flexing and extending my foot to ensure the ribbons don't interfere with muscles or tendons or rub irritatingly against the bone. Hannah, whom I'm sitting beside, stands up and performs her own ritual, rising onto the balls of her feet a few times and shifting her balance from one foot to the other. Watching her, I admire the beauty of the slender satin slipper encasing her foot and vividly remember how they feel on the foot: rigid, at times painful, constraining and pressing uncomfortably on the bones and flesh, but in so doing enabling the dancer to rise onto their toes. In contrast my leather slippers accommodate every curve and contour of my foot but leave me unable to perform quintessential ballet movements. I feel slightly amateurish in them. As if reading my thoughts, Hannah muses, "There's something about [pointe shoes], isn't there? I feel like a real dancer in them. . . . They're not just pretty; they make me feel like I can do anything a prima ballerina can."

In this encounter, the boundaries between matter and meaning, my body and Hannah's body, satin and flesh were blurred. As Hannah tied her ribbons, my fingers felt the smooth satin sliding between them, and my ankles, muscles, and tendons recalled and responded to an absent pressure. My history as a dancer and previous embodied experiences of wearing pointe shoes became entangled with the bodies (human and nonhuman) acting at the site of inquiry, prompting physical and emotional responses. In keeping with Barad's (2007) ontology, I sought to resist the seductive trap of reflexivity when revisiting this encounter but to somehow put these responses to work, to acknowledge the flows of sensations and relations that emerged in the encounter as well as the materiality of the bodies involved. To think with new materialist thought, Lenz Taguchi (2012) suggests researchers attend to those

> bodymind faculties that register the flows of smell and the intensities of touch, level, temperatures, pressure tension and force in the interconnections that

> emerge in the engagement. We try to register how the data *interferes* with the sensibilities of our bodyminds and what this brings to the event of reading the data. (p. 272, italics in original)

Therefore, I paid attention to the sensory and affective responses prompted through the entanglements with the bodies around me and attempted to think through them, rather than to describe or interpret them. I attended to the blurring of boundaries between my body and Hannah's body by wondering what might emerge when we think of bodies not as discrete entities with clear borders and instead "think in terms of processes of entanglements" (Lenz Taguchi, 2012, p. 271). As a result, there ceased to be any identifiable starting point for meaning; rereading this encounter was not about producing knowledge about my body, Hannah's body, the pointe shoe or establishing its discursive meanings. Rather, it became a practice of knowing in being where knowing is "a matter of part of the world making itself intelligible to another part of the world" (Barad, 2007, p. 185).

This shift allowed for bodies, pointe shoes, and discursive practices to emerge as mutually constitutive performative agents (Barad, 2007). Matter pressed upon matter to create new possibilities for movement. Satin ribbons pressed upon muscles and tendons to transform corporeal capacities. The boundaries between subject and object, matter and meaning, human and nonhuman bodies softened, became porous, giving way to what was previously unthought. This softening and blurring of boundaries allowed pointe shoes and ballet bodies to emerge as dynamic entities with the capacity to evoke responses and prompt change in the other bodies they encountered. Through such encounters, bodies, practices, and subjectivities were transformed.

Take, for example, Hannah's body that performs and approaches dance class differently in pointe shoes than in leather ballet slippers. She explains that pointe shoes make her "feel like a real dancer," they enable particular bodily practices and movements that align with discursive understandings of what ballet bodies look like and how they behave. Yet they cannot be understood as simply artifacts that appear on the discursively constructed ballet body. This understanding does not capture what they *do*. In this example, the pointe shoes produce a response in Hannah and transform Hannah's body, giving way to a different comportment as she moves in the change room and studio. They produce a sense in Hannah and others around her (including myself) that she belongs in and inhabits this space in a particular way, as a "real dancer" with the capacity to move and perform as such. Therefore, Hannah's body, the pointe shoes, and the studio space work together to produce Hannah as a material-discursive force that continues to become through future intra-actions and movements.

CONCLUSION: INSEPARABLE BODIES, TRANSFORMATIVE BODIES

The analyses above illuminate how intricately (ballet) bodies are entangled with the material and discursive forces around them. My initial curiosities centered on the ballet body, but thinking with Barad yielded a diffractive, performative process of knowing in which ballet bodies, pointe shoes, ribbons, bobby pins, and discourse came to bear. Reimagining the ballet body through Barad therefore illuminated the connections continually forged with other material-discursive bodies and produced the body as a collection of forces rather than a discrete entity. As Grosz (1994) explains, "The body does not hide or reveal an otherwise unrepresented latency or depth but is a set of operational linkages and connections with other things, other bodies" (p. 120).

Therefore, reimagining the ballet body through new materialism meant reconfiguring or dissolving its assumed boundaries and thinking in the *in-between* of its connections and intra-actions. This blurring of boundaries aligns with the concept of inseparability, or entanglement, so central to Barad's (2003, 2007) ontology. As Barad (2007) notes, "To be entangled is not simply to be intertwined with one another, as in the joining of separate entities, but to lack an independent, self-contained existence" (p. ix). This means entanglement insists on ontological inseparability, which in turn disrupts deeply sedimented ways of knowing that privilege binaries, dualisms, and categories. It is when these systems are shaken up enough that we can begin to imagine thinking anew.

Indeed, knowing and thinking about ballet bodies through Barad's onto-epistemological framework allowed me to understand ballet bodies as more than discursive constructions and accommodated the liveliness I sensed in my original dissertation study but was unable to adequately theorize. Plugging into Barad illuminated the capacity of ballet bodies to create new ways of becoming. As ballet bodies become through their movements and relationships with other bodies, they created flows of intensities that moved in myriad directions, never locatable or predictable, but always giving way to new intra-actions.

However, thinking anew required a rethinking of how to think and know. My own researching bodymind was continuously transformed through processes of knowing. To know ballet bodies differently required inserting myself into the "threshold" of data events in order to animate theoretical concepts. But what does one do when in the threshold? This is what needed to be questioned continuously and remains the challenge for inquiry through a new materialist lens. As I installed myself into the inquiry, the boundaries between my body and participants' bodies blurred, and my corporeal memories and performances became entangled with other dancing bodies and their doings, with the leather slippers on my feet, and with discursive meanings of pointe shoes and ballet bodies. At times, this was disorienting. These responses were powerful, and we are well trained to locate meaning in our own or our participants' experiences and voices or to provide interpretations that are grounded in some certainty. Yet new materialisms abandon such linearity for multiplicity and diffractive thinking and eschew certainty for processes of knowing that may (or may not) produce that which was previously unthought. Therefore, I sought to examine what emerged in between these blurred boundaries and relinquished my desire for direction and certainty. In turn, this allowed me to extend my analyses beyond critique and representation. New questions have emerged for me, and many remained unanswered. There is much left to know about ballet bodies and what they do. Consequently, further and ongoing inquiries are needed that engage new materialisms and seek to theorize through performative processes of knowing.

When dancers perform warm-up exercises at the barre, the same movement is repeated multiple times as muscles and joints warm up and loosen and technique is refined. However, no movement ever feels, looks, or *is* the same. The relationship between foot and floor, for example, always shifts, always changes. The leg may extend farther out from the body or sail higher in the air next time. There is always that possibility; the limits are never really known. Similarly, new materialist inquiry can be engaged in over and over again, and the processes of knowledge can be likened to the performative practice of ballet: always yielding something new, something unthought, reaching toward ever-expanding limits to what can be known.

REFERENCES

Aalten, A. (2004). "The moment when it all comes together": Embodied experiences in ballet. *European Journal of Women's Studies, 11*(3), 263–276.

Alaimo, S., & Hekman, S. (2008). Introduction: Emerging models of materiality in feminist theory. In S. Alaimo & S. Hekman (eds.), *Material feminisms* (pp. 1–19). Bloomington: Indiana University Press.

Alexias, G., & Dimitropoulou, E. (2011). The body as a tool: Professional classical ballet dancers' embodiment. *Research in Dance Education, 12,* 87–104.

Barad, K. (2003). Posthumanist performativity: Toward an understanding of how matter comes to matter. *Signs: Journal of Women in Culture and Society, 28*(3), 801–831.

Barad, K. (2007). *Meeting the universe halfway: Quantum physics and the entanglement of matter and meaning.* Durham, NC: Duke University Press.

Barad, K. (2012). Intra-actions. *Mousse, 34,* 76–81.

Bennett, J. (2010). *Vibrant matter: A political ecology of things.* Raleigh, NC: Duke University Press.

Bowerman, E. A., Whatman, C., Harris, N., & Bradshaw, E. (2015). A review of the risk factors for lower extremity overuse injuries in young elite female ballet dancers. *Journal of Dance Medicine & Science, 19*(2), 51–56.

Caine, D., Goodwin, B. J., Caine, C. G., & Bergeron, G. (2015). Epidemiological review of injury in pre-professional ballet dancers. *Journal of Dance Medicine & Science, 19*(4), 140–148.

Campoy, F. A. S., de Oliveira Coelho, L. R., Bastos, F. N., Júnior, J. N., Vanderlei, L. C. M., Monteiro, H. L., . . . Pastre, C. M. (2011). Investigation of risk factors and characteristics of dance injuries. *Clinical Journal of Sport Medicine, 21*(6), 493–498.

Clark, M., Spence, J. C., & Holt, N. L. (2011). In the shoes of young adolescent girls: Understanding physical activity experiences through interpretive description. *Qualitative Research in Sport and Exercise, 3,* 193–210.

Clark, M. I. (2017). Multiple bodies: In the ballet studio with adolescent dancers. In P. Markula & M. I. Clark (eds.), *The evolving feminine ballet body* (pp. 77–100). Edmonton: University of Alberta Press.

Clark, M. I., & Markula, P. (2017). Foucault at the barre and other surprises: Examining adolescent girls experience of discipline and docility in the ballet studio. *Qualitative Research in Sport, Exercise and Health, 9,* 435–452.

Coole, D., & Frost, S. (eds.). (2010). *New materialisms: Ontology, agency, and politics.* Durham, NC: Duke University Press.

Dolphijn, R., & Van der Tuin, I. (2012). *New materialism: Interviews & cartographies.* Ann Arbor, MI: Open Humanities Press.

Dryburgh, A., & Fortin, S. (2010). Weighing in on surveillance: Perception of the impact of surveillance on female ballet dancers' health. *Research in Dance Education, 11,* 95–108.

Dowda, M., Dishman, R. K., Porter, D., Saunders, R. P., & Pate, R. R. (2009). Commercial facilities, social cognitive variables, and physical activity of 12th grade girls. *Annals of Behavioral Medicine: A Publication of the Society of Behavioral Medicine, 37,* 77–87.

Foster, S. L. (1997). Dancing bodies. In J. Desmond (ed.), *Meaning and motion: New cultural studies of dance* (pp. 235–257). Durham, NC: Duke University Press.

Foucault, M. (1972). *The archaeology of knowledge.* London, England: Tavistock.

Foucault, M. (1979). *Discipline and punish: The birth of the prison* [Surveiller et punir] (2nd ed.). New York, NY: Vintage.

Fullagar, S. (2017). Post-qualitative inquiry and the new materialist turn: Implications for sport, health and physical culture research. *Qualitative Research in Sport, Exercise and Health, 9*(2), 247–257.

Gamboa, J. M., Roberts, L. A., Maring, J., & Fergus, A. (2008). Injury patterns in elite preprofessional ballet dancers and the utility of screening programs to identify risk characteristics. *Journal of Orthopaedic & Sports Physical Therapy, 38*(3), 126–136.

Green, J. (1999). Somatic authority and the myth of the ideal body in dance education. *Dance Research Journal, 31,* 80–89.

Green, J. (2001). Socially constructed bodies in American dance classrooms. *Research in Dance Education, 2,* 155–173.

Green, J. (2003). Foucault and the training of docile bodies in dance education. *Journal of the Arts and Learning Special Interest Group of the American Education Research Association, 19,* 99–125.

Grieser, M., Vu, M. B., Bedimo-Rung, A. L., Neumark-Sztainer, D., Moody, J., Young, D. R., & Moe, S. G. (2006). Physical activity attitudes, preferences, and practices in African American, Hispanic, and Caucasian girls. *Health Education and Behavior, 33,* 40–51.

Grosz, E. (1994). *Volatile bodies: Toward a corporeal feminism.* Bloomington: Indiana University Press.

Grosz, E. (2005). *Time travels: Feminism, nature, power.* Durham, NC: Duke University Press.

Haraway, D. (1991). A cyborg manifesto: Science, technology, and socialist feminism in the late twentieth century. In D. Haraway (ed.), *Simians, cyborgs and women: The reinvention of nature* (pp. 149–181). New York, NY: Routledge.

Haraway, D. (2003). *The companion species manifesto: Dogs, people, and significant otherness.* Chicago, IL: Prickly Paradigm Press.

Hekman, S. (2010). *The material of knowledge: Feminist disclosures.* Bloomington: Indiana University Press.

Hultman, K., & Lenz Taguchi, H. (2010). Challenging anthropocentric analysis of visual data: A relational materialist methodological approach to educational research. *International Journal of Qualitative Studies in Education, 23*(5), 525–542.

Jackson A. Y., & Mazzei, L. A. (2012). *Thinking with theory in qualitative research: Viewing data across multiple perspectives.* Abingdon, England: Routledge.

Kleiner, S. (2009). Thinking with the mind, synching with the body: Ballet as symbolic and nonsymbolic interaction. *Symbolic Interaction, 32*, 236–259.

Kuo, J., Schmitz, K. H., Evenson, K. R., McKenzie, T. L., Jobe, J. B., Rung, A. L., . . . Pate, R. R. (2009). Physical and social contexts of physical activities among adolescent girls. *Journal of Physical Activity and Health, 6*, 144–152.

Lenz Taguchi, H. (2011). Investigating learning, participation and becoming in early childhood practices with a relational materialist approach. *Global Studies of Childhood, 1*(1), 36–50.

Lenz Taguchi, H. (2012). A diffractive and Deleuzian approach to analysing interview data. *Feminist Theory, 13*(3), 265–281.

Lenz Taguchi, H. (2013). Images of thinking in feminist materialisms: Ontological divergences and the production of researcher subjectivities. *International Journal of Qualitative Studies in Education, 26*, 706–716.

Markula, P., & Clark, M. I. (2017). Dance practices as physical culture. In M. Silk, D. L. Andrews, & H. Thorpe (eds.), *The Routledge handbook of physical cultural studies.* New York NY: Routledge International.

McEwen, K., & Young, K. (2011). Ballet and pain: Reflections on a risk-dance culture. *Qualitative Research in Sport, Exercise and Health, 3*, 152–173.

Merrell, F. (2003). *Sensing corporeally: Toward a posthuman understanding.* Toronto, Canada: University of Toronto Press.

Mol, A. (2002). *The body multiple: Ontology in medical practice.* Durham, NC: Duke University Press.

O'Flynn, G., Pryor, Z., & Gray, T. (2013). Embodied subjectivities: Nine young women talking dance. *Journal of Dance Education, 13*(4), 130–138.

Pickard, A. (2012). Schooling the dancer: The evolution of an identity as a ballet dancer. *Research in Dance Education, 13*, 25–46.

Pickard, A. (2013). Ballet body belief: Perceptions of an ideal ballet body from young ballet dancers. *Research in Dance Education, 14*, 3–19.

Pickard, A. (2015). *Ballet body narratives: Pain, pleasure and perfection in embodied identity.* Bern, Switzerland: Peter Lang.

Probyn, E. (2004). Teaching bodies: Affects in the classroom. *Body and Society, 10*, 21–43.

Ringham, R., Klump, K., Kaye, W., Stone, D., Libman, S., Stowe, S., & Marcus M. (2006). Eating disorder symptomatology among ballet dancers. *International Journal of Eating Disorders, 39*, 503–508.

Ritenburg, H. M. (2010). Frozen landscapes: A Foucauldian genealogy of the ideal ballet dancer's body. *Research in Dance Education, 11*(1), 71–85.

Solutions Research Group. (2014). *Massive competition in pursuit of the $5.7 billion Canadian youth sports market*. Retrieved from http://www.srgnet.com/2014/06/10/massive -competition-in-pursuit-of-the-5-7-billion-canadian-youth-sports-market

Thorpe, H. (2016). Athletic women's experiences of amenorrhea: Biomedical technologies, somatic ethics and embodied subjectivities. *Sociology of Sport Journal, 33*(1), 1–13.

Tuana, N. (2008). Viscous porosity: Witnessing Katrina. In S. Alaimo & S. Hekman (eds.), *Material feminisms* (pp. 188–201). Durham, NC: Duke University Press.

Wainwright, S. P., & Turner, B. S. (2004). Epiphanies of embodiment: Injury, identity and the balletic body. *Qualitative Research, 4,* 311–337.

Wainwright, S. P., Williams, C., & Turner, B. S. (2005). Fractured identities: Injury and the balletic body. *Health: An Interdisciplinary Journal for the Social Study of Health, Illness and Medicine, 9*(1), 49–66.

Wellard, I., Pickard, A., & Bailey, R. (2007). "A shock of electricity just sort of goes through my body": Physical activity and embodied reflexive practices in young female ballet dancers. *Gender and Education, 19*(1), 79–91.

Wulff, H. (2008). Ethereal expression: Paradoxes of ballet as a global physical culture. *Ethnography, 9*(4), 518–535. doi:10.1177/1466138108096990.

PART III BODY ECOLOGIES

10 · REASSEMBLING "SPORT FOR DEVELOPMENT AND PEACE" THROUGH ACTOR-NETWORK THEORY

SIMON C. DARNELL

INTRODUCTION: THE BECOMING OF "SPORT FOR DEVELOPMENT AND PEACE"

In September 2015, the United Nations General Assembly adopted the *2030 Agenda for Sustainable Development*, outlining its vision for sustainable development in the 21st century. At the core of the agenda are the 17 Sustainable Development Goals (SDGs) and their associated targets, which are intended to focus attention on the most important development issues—such as poverty reduction, gender equality, and action on climate change—while striking a balance between social, economic, and environmental priorities.

Included in the 2030 agenda is a specific recognition of the potential for sport to contribute to achieving the SDGs. Article 37 of the 2030 agenda reads,

> Sport is also an important enabler of sustainable development. We recognize the growing contribution of sport to the realization of development and peace in its promotion of tolerance and respect and the contributions it makes to the empowerment of women and of young people, individuals and communities as well as to health, education and social inclusion objectives.

For those who have worked in the area of sport and international development, article 37 was "an important milestone" (Lindsey & Chapman, 2017, p. x) in that it drew more formal attention and recognition to sport's development possibilities and in turn suggested "an opportunity to build on previous commitments and progress made by the Commonwealth, the United Nations, and other international, national and local stakeholders" in regards to sport for development (Lindsey & Chapman, 2017, p. x). These previous commitments include the naming of a special adviser to the United Nations Secretary-General on Sport for Development and Peace (currently Wilfried Lemke), the establishment of the United Nations Office on Sport for Development and Peace (UNOSDP),[1] and the emergence of hundreds of organizations (governmental, nongovernmental, corporate, charitable, etc.) that now mobilize sport to meet international development goals.

Overall, then, the inclusion of sport in the SDGs offers one of the latest illustrations of the establishment and institutionalization of sport for development and the Sport for Development and Peace (SDP) sector in recent years. Thus it is possible to identify the emergence of the SDP sector even if its strict boundaries remain difficult to ascertain (Svensson, 2017).

Accompanying this institutionalization has been an increase in research and scholarly analyses of SDP, including typologies of SDP stakeholders (Giulianotti, 2011a), empirical assessments of its outcomes (Coalter, 2013), and even attempts to theorize new and more progressive ways in which to mobilize sport toward development (Hayhurst, Kay, & Chawansky, 2016). In this chapter, however, I take a different approach to the study of sport for development (as a concept and practice) and the SDP sector (as an entity made up of different organizations and individuals). Rather than continuing critical sociological analyses that tend to focus on the outcomes or political logic of SDP, I use the insights of actor-network theory (ANT), and particularly the work of Bruno Latour, to discuss some of the ways in which to examine SDP as a social assemblage. I draw attention to the variety of "actants" involved in SDP and the processes by which it coheres. In so doing, I am not suggesting that recent SDP literature is necessarily lacking, but I do posit that bringing an ANT perspective to SDP both broadens and focuses its analysis by encouraging a wider understanding of the forces at play in SDP as well as more specific and particular conceptualizations and assessments of the formations, power, agency, and even truth that facilitate it. In this way, I suggest that ANT encourages a less deterministic understanding of sport for development and SDP while productively complicating critical analyses offered to date. Overall, I submit that ANT can help critical scholars formulate answers to the following questions: How did/does something like SDP become a "thing"? In turn, what can we—as social scientists—learn about the world of sport, development, and SDP through the study of this process of SDP's becoming?

With this in mind, the remainder of the chapter proceeds in four parts. In the next section, I discuss some recent trends in the study of sport for development and SDP, paying particular attention to issues of theory and method that I suggest demonstrate the timeliness and importance of an ANT analysis. This is followed by an overview of ANT and some of its main tenets. The penultimate section discusses some of the applications to, and implications of, ANT in the study of SDP. The final section offers some concluding thoughts and questions for future research.

TRENDS IN SDP RESEARCH

As mentioned, the institutionalization of SDP has been accompanied by an increase in SDP research—research that increasingly varies in terms of theory and method. For example, perspectives from the fields of sociology (e.g., Hayhurst, 2013), management (e.g., Welty-Peachey, Borland, Lobpries, & Cohen, 2015), and social psychology (e.g., Guest, 2013) have all been deployed, as have different methodologies, including ethnographic and qualitative (Forde, 2015) as well as quantitative assessments based on pre- and posttests with the targeted beneficiaries of SDP programs (Coalter, 2013). Indeed, even randomized controlled trials (RCTs) that attempt to measure the outcomes of SDP programs (McFarlane et al., 2017) have emerged recently, which in some ways stand as further evidence of the institutionalization of SDP within the mainstream scientific research community.

That said, tensions are evident within this growing body of literature, and two are particularly useful for framing the discussion in this chapter. The first is the division between what can be labeled *positivist* and *critical* analyses. The positivist approach tends to analyze SDP initiatives through assessments of what sport can or cannot do, or does or does not do, for targeted populations and user groups. It is in this approach that surveys, pre- and posttests, and RCTs are often the preferred methodologies (see Coalter, 2013). Research that investigates the various processes of achieving development and/or peace through sport has also been conducted in the hope that best practices might be implemented or replicated in other contexts (e.g., Schulenkorf, 2012). These approaches lend important insights to the literature but often stop short of critical analysis informed by historical, social, or political forces.

In contrast to the positivist approach is a growing body of critical research that tends to analyze SDP programs in relation to their historical, material, political, social, and/or discursive relations of power. These analyses often focus on issues of subjectivity, colonization, politics, agency, and knowledge production and tend to pose critical questions about the extent to which SDP programs support social change and/or secure relations of global dominance and authority (see Forde & Frisby, 2015; Nicholls, Giles, & Sethna, 2010; Hayhurst, 2009, among others). Such approaches do not necessarily argue against the efficacy of SDP but do tend to question some of the conventional wisdom regarding sport's development utility.

At times, these two approaches to SDP research do not sit well together. For example, Coalter (2013) has argued that critical analyses of SDP employing postcolonial perspectives and/or participatory methodologies may be based on ideologies or first principles rather than data or insights from the field.[2] From this perspective, critical scholars of SDP are seen to overlook the benefits (or even the limitations) of SDP programs by committing to ideological critiques prior to data collection. Similarly, critical scholars have called attention to the limits of positivist research—namely, that pre- and posttests or surveys do little to capture the diversity and complexity of SDP programs and their participants while failing to illuminate the ideologies, politics, or funding structures that underpin such programs, particularly on a global scale (Darnell & Hayhurst, 2011). As a result, critical scholars have called attention to the importance of democratic approaches to SDP research that respect the agency of SDP's presumed beneficiaries and resist the neocolonization of marginalized groups (see Kay, 2009).

A second tension evident within SDP research exists around the sector's position within an increasingly complex and integrated global society. While important attention has been drawn to local agency and diversity (see Kay, 2009; Lindsey, Kay, Jeanes, & Banda, 2017), Giulianotti (2011b) also puts forth the argument that as a body of literature, studies of SDP have

> yet to engage fully with the transnational complexity of the SDP sector . . . and have not examined adequately the different scales, tiers, and professional networks which have rapidly grown up across local, national, and transnational levels within the SDP sector. (p. 51)

In response, there is a need to situate the SDP sector within global civil society and understand SDP's practices and effects in relation to the range of institutional actors and political agendas by and through which it is constituted (see Giulianotti, 2011b). Hayhurst (2016), in particular, argues that SDP programs should be understood as sociopolitical practices that are (re)constructed, implemented, and negotiated within a networked set of dynamic relationships and connections. She shows that SDP practices are fundamentally relational and shaped "through the nature of (global) connections, nodes and links among NGOs, donor staff, their allies, adversaries and dominant institutions" (p. 426). From this perspective, SDP needs to be viewed as co-constructed by the increasingly broad range of stakeholders (including nonhuman actors) that are necessary for its coherence and stability. As Hayhurst writes, the SDP sector—or network—is not simply a repository for the implementation of sport programs but rather constitutes "a set of relationships that are always fluid and changing interactions with a variety of supporters and intermediaries, whether individuals, organizations, discourses, or other social structures" (p. 429)

Adopting such a perspective means that it is insufficient to study SDP programs in sociopolitical isolation or only in terms of their effects or results. Instead, the complexity of interactions between various SDP actors (understood in ANT terms) needs to be accounted for. Commensurate with this is the need to consider sport for development as a material reality. For example, Douglas Hartmann (2016) concludes his monograph on midnight basketball programs in the United States by appealing for analyses that consider processes not only of social construction but also of reality. That is, instead of insulating critical analyses in conversations with like-minded thinkers or eschewing the concept or significance of "reality" altogether, Hartmann (2016) argues that

> the central task and challenge for critically oriented, culturalist social scientists is to stake a claim to "the real" as it exists in the cultural common sense and is represented with mainstream, empiricist social scientific methods and techniques. (p. 223)

From this perspective, there is arguably a gap in SDP research given how little has been asked and explored regarding the constitution and solidification of its reality. Following Hartmann, investigating this reality of SDP need not be inherently positivist or deterministic, but neither does it have to be merely relativistic. Rather, there is space for the study of SDP that focuses on its ontology and the organization of society necessary for SDP to cohere. In other words, in addition to analyses of SDP's outcomes and effects, the research literature would benefit from assessments of how SDP came to be and the ways in which the reality of SDP is organized and maintained. To pursue understandings of the truth of SDP in this way, the insights of actor-network theory are useful, and these insights are discussed in the next section.

THE TENETS OF ACTOR-NETWORK THEORY

Actor-network theory is most often associated with the writings of Latour as well as Michel Callon, John Law, and Steve Woolgar (Law & Lodge, 1984; Cressman, 2009). It is best known for its application in science and technology studies but is versatile enough to have been used across the social sciences. Since actor-network theory does not describe a particular and specific understanding of how the world is built or operates, it is something of a misnomer to label it a theory.

Instead, it is more accurate to describe ANT as a methodology and/or a treatise on ontology. Law's (2009) definition is useful in this regard:

> Actor-network theory is a disparate family of material-semiotic tools, sensibilities and methods of analysis that treat everything in the social and natural worlds as a continuously generated effect of the webs of relations within which they are located. (p. 141)

An ANT approach focuses its analytical attention less on the social world as somehow already made and therefore waiting to be analyzed and more on the process of its construction, organization, and stability, or its "becoming." In this sense, ANT proceeds from the perspective that social order is not to be expected, nor is it the result of any set of universal rules. Instead, social order and organization are exceptions that call for explanation (Latour, 1996).

One of the challenges of using ANT (and one of the reasons for its misuse) is the struggle to clarify the terminology it deploys. Latour (1996) has gone to some lengths to explain that a network in ANT is *not* a technical network—like a sewer or a subway—nor is it a social network that connects rational, motivated individuals. Indeed, rather than denoting something socially produced, a network in ANT is used in reference to the ontology of the (social) world. In Latour's (1996) words, "ANT aims at accounting for the very essence of societies and natures. It does not wish to add social networks to social theory, but to rebuild social theory out of networks" (p. 369).

For example, rather than investigating connections between the (already constituted) scales of local and global, an ANT perspective sees global entities as continuously local. Or, put differently, no place is "bigger" than any other, but some places enjoy connections with many more places (Latour, 2005, p. 176). Indeed, metaphors of shape, scale, and proximity are recurrent in Latour's explanations of ANT, particularly in illustrating the importance of networks. For example, the physical or geographic proximity of actants say very little about their connections, because the key characteristic of networks is not whether they are closer or bigger than others but whether they are "more intensely connected" (Latour, 1996, p. 371). Similarly, a network in ANT has no inside or outside and conceives no space to be filled; it is all boundary. This means that the point of ANT research is not to examine what is outside the network but rather to explain and account for its existence. Put differently, the point of these metaphors is not to suggest that the global, the macro, or the exterior do not exist but rather to draw attention to the importance of investigating what is necessary for these effects to come into being—that is, "in order to obtain the *effects* of distance, proximity, hierarchies, connectedness, outsiderness and surfaces an enormous *supplementary* work has to be done" (Latour, 1996, p. 372, emphases in original), and it is this supplementary work to which the ANT perspective draws attention.

So what, then, is a network, or an actor-network, and how is it defined? In distinction from the rational, humanist, and sociologically conventional notions of structure/agency or objectivity/subjectivity, ANT sees actors and networks as overlapping and connected by their abilities to produce and stabilize. An actor-network, therefore, cannot be reduced to either an actor or a network. Rather, "everything can be considered both an actor and a network" (Cressman, 2009, p. 3). Or, in other words, "an actor-network is simultaneously an actor whose activity is networking heterogeneous elements and a network that is able to redefine and transform what it is made of" (Callon, 1987, p. 93).

In the study of science and technology, ANT has often been used in micro-level assessments focused on the specific places where science and technology come to be, such as in "labs, institutes, government departments, boardrooms and funding agencies" (Cressman, 2009, p. 2). In such cases, ANT researchers tend to trace actors and actions through the various processes of network formation that are necessary for science and technology to accrue. These networks are likely to be composed of a range of actors, including scientists, but also things (like money), institutions (like government), or even ideas themselves (Cressman, 2009). The ANT researcher should treat any of these entities—or actants—in a similar manner so long as they play some role in the process of becoming. An actant, therefore, is understood to be "something that acts or to which activity is granted by another. . . . An actant can literally be anything provided it is granted to be the source of action" (Latour, 1996, p. 373).

Notably for its application to the study of SDP, ANT has been considered in relation to international development studies as well as the critical study of sport. In the former, Donovan (2014) has argued that whereas critical development studies—and particularly the critical school of postdevelopment—tended to reduce development to a continuation of Western colonialism or to presume a homogeneous subject in the developing world who either accepts or resists the first-world incursions of development interventions, ANT instead attempts "to trace the activity by which locals become included in an intervention, [while also] trusting in their agency and right to be convinced that becoming enrolled in 'development' is in their interest" (p. 873). In the latter, Weedon (2015) has used ANT to study the phenomenon of adventure or obstacle racing—in particular, the Tough Mudder, in which the camaraderie between participants is literally produced, at least in part, by mud itself. Mud binds the shared experiences of participants, and "the material weight of mud, its adhesive capacity to form socio-natural bonds among diverse beings, human and otherwise, establishes it as a constitutive actor in Tough Mudder's characteristic ethos of togetherness" (Weedon, 2015, p. 442). Here, nonhuman actors are shown to play an important part in the constitution of a physical activity and social experience. Conversely, to ignore the role of mud would be to omit an important element of the constitution of the social.

What connects all of these viewpoints and topics—and constitutes a central tenet of the ANT approach to social science theory/method—is that the analyst or researcher becomes responsible for investigating and reconciling the social from the perspective of social actors, but *not* through the deployment of social structures or categories that are already presumed to exist. Social order has to be accounted for, not presupposed, and power constitutes that which requires explanation instead of a tool in and of itself that can be used to explain society and its machinations. As Latour (2005) writes, "Power, like society, is the final result of a process and not a reservoir, a stock, or a capital that will automatically provide an explanation. Power and domination have to be produced, made up, composed" (p. 64).

Here, the processes and methodologies of research are key. Specifically, in recognizing and building on the importance and significance of discourse as not simply communication but indeed meaning making, ANT shifts the methodological gaze to describing the deployment of associations and tracing the path(s) of generative social processes. ANT does not presume to know the outcome of any analysis but instead "places the burden of theory on the *recording*, not on the specific shape that is recorded" (Latour, 1996, p. 374, emphasis in original).

The implications for theorizing the social in this way are significant. Whereas in more traditional sociology, social structures and hierarchies are often presumed from the outset and then used to explain and account for the organization of the world and the experiences of individuals, in an ANT perspective, the social is the result itself of various associations. As Latour (2005) writes, "Social does not designate a thing among other things . . . but a *type of connection* between things that are not themselves social" (p. 5; italics in original). When new ideas or movements are formed (such as the idea and practice of organizing sport in the service of international development and peace building), analysts are tasked with "reassembling" the pieces of these new connections. Again, ANT sees stability in the social world as rare, even exceptional, and the result of the efforts of actors who built it. It therefore seeks to identify "the controversies and disputes that must be settled for consensus to be reached" (Donovan, 2014, p. 871).

Thus the definition and ordering of the social is the purview not of the analyst but of social actors themselves. One way to assess the quality of an ANT analysis, therefore, is to ask, "Are the concepts of the actors allowed to be *stronger* than that of the analysts, or is it the analyst who is doing all the talking?" (Latour, 2005, p. 30). In focusing on the formation of groups rather than the existence of the groups themselves, an ANT researcher would follow the controversies that trail from the formation of groups and ideas. Indeed, a key source of uncertainty or instability in Latour's description of ANT is that there are no stable groups, only group formations, and understanding these formations should be central to social science analysis.

A second source of uncertainty relates to action and agency. Whereas more traditional social science theory and method see action as something undertaken consciously, ANT views action as "overtaken" by various other agentic forces. This is not the same as saying that social forces determine action or that people simply act unconsciously but more to suggest that a whole range of agentic forces lead to social action and organization. Thus the actor in ANT is "not the source of an action but the moving target of a vast array of entities moving towards it" (Latour, 2005, p. 46). The social scientist, in turn, should understand these data on their own terms rather than interpreting them into social abstraction. In this sense, ANT scholars should, as much as possible, avoid the tendency "to substitute an unknown expression for a well-known one" (Latour, 2005, p. 48). The goal is to explain what can be seen, not expose what is hidden.

A third source of uncertainty for Latour is arguably the most controversial and radical—namely, that understandings of the types of agentic actors should be expanded to include nonhumans. If, as ANT advocates, mere references to "society" or "social forces" are insufficient to explain power, dominance, hierarchies, or indeed the durability of the social world, then it becomes reasonable to expand the notion of agency beyond intentional human action. As a result, from an ANT perspective, "anything that does modify a state of affairs by making a difference is an actor" (Latour, 2005, p. 71). This is not to say that "things" have agency in terms of sentience, nor do things simply become the cause of action in a reversal of influence. However, social scientists should be on the lookout for agents that make a difference in the actions of other agents. Such influence constitutes an important form of agency, and ANT seeks to understand these agentic forces as they contribute to a durable social whole. Therefore, the interrelationships between humans and nonhumans are a crucial part of the network story that should not be overlooked. As Latour (2005) argues,

> One cannot call oneself a social scientist and pursue only some links—the moral, legal and symbolic ones—and stop as soon as there is some physical relation interspersed in between the others. That would render any enquiry impossible. (p. 78)

The discussion of ANT in this section illustrates some of the important ways in which it differs from more conventional approaches to social science. With this in mind, I turn next to the application of ANT to the sociological study of SDP.

CONSIDERING THE INSIGHTS
OF ANT IN THE STUDY OF SDP

Considering the overview of ANT above, several questions emerge for SDP scholars. What do the insights of ANT suggest for the study of SDP? What would an ANT analysis of SDP look like? And how would this be similar to or different from the SDP research that has been conducted to date? In this section, I consider these questions, and in so doing, I build on previous efforts to connect ANT to the critical study of international development. As discussed above, Donovan (2014) has argued that an ANT approach would encourage critical scholars of development to think about their work differently in several respects. Rather than championing and even protecting marginalized people from top-down, neocolonial development, as it is seen to be thrust upon them, by contrast "an ANT approach to development would give attention to how globally dispersed norms and practices come to a local development initiative" (Donovan, 2014, p. 876). This is particularly the case because in an ANT approach, *emancipation* means not "freed from bonds" but rather "well attached." Similarly, while it is possible to view funding agencies and international organizations in development as having macrolevel power in a de facto sense and using this power to influence local practice, in an ANT approach, no place is nonlocal, meaning that the macro (such as it is) must be understood in the specific places it is operationalized (Donovan, 2014). The ANT method for achieving such understanding is to trace the networks between actors that are not inherently hierarchical but only more or less connected. Drawing on these insights into what ANT can do for development studies, I discuss the possibilities and benefits of ANT for the study of SDP.

First and foremost, an ANT approach to SDP would necessitate the inclusion of and accounting for all the various actants that play a part in its formation. To date, SDP scholars have paid attention to power, subjectivity, knowledge production, policy, politics, and even results and impact. Rarely, however, have the productive interactions of these various actants been considered, particularly for the ways in which they facilitate and solidify the becoming of SDP. In turn, the influence or agentic capacities of things (such as funding) and/or concepts (such as sport's universal appeal) have received little to no attention at all. This is not to suggest that the political or cultural economies of SDP have not been analyzed but rather that the tendency has been to think of funding and concepts as tools wielded by rational actors with relative amounts of power. By contrast, ANT encourages an analysis of the extent to which such nonhuman actors actually "modify a state of affairs by making a difference" (Latour, 2005, p. 71). This is a qualitatively different approach to SDP research, one that would be less concerned with revealing forms of subjectivity and power that are presumed to exist and more interested in explaining the conditions by and through which such social effects are achieved. Put differently, while it may remain useful to think of SDP as "socially constructed" (i.e., the extent to which the notion of sport's development utility is a social construct), this approach is insufficient from the perspective of ANT. What is needed is an analysis that reveals the various mechanisms by which the social construction of SDP is in fact accomplished.

Specifically, this means embracing the range of actants implicated in SDP and examining the ways in which these actants constitute SDP's formation. Fortunately, there are examples from research that are useful in this regard. For example, Jeanes and Lindsey (2014) have analyzed the role of monitoring and evaluation (M&E) and funding in the everyday processes of SDP. Their research illustrates how local employees hired by SDP NGOs with funds obtained from international donors often have their livelihoods at stake when conducting ostensibly objective analyses of SDP programs and their outcomes. In Jeanes and Lindsey's (2014) words,

> Inevitably, staff became very nervous of communicating information that may suggest they were not meeting the ambitious outcomes of the funding agency and this has considerable implications for how they will assist external evaluators. (p. 207)

As a result, staff members may be willing to pursue M&E in such a way as to satisfy evaluators and maintain programs. The agency of staff members in this scenario is unquestionably important. In addition, though, and from an ANT perspective, the significance of M&E as a concept and practice and funds as both material and symbolic hold agentic capacity to shape the organization and practice of SDP and therefore also need to be acknowledged. M&E and funding are not simply by-products of human action but rather important elements of the assemblage of SDP and therefore should be considered in relation to human actors but not subjugated by them.

Similarly, the notion of sport itself can and should be considered an actant in SDP, particularly the ways in which it has been presented as a universal language with inherently positive characteristics and social effects. The history of the United Nations, for example, shows that notions of sport have been instrumental to the ways the organization has made and remade itself by aligning sport with the broader landscape of international development (Millington, 2015). Rather than relegating sport to a tool used or wielded by the United Nations, ANT encourages a recognition of sport's agentic capacity and its role in the process whereby the UN has become an organization interested in sport and has committed itself to sport for development in different ways at different times. Sport is "real" in the sense that it has demonstrated an ability to shape the UN in its becoming an SDP organization, which was itself necessary for the broader SDP sector to stabilize socially.

An important implication of these assessments is that when considered through the insights of ANT, empirical analyses (of the kind likely to be favored by the advocates of positivism) need not be divorced from critical approaches. Indeed, just the opposite is true; critical assessments are likely to be more robust and effective when grounded in the empirical, provided that the treatment of the data is dynamic and nuanced, not predetermined or ideological. Thus ANT likely goes some way toward reconciling the first tension in SDP research discussed above.

In turn, ANT can also make a positive contribution in response to the second tension in the SDP literature cited—namely, the place of SDP within a complex and increasingly integrated global society. By refusing to presume that there are fundamental social structures of scale, proximity, or distance, an ANT approach would resist attempts to situate SDP within a micro/macro divide from the outset. Instead, by analyzing networks and, more specifically, the social formations necessary for the effects of scale and the micro/macro to be seen and felt within SDP, ANT would help illustrate and document the complex interactions of global SDP processes that scholars like Giulianotti and Hayhurst have argued are currently underdeveloped in the literature. Specifically, this would mean analyzing how the distance between the macro and micro in SDP is maintained and/or challenged by, for example, the agentic characteristics of "things" like funding or curricula that connect various stakeholders within the burgeoning global SDP sector. Again, following Hayhurst (2016), the SDP sector is constitutive of "fluid and changing interactions with a variety of supporters and intermediaries" (p. 429). The point is that an ANT approach to the study of SDP would have to show how micro/macro relations come to be solidified amid changing interactions rather than using the micro/macro divide to explain SDP and its social effects.

From these perspectives, ANT presents a challenge to the burgeoning orthodoxy of critical sociological analyses of SDP (an orthodoxy to which my own work has contributed) by questioning the extent to which power in SDP can be understood or known *a priori*. Instead, critical social science studies of SDP through ANT would have to be able to account for power at the conclusion of an analysis by showing how it has operated by and through actants. For example, rather than presuming that top-down policies or practices of SDP that flow from the Global North to the Global South illustrate and confirm neocolonial power relations, ANT would ask SDP scholars to explore such processes and explain and account for their implications. While it may very well be that such processes are indeed neocolonial as many SDP researchers have concluded, it is the researcher's task to show how and why this has come to be rather than drawing conclusions that might confirm his or her own suspicions. This is what Donovan (2014) speaks to when he refers to the importance of trusting the agency of local people within development initiatives; in addition, this kind of research needs to eschew the binary of structure versus agency, an issue that has preoccupied recent SDP thinking and debate (see Lindsey & Gratton, 2012; Darnell & Hayhurst, 2012).

Indeed, recent discussions in the sociology of sport literature more broadly have drawn attention to the limits of critique as the basis for social science research and suggested that the epistemological privileging of the qualitative, subjective, and critical over the quantitative and traditionally scientific has limited the field's ability to affect change because it offers a limited tool kit to researchers (Pringle & Falcous, 2018). In response, Pringle and Falcous argue for the importance of at least considering the possibilities presented by quantitative and/or interdisciplinary approaches—approaches that might help critical researchers connect their work to the broader scholarly and political fields. Implicit in their invocation is that the research process—and by extension, the results produced through research and their political effects—is limited and incomplete if the choices of theory/method made by the researchers preclude significant aspects of the social assemblages under analysis. In this sense, even those critical scholars who are motivated or determined to facilitate change or to see their work have political impact would do well to consider the possibilities for new and radical results that ANT may encourage, particularly through its insistence that all actants are analyzed and accounted for in any social formation. The point is that while the meaningful, rigorous, and impactful study of SDP should include qualitative, political critique, it needs to be more than this as well. SDP research should *also* seek to understand how the SDP field is constructed, what its component parts are, and what these assemblages mean in relation to people and cultures, as well as things and ideas.

With this in mind, I offer some final examples in the form of self-reflection. My previous research into SDP, similar to other critical social science research on the topic, has tended to proceed from a theoretical framework of power (such as hegemony, biopower, or Orientalism) and to use these frameworks to explain how SDP practices are conceived, constructed, and/or implemented by various stakeholders, such as international volunteers, program officials, and funders or even journalists. This work has allowed me to develop critiques of the sociological and political implications of SDP, but in hindsight, it has left important questions unanswered, such as, How has the idea of sport-for-development and the existence of SDP programs actually come to be? What controversies needed to be reconciled for this to occur? And what actants served this process of SDP's becoming? In this sense, instead of using Gramscian theory to investigate the building of consent for SDP among various subjects, an ANT approach would trace, analyze, and document how and why the specific idea of sport's positive social contribution has in fact contributed to the solidification of the SDP sector. Or, instead of examining the extent to which SDP is a process of governmentality, ANT would encourage an examination of the "ability" of things or concepts, such as funding or best practices, to facilitate networks in and of SDP, even across significant geographic spans. Crucially, an ANT perspective would also ask us to reserve our judgment about the operation of power—and any conclusions about subjugation or inequality—until we had an empirical basis from which to do so. This would require tracing the processes by which actors become involved in SDP and the reasons they do so before drawing any political or ethical conclusions about such processes. In sum, ANT asks us to paint a clear picture of SDP and its constitutive parts before we presume to know its social impact.

CONCLUDING THOUGHTS

In this chapter, I have offered an overview of actor-network theory and the work of Bruno Latour and have made some suggestions about its possible applications to the study of SDP as well as some of the benefits of taking such an approach. In concluding this discussion, I offer some reflections on these arguments, particularly with respect to the limitations or criticisms aimed at ANT.

A major criticism of ANT is that through its commitment to radical empiricism and its refusal to theorize power *a priori* or to presume forces at work that are larger than actants themselves, it abandons the critical project or search for social justice that has become a hallmark of the social sciences. Whittle and Spicer (2008), for example, claim that ANT can do little to contribute to critical analyses because it minimizes, or even ignores, context and situated constructions in its rush to understand the effects of the "nonhuman" world. In essence, they suggest that ANT is a depoliticizing framework that problematically claims that "repetitive action of an automated system is of the same status as political action" and one in which trivial events of everyday life are equated with the comparatively consequential struggles for political and social justice (Whittle & Spicer, 2008, p. 620).

Such criticisms of ANT are likely to be significant to critical social scientists studying SDP who wish for their work to inform practice and policy while making a positive contribution toward equality and justice. How then to connect ANT and its foci to practice and/or policy? And how do we avoid reducing SDP research to only esoteric truth claims rather than critique and legitimate knowledge production?

One possible response lies in the understanding that the methodological commitment to the empirical, which is a hallmark of ANT, is compatible with the goal of understanding the politics of power and inequality that are rightly important to critical SDP scholars. In other words, far from eschewing politics, an ANT approach to SDP can help illuminate and explain politics, making a stronger case to program officials, funders, and policy makers about the everyday experiences and social formations around and through which SDP coheres. That is, using ANT can help SDP researchers move toward fuller and richer analyses that include the array of forces and actors (human and nonhuman) that influence the organization and deployment of SDP. In turn, and when done well, an ANT analysis leaves few stones unturned, meaning that the results—and their political implications—are likely to be more compelling to audiences with the authority to respond.

Importantly, such rich assessments of SDP and its various agentic forces need not be depoliticizing but in many ways can be just the opposite. As Weedon (2015) writes, "Politics is also a matter of apprehending who and what counts as an active subject in a given social configuration" (p. 449). The ANT framework can serve as a reminder that political experiences and analyses of power in SDP are not (only) the purview of the (critical) social scientist but also rooted in the experiences of various stakeholders who take part in SDP activity regularly. In turn, it is more likely that these everyday politics and struggles of SDP will be illuminated in and through ANT, particularly compared to a research framework that presumes that critique begins with the analyst. This of course, requires the researcher to be open to politics within a variety of associations, to account for nonhuman actors, and to respect the agency of humans who seek alliances that might sit uncomfortably with a predetermined theoretical framework of social power.

Overall, then, ANT's political world view asks for recognition that "there is strength in dependency, in fostering alliances, as well as the potentially ethical recognition of precarity and responsibility" (Weedon, 2015, p. 450). From this perspective, SDP researchers are charged with understanding a range of formations—dependency, alliances, precarity, and responsibility, among others—as they are assembled through *all* the constitutive parts of the SDP field. This may be something of a radical shift, but the benefit of doing so is a more detailed, empirical, and respectful assessment and understanding of the place of sport within ongoing struggles for development and peace.

NOTES

1. In May 2017, UN Secretary-General Antonio Guterras announced a new direct partnership between the United Nations and the International Olympic Committee, coupled with the closing of the UNOSDP in order to avoid "parallel work" between the organizations (Wickstrom, 2017).

2. See Darnell (2015) for a discussion of the strengths and weaknesses of Coalter's arguments.

REFERENCES

Callon, M. (1987). Society in the making: The study of technology as a tool for sociological analysis. In W. Bijker, T. Hughes, & T. Pinch (eds.), *The social construction of technological systems* (pp. 77–98). Cambridge, MA: MIT Press.

Coalter, F. (2013). *Sport for development: What game are we playing?* London, England: Routledge.

Cressman, D. (2009). A brief overview of actor-network theory: Punctualization, heterogeneous engineering & translation. Centre for Policy Research on Science and Technology. *Simon Fraser University Institutional Repository.* Retrieved from http://summit.sfu .ca/item/13593

Darnell, S. C., & Hayhurst, L. (2012). Hegemony, postcolonialism and sport-for-development: A response to Lindsey and Grattan. *International Journal of Sport Policy and Politics, 4*(1), 111–124.

Darnell, S. C., & Hayhurst, L. M. (2011). Sport for decolonization: Exploring a new praxis of sport for development. *Progress in Development Studies, 11*(3), 183–196.

Darnell, S. (2015). Review: Sport for development: What game are we playing? by Fred Coalter. *International Journal of Sport Policy and Politics, 7*(2), 315–318.

Donovan, K. P. (2014). "Development" as if we have never been modern: Fragments of a Latourian development studies. *Development and Change, 45*(5), 869–894.

Forde, S. D. (2015). Fear and loathing in Lesotho: An autoethno-graphic analysis of sport for development and peace. *International Review for the Sociology of Sport, 50*(8), 958–973.

Forde, S. D., & Frisby, W. (2015). Just be empowered: How girls are represented in a sport for development and peace HIV/AIDS prevention manual. *Sport in Society, 18*(8), 882–894.

Giulianotti, R. (2011a). The sport, development and peace sector: A model of four social policy domains. *Journal of Social Policy, 40*(4), 757–776.

Giulianotti, R. (2011b). Sport, transnational peacemaking, and global civil society: Exploring the reflective discourses of "sport, development, and peace" project officials. *Journal of Sport & Social Issues, 35*(1), 50–71.

Guest, A. M. (2013). Sport psychology for development and peace? Critical reflections and constructive suggestions. *Journal of Sport Psychology in Action, 4*(3), 169–180.

Hartmann, D. (2016). *Midnight basketball: Race, sports, and neoliberal social policy.* Chicago, IL: University of Chicago Press.

Hayhurst, L. M. (2009). The power to shape policy: Charting sport for development and peace policy discourses. *International Journal of Sport Policy, 1*(2), 203–227.

Hayhurst, L. M. (2013). Girls as the "new" agents of social change? Exploring the "girl effect" through sport, gender and development programs in Uganda. *Sociological Research Online, 18*(2), 8.

Hayhurst, L. M. (2016). Sport for development and peace: A call for transnational, multi-sited, postcolonial feminist research. *Qualitative Research in Sport, Exercise and Health, 8*(5), 424–443.

Hayhurst, L. M., Kay, T., & Chawansky, M. (eds.). (2016). *Beyond sport for development and peace: Transnational perspectives on theory, policy and practice.* Abingdon, England: Routledge.

Jeanes, R., & Lindsey, I. (2014). Where's the "evidence?" Reflecting on monitoring and evaluation within sport-for-development. *Research in the Sociology of Sport: Vol. 8. Sport, Social Development and Peace* (pp. 197–217). Emerald Group Publishing.

Kay, T. (2009). Developing through sport: Evidencing sport impacts on young people. *Sport in Society, 12*(9), 1177–1191.

Latour, B. (1996). On actor-network theory: A few clarifications. *Soziale Welt, 47* (4), 369–381.

Latour, B. (2005). *Reassembling the social: An introduction to actor-network theory.* Oxford, England: Oxford University Press.

Law, J. (2009). Actor network theory and material semiotics. In B. Turner (ed.), *The new Blackwell companion to social theory* (pp. 141–158). Chichester, England: Wiley-Blackwell.

Law, J., & Lodge, P. (1984). *Science for social scientists.* London, England: Macmillan.

Lindsey, I., & Chapman, T. (2017). *Enhancing the contribution of sport to the sustainable development goals.* London, England: Commonwealth Secretariat.

Lindsey, I., & Grattan, A. (2012). An "international movement"? Decentring sport-for-development within Zambian communities. *International Journal of Sport Policy and Politics, 4*(1), 91–110.

Lindsey, I., Kay, T., Jeanes, R., & Banda, D. (2017). *Localizing global sport for development.* Manchester, England: Manchester University Press.

McFarlane, J., Karmaliani, R., Maqbool Ahmed Khuwaja, H., Gulzar, S., Somani, R., Saeed Ali, T., & Jewkes, R. (2017). Preventing peer violence against children: Methods and baseline data of a cluster randomized controlled trial in Pakistan. *Global Health, Science and Practice, 5*(1), 115–137.

Millington, R. (2015). *The United Nations and sport for development and peace: A critical history* (Unpublished doctoral dissertation). Queen's University.

Nicholls, S., Giles, A. R., & Sethna, C. (2010). Perpetuating the "lack of evidence" discourse in sport for development: Privileged voices, unheard stories and subjugated knowledge. *International Review for the Sociology of Sport, 46*(3), 249–264.

Welty-Peachey, J., Borland, J., Lobpries, J., & Cohen, A. (2015). Managing impact: Leveraging sacred spaces and community celebration to maximize social capital at a sport-for-development event. *Sport Management Review, 18*(1), 86–98.

Pringle, R., & Falcous, M. (2018). Transformative research and epistemological hierarchies: Ruminating on how the sociology of the sport field could make more of a difference. *International Review for the Sociology of Sport, 53*(3), 261–277.

Schulenkorf, N. (2012). Sustainable community development through sport and events: A conceptual framework for sport-for-development projects. *Sport Management Review, 15*(1), 1–12.

Svensson, P. G. (2017). Organizational hybridity: A conceptualization of how sport for development and peace organizations respond to divergent institutional demands. *Sport Management Review, 20*(5), 443–454.

Weedon, G. (2015). Camaraderie reincorporated: Tough Mudder and the extended distribution of the social. *Journal of Sport & Social Issues, 39*(6), 431–454.

Whittle, A., & Spicer, A. (2008). Is actor network theory critique? *Organization Studies, 29*(4), 611–629.

Wickstrom, M. (2017, May 11). UN secretary-general closes UNOSDP. *Play the Game.* http://www.playthegame.org/news/news-articles/2017/0309_un-secretary-general -closes-unosdp/.

11 · ENTANGLING CORPOREAL MATTER AND GEOMATTER

Making and Remaking the Beach

DOUGLAS BOOTH

Urban beaches around the world are hives of activity. "No matter what time you go to the beach," writes Meg Stewart (1984) with reference to the well-known Australian playground of Bondi, you see physical activity: "Breathing, stretching, yawning, yogaing, tai-chiing, wet sand jogging, punitively pounding along the promenade or slipping in soft sand, the whole beach is on the move" (p. 36). Activity is not confined to humans. The earth, too, is on the move: Tides ebb and flow, waves crash, winds blow, storms brew, the sun and the moon rise and set, light brightens and fades, the sun's rays change angles, temperatures climb and fall, and clouds change shape, size, and color. Traditionally, we separate these two forms of activity into culture and nature, or meaning and matter, and locate humans and culture "outside nature." More recently, under the influence of new materialism, all movement is considered "*of* the world" (Schouwenburg, 2015, p. 64). Grounded in an interactionist ontology (Tuana, 2008), new materialism gives a fresh impetus to the "co-constitutive" relationships between human (corporeal) and nonhuman (bio-, geo-, and manufactured) matter (Schouwenburg, 2015, p. 64). Examples of co-constitution abound: A shining sun in a blue sky (geomatter) awakens and arouses human spirits, emotions, and feelings (corporeal matter); on a golden beach, these conditions affect and prompt activity—to disrobe and feel, wander over, run along, lie on, and dig in the sand.

In this chapter, I explore the interactions and entanglements between corporeal matter and geomatter (sun, sand, surf, wind, rain, geomorphology) at the beach. My specific reference is Bondi Beach, which is synonymous with some of Australia's boldest archetypes of physical culture: the surfbather, the sunbather, the surf lifesaver, the beach belle, the beach bum, and the surfer (rider of waves). I am interested in what the sociologist of science Andrew Pickering (2005) calls the third intellectual space, which resides between studies of the natural material environment and constructionist studies of the social world and which "couple[s] . . . the human and the nonhuman" (p. 35). I propose that interactions between corporeal matter and geomatter (e.g., bodies and waves) laid the foundations for new physical cultures (e.g., surfing) that transformed the form and conceptualization of Bondi Beach.

This chapter comprises three substantive sections. In the first section, I provide an overview of social constructionist approaches to understanding the beach. Social constructionism refers to individuals and groups constructing (and discursively representing) social and cultural relationships largely independent of the material world. Today, we receive Bondi through a myriad of representations proffered by real estate agents, local councilors, artists, and celebrities. These typically present geomatter as little more than an "infrastructure" of side interest (Law, 2009, p. 146) or a "background object" upon which human designs have been "imposed at will" (Whatmore & Hinchliffe, 2010, p. 449).

In the second section, I introduce geomatter as an active agent that "interacts with, resists and co-shapes" corporeal matter (Schouwenburg, 2015, p. 65). Just as warm temperatures and clear blue skies draw people to Bondi, so southerly storms send them scurrying and blanket their amenities in sand; just as the sun's rays tan skin, so too ultraviolet radiation prematurely ages it; just as cool waters and gentle swells relieve bathers from the effects of scorching northwesterly winds, so rip currents unceremoniously drag them seaward—occasionally with deadly consequences. In this section, I am particularly interested in the early interactions between corporeal matter and geomatter that precipitated the physical cultures of surfbathing and surfing (riding waves with and without a board).

In the third section, I argue that surfbathing and surfing transformed the perceptions and forms of corporeal and geomatter at Bondi. The surf lifesaver, for example, engendered a notion of human agents domesticating the surf. But as events repeatedly demonstrate, domestication expresses an unwarranted confidence in human agency to conquer other forms of material agency (Tuana, 2008, p. 197). Herein lies the rationale for this chapter. Physical cultures that celebrate the power of human agency over all other forms of matter produce and maintain an "ignorance" of dynamic geo environments (Tuana, 2008, p. 203), which at the beach can have fatal consequences (Booth, 2015). For this reason alone, we should "take the nonhuman world" and the agency of geomatter "seriously" (Schouwenburg, 2015, p. 65).

CONSTRUCTING BONDI

Today, we receive urban beaches as socially constructed features. In a particularly rich history of the domestication of the urban beach, Alain Corbin (1994) traces the origins of play, exercise, and bathing at the beach to ideas associated with Romanticism and to Edmund Burke and Immanuel Kant's aesthetic theories of the sublime, in which "standing on the shore . . . confronting the elements . . . stirred the self in a special way" (p. 64). Renowned poets (e.g., Byron, Keats, Shelley) and painters (e.g., Turner, Friedrich) of the age presented the sea and the beach as special places where people could discover their spirituality. Ideas about the therapeutic value of fresh air and bathing in cold water accompanied these aesthetic and spiritual notions: Physical activity at the beach became a panacea to temper melancholy and anxiety, maintain the regular rhythm of the lungs, restore energy and dynamism, eradicate pale complexions, prepare children for the pains of puberty and women for the sufferings of childbirth, and improve the muscle tone and strength of boys and men (Corbin, 1994). Construction of the built environment followed. Boardwalks, promenades, piers, parklands, changing rooms, benches, shelter sheds, clubhouses, refreshment rooms, vendors' kiosks, regulatory signs (no dogs, no vehicles, no littering), litter bins, and safety equipment transformed beaches. These facilities and resources, which emerged from social thoughts and behaviors and are dependent on them, further popularized interactions with the natural elements and defined the rules of conduct in this emergent space.

Social constructionist approaches can contain evocative, detailed, and even dramatic descriptions of geomatter. Stewart (1984) eloquently refers to Bondi's "stretch and spread of white sand, blue sky and a sea from palest turquoise to deep cobalt" (p. 28). She includes rich detail: Under the waves "sunlight patches reflect . . . on the sandy bottom like leopard skin" (p. 28). And she captures the dramatic: A strong wind "whipped up . . . sheets of . . . salt and sand" and "veiled" the beach "in dull grey, unending desert nothingness." Later, "the sun came out to set clearly. The surf resumed its natural effervescence, the last of the sun's rays jewel glinting on foam white, dark blue water shining gold, perfection" (p. 34). But such descriptions are merely annotations in anthropocentric accounts that privilege and elevate human thought, perspective, design, and agency above all else. Critically orientated streams of social constructionism undoubtedly offer important insights into power struggles over the legitimate use of space and the body at the beach. Bondi, for example, was a key site in debates over the presentation of revealed bodies and laws around bathing costumes in Australia. In this contest, which spanned nearly a century, "every square inch of epidermis became hotly contested ground, with each advance toward nakedness decried as a plunge into social anarchy and moral turpitude" (Lenček & Bosker, 1998,

p. 194). Nonetheless, anthropocentric approaches exclude consideration of equally important nonhuman agents.

Underpinning the socially constructed beach is a Cartesian ontological model in which the world comprises matter—"solid, bounded objects that occupy space and whose movements or behaviors are predictable, controllable, and replicable because they obey fundamental and invariable laws of motion"— and nonmatter, or human subjects and agents (Coole & Frost, 2010, pp. 7–8). This dichotomized model "yields a conceptual and practical domination of nature" by human agents (Coole & Frost, 2010, p. 8) for whom "the natural is a cultural product" and "nature exists only as a conceptual opposition to culture" (Fiske, 1983, p. 119; see also Flynn, 1987). Significantly, matter may hinder human agency, but it never generates human action (Whatmore & Hinchliffe, 2010, p. 442).

Cartesian assumptions shine through histories of Bondi, a beach that emerged in "the consciousness of European Australia" in the late 19th century and developed into a national icon over the course of the following century (Taylor, 2005, p. 2). In Richard Taylor's (2005) version, Bondi's "near-kilometer of golden sand between twin craggy headlands" contributes to its iconic status (p. 10). But ultimately, these "natural attractions" are nothing more than the props of cultural agents and events. Taylor's cultural agents are property developers, painters, writers and photographers, film and television workers, community activists, music and media stars, young tycoons, and above all, surf lifesavers. At the turn of the 20th century, property developers with "no partiality to wealth" opened the "pleasures of the sun, sea and surf . . . to all," including the working classes (p. 3). In the interwar years, a generation of painters, writers and photographers re-presented "the idea of the beach" as an Australian playground and lifesavers as the heroes of the surf (p. 4). Events corroborate these images. During the Great Depression, "armies of unemployed men and women flocked to Bondi to spend the day"; in so doing, they reinforced the idea of the beach as a bastion of egalitarianism (p. 5). On Sunday, February 6, 1938, Bondi lifesavers performed a dramatic mass rescue when they pulled from the surf more than 200 swimmers who had been swept seaward. Although 5 bathers drowned, in the absence of lifesavers, the toll would have been much higher (p. 7), and the incident enshrined the surf lifesaver at the center of the national beach mythos. In the 1960s, Bondi became synonymous with the Australian beach as "a place of physical pleasure"; since the 1970s, it has been the "default media beach, the almost inevitable backdrop for any artistic or commercial project" set at the beach (p. 8). Gentrification precipitated local activism.

In the 1970s, community activists challenged plans for commercial development. Apartment blocks constructed in the 1920s and 1930s and previously "labelled eyesores" became objects of celebration, with their "modest facades" conveying "a lingering aura of democracy" (Taylor, 2005, p. 9). Taylor (2005) claims that Bondi "made an emphatic statement of [its] status as Australia's iconic summer landscape" at the Sydney Olympic games. Bondi hosted the beach volleyball tournament, and its lifesavers carried a giant reel-and-line (iconic surf-rescue equipment) around the arena during the closing ceremony. Today, Bondi "looks better . . . than it has for 80 years. The buildings that hug the bay are scrubbed and bright and the water is among the cleanest along the coast. On a hot summer day the beach entertains tens of thousands of local, city, national and international visitors" (p. 10).

Social constructionism challenged the assumptions of powerful paradigms emanating from the biological sciences that ranked the human species and rendered whole groups inferior and unworthy. In the second half of the 20th century, a range of "minority" groups, including women and people of color, embraced social constructionism and argued that knowledge is constituted as much by culture, language, and semiotics as by pure logic and reason (Frost, 2011). In this sense, social constructionism was intellectually and politically liberating. However, in putting human subjectivity and agency at the center of analysis, social constructionism ignores the "peculiar and distinctive" agency of matter that is informed and compelled by forces beyond "human intentionality" and that has "its own impetus and trajectory" (Frost, 2011, p. 70). In the remainder of this chapter, I operationalize new materialism as it "breaks through" the dualisms of Cartesian thought—organism and environment, living and inert, nature and culture, human and nonhuman—and reveals the co-constitutive relationships between corporeal matter and geomatter at the beach. Using surfbathing and surfing as examples, I propose that geomatter acts directly on, and inscribes itself in, corporeal matter to produce effects and sensations. Surfbathers and surfers convert these into discourses, demeanors, attitudes, styles of dress, economies, and social relationships that constitute their respective cultures.

MATERIALIZING BONDI

New materialism foregrounds the agency—productive and generative powers and forces that operate irrespective of intentionality—of human and nonhuman matter that exists in a myriad of forms, each of which is capable of shaping, resisting, and coshaping other forms (Dolphijn & Van der Tuin, 2012, pp. 100–101). In this section, I examine a set of interactions between geomatter and corporeal matter that manifested in surfbathing and surfing that I call corporeally creative physical cultures. Two interrelated aspects of these interactions warrant comment. First, creativity is "never simply the property of a masterful agent"— geomatter or corporeal matter. Creativity flows "through and between agents" without being reduced to a property of one or the other (Connolly, 2013,

p. 407). Second, corporeal creativity is often contingent and involves spontaneous improvisation and experimentation (Coole & Frost, 2010, p. 20). Corporeal creativity is not a product of rational, reasoned thought or of "pre-existing visions and designs" (Whatmore & Hinchliffe, 2010, p. 457). In contradistinction to social constructionism, which searches for origins and trends—such as British bathing practices and foreign role models in the case of surfbathing and surfing, respectively, in Sydney—new materialism emphasizes relational engagements between matter, particularly at the points at which they emerge and are first experienced "before they are made" into practices (Whatmore & Hinchliffe, 2010, p. 457). In contradistinction to social constructionism, which paints prognoses, new materialism accentuates open and indeterminate futures. For example, no one could have predicted that the early interactions between bathers and the surf would so radically transform Bondi.

Bathing developed in Australia in the 19th century despite laws that restricted the practice. These laws attempted to preserve morality by shielding the public from indecently exposed bathing bodies. Reports from the young colony describe bathing in the harbor as a "favorite recreation" (Booth, 2016, p. 25). The effects of water on corporeal matter undoubtedly contributed to this popularity. Marine biologist and advocate for the ocean Wallace Nichols (2014) notes that "our bodily senses—touch, pressure, temperature, motions, position, balance, weight, vibration—are truly alive" in water (p. 101). Nichols (2014) traces corporeal connections with water to neurochemicals and physiological functions that have positive effects on the brain. For example, the weightlessness of the body in water can calm the mind by slowing down brain waves. Floatation, Nichols (2014) says, allows the brain to

> transition from the waking state (beta waves) through the state of wakeful relaxation (alpha waves) and ultimately to the state of deep meditative consciousness akin to the moment between waking and sleeping (theta [waves]). In this altered state the mind settles into nothingness, the inner voice is silent, and often a feeling of oneness and bliss occurs. (p. 104)

Bathers' accounts support these scientific claims. Game and Metcalf (2011, p. 38) cite one bather who described a sense of total relaxation while floating in Bondi Bay and reported the boundaries between their body and the sea and air evaporating as they merged with the ocean and sky.

Bathing in the calm waters of Sydney Harbor was, of course, a different practice compared to bathing in the surf at nearby ocean beaches. But even in the harbor, bathers could learn skills. In response to immersion, for example, bathers had to learn how to find the surface and "level off," to tread water using the legs to generate lift forces and keep the body upright and airways clear (Schnitzler,

Bondi circa 1870 (American and Australasian Photographic Company, State Library of New South Wales, Reference Code 63476).

Button, Croft, & Seifert, 2015), and to coordinate breathing with arm and leg actions—that is, swim. (Prior to its codification as a sport in the late 19th century, swimming incorporated a wide variety of movements, from breaststroke to various forms of overarm propulsion, using either one arm or two, and different styles of kicking, including frog, scissors, and vertical beats.)

The acclaimed Australian poet Agnes Storrie (1905) doubted whether any "jaded" Sydney-sider could "resist" the sea:

> The colours . . . hold out . . . a perpetual invitation. . . . Ultramarine, gobelin-blue, emerald, eau de nil, topaz—what a glory of intermerging hues, rich with the royal opulence of the sun. Each wave as it rises in the air catches the light in prismatic flashes, holds it imprisoned—a gigantic goblet of liquefied-sunshine—then curls over and breaks into "shallows sheeted with flaming foam" upon the sand.

Arthur Lowe (1958), the self-proclaimed pioneer of surfbathing in Sydney, concurred. "[I] looked at the big, white rollers, chasing each other into the beach," he wrote, and "the more I watched them, the more I wanted to watch them. They

Bondi 2014 (Photographer Franck Gazzola, published with permission).

seemed to recall something long past. What it was I did not know. But a longing to go in amongst those waves took hold of me" (p. 19). Both descriptions suggest that surfbathing began before cognition, in moments of experiential spontaneity and creative contingency, and in those moments that "precede" and are "separate" from and "independent" of thought (Meillassoux, 2012, p. 79). The historical record of interactions with the surf around Sydney, which begin in the mid-19th century, supports this argument, as well as the view that these interactions continued in a process of affected learning in which the sustained effects of geomatter—waves, currents, winds—on corporeal matter were slowly "mapped into knowledge" (Whatmore & Hinchliffe, 2010, p. 451).

The first surfbathers were mostly cautious, tentative, and hesitant. Initial interactions with the sea often occurred in rock pools, colloquially known as bogey holes (i.e., a place to swim or bathe; Booth, 2016). In the late 19th century, the state developed numerous bogey holes around Sydney into larger pools by excavating the relatively soft sandstone rock. These pools, such as the one at the south end of Bondi, enabled the state to preserve traditional moralities and exert some control over public representations of bodies. But the pools enable bathers to acquire "embodied knowledges of the ocean and the rocky shore" and to cultivate "convivial relations" with geomatter (McDermott, 2012, p. 6).

Notwithstanding the allure of the surf, a fear of "being battered" by big waves (Lowe, 1958, p. 20) or an inability to float or swim (Jaggard, 2007, pp. 90–91) confines most surfbathers to the water's edge. Here they find their

Bondi Baths circa 1900. Tidal pools such as the Bondi Baths were important halfway houses in learning to interact and develop intimate relationships with the surf (Photographer Michael Drew, State Library of Victoria, Accession no H2012.171/29).

pleasure "tumbling" in the shallows and feeling the "friction" and "swirl" of the water around their torsos and limbs (Maxwell, 1949, p. 8; Lowe, 1958, p. 24). Fuller engagement means a long process of learning to read the surf, avoiding areas "where sand and water mix . . . in bubble and confusion," and evading rip currents—seemingly "quiet sections" where the water travels at high speed back to sea, sometimes for hundreds of meters (Maxwell, 1949, p. 11). "Experience has taught us," said Fred Williams (1907, p. 154), pioneer surfbather and long-serving councilor on the New South Wales Amateur Swimming Association, that it is "fruitless" to fight a rip. The bather must maintain "presence of mind" and "swim in the same direction as the current, but bearing left or right." "Treading water" should be "the only exertion" and "floating" will give the bather rest (p. 154).

Riding waves expanded the corporeal dimensions of surfbathing. Lowe (1958, p. 24) claimed there is an intrinsic appeal to "plunging" with a wave "as it roll[s] to the shore." Yet while it is easy enough to "spring off the sand" into a "billow" (Mills, 1906a) and merge with it like "a cataract that is half human and half [champagne] froth" (Russell, 1910, p. 257), the best bodysurfers (initially "surfshooters") catch the waves before they break. They also position themselves on the face of the breaking wall in front of the whitewater; this requires a delicate combination of "timing," "judgement," and "speed" (Jaggard, 2007), as well as the ability to distinguish a "hollow-fronted 'dumper'" from a "beacher." The former thrusts the bodysurfer to "the bottom of . . . a pile of water" that detonates on a sandbank with enough power to dislocate shoulders and fracture necks (Maxwell, 1949, p. 10); the latter lifts the bodysurfer high until their "heart is in their mouth," and then sends them "skimming along" the sloping face of the wave "like a human torpedo" (Russell, 1910, pp. 255–256). Roger Caillois (1961/2000) calls this moment *ilinx*. One of four types of play, *ilinx*—from the "Greek term for whirlpool" and "from which is also derived the Greek word for vertigo, *ilingos*"—involves moments that "destroy the stability of perception and inflict a kind of voluptuous panic upon an otherwise lucid mind" (Caillois, 1961/2000, p. 23). Critically, *ilinx* contains a physiological dimension that involves increased concentrations of neurotransmitters—notably epinephrine, norepinephrine and dopamine (Gill & Beaven, 2007).

Of course, neurophysiology did not solely determine bathers' engagements with waves. The full relationship is probably best analyzed as a complex entanglement that includes the language used by bathers to articulate and explain their experiences, which subsequently became the discourses adopted by their followers. Journalist Samuel Mills (1906b), for example, identified the beach as a "mission for mankind": "We bathe in [its waters] and acquire its strength; we inhale [its air] and receive the breath of life. We doze in its presence and the salt-laden rays of its sun go clean through our bodies, purifying our blood" (p. 634). Such lyrical waxing was part of the discourse of health that helped construct the modern Australian beach. Nonetheless, my focus in this entanglement is the agency of geomatter.

Surfboards heightened the sensations of *ilinx* in riding waves and "made body-shooting feel like dog-paddling" (Foran, 1928). Learning to ride a surfboard, however, is more difficult than learning to bodysurf. According to Maxwell (1949, p. 235), the early riders in Sydney "all but turned themselves inside out and upside down" in their efforts to master the board. "But . . . their ideas were wholly wrong": "they attempted only broken waves at low tide . . . and their only reward was to see the board dip its nose, dig into the sand, and come back, bang over tip, to slam them hard in the ribs" (p. 235). In the tradition of social constructionism, Maxwell (1949) traces the origins of boardriding in Australia to a visit by the Hawaiian waterman Duke Kahanamoku in the summer of 1914–15. But Frank Foran (1928), a member of the North Bondi Surf Life Saving Club and Bondi's "star board shooter" in the interwar years, maintains that not even Kahanamoku could teach the intricacies of boardriding. "I watched Duke," "studied his methods," and analyzed his "tricks," recalled Foran (1928). "He was superbly skilful," but there must have been something amiss in the manner in which I sized him up. . . . "Bluey" and I . . . obtained boards, and after a little imitation of the adroitness of the Duke, "Bluey" was bluer than ever with bruises, and I ached from the tip of my nose to the extremities of my toes. Naturally, we realised we were doing it the wrong way so we tried different tactics, and gradually improved. (Foran, 1928, p. 25)

Surfbathing and surfing emerged from a myriad of interactions between geomatter and corporeal matter. These interactions, many of which began as spontaneous and exploratory play among waves, transported surfbathers, and especially surfers, into a new realm. The transformative power of these interactions should not be underestimated. In the words of Egbert Russell (1910), an early observer of the surfbathing scene in Sydney, the surf "represents a readjustment of all the classifications that history and politics and social conditions ever brought about" (p. 262). Interactions with the surf were neither static nor unidirectional. Rather, they spawned ongoing dynamic relationships (Tuana, 2008) that at Bondi involved enculturating corporeal matter and transforming the material content of the beach. In the following section, I explore these processes with a particular focus on the surf lifesaver, who played a key role in transforming social perceptions of the surf as a hazard. But if surf lifesavers transformed

Bodysurfing transformed corporeal relationships with geomatter (Photographer Franck Gazzola, published with permission).

Surfboards further heightened corporeal relationships with geomatter. Technological developments enabled surfers to explore every dimension of the wave and to redefine a surfable wave. But regardless of their skill and creativity, surfers are always "reacting to what the wave will let [them] do" (Potter, 2017). (Photographer Franck Gazzola, published with permission).

perceptions, they could not eliminate biohazards (e.g., sharks, bluebottles[1]) or geohazards (e.g., rip currents—a major cause of drowning [Brander, 2010]); these forms cannot be predicted or calculated, let alone managed (Brander, 2015; Gershwin, 2015; Neff, 2012).

ENCULTURATING BONDI

Sydney's beaches became a new cultural scene early in the 20th century. At the center of this scene was the "surfer," a portmanteau term for surfbathers, body-surfers, boardriders, and surf lifesavers—essentially anyone who displayed corporeal capital on the sand and physical skill in the waves.[2] In these initial years, tanned skin was the critical marker of surfers and distinguished them from nonsurfers. Only those with a "brown tint" acquired by "lying in the sun" could achieve "full citizenship" on the beach, Russell (1910, p. 263) reported, adding that white bodies without "the coat of tan" were scorned and abused. Corporeal exteriors revealed even finer status distinctions among surfers. Mills (1906a) referred to "sun-boys," "whose hair is full of sand"; "limber ones," "who buckle in the water like well-oiled leather"; "pets," "whose skins have been coloured up like a meerschaum"; and "talent," who possessed a "brown hide with a real good outside cake of salt." Such distinctions, of course, are in constant flux and mostly based on tacit and visceral classifications and judgments that do not necessarily "pass through conscious awareness" (Coole & Frost, 2010, p. 20).

While commentators as divergent as Mills (1906b) and Fiske (1983) argue that the pursuit of tanned skin is a quest for social status, the tan should not be reduced to a social construction. Biological research shows that tanning releases endorphins, neurotransmitters that generate feelings of well-being (Fell, Robinson, Mao, Woolf, & Fisher, 2014). Thus sunbathing at the beach is not just a search for sun to inscribe social meaning onto corporeal matter. It also expresses the agency of geomatter, over which corporeal matter does not have total control. The line between tanned and damaged skin is extremely fine. Ultraviolet light not only changes the color of skin; it burns and prematurely wrinkles the skin as well as stimulates irregular pigmentation, melasma (chloasma), solar lentigines, solar keratosis, carcinomas, and melanomas (Hayward, 2003). Indeed, if skin is a boundary between corporeal matter and geomatter, it is highly porous with regard to the effects of the sun. The sun socially enhances and disfigures corporeal matter and extinguishes bodies (Tuana, 2008).

Sydney's beaches became immensely popular as increasing numbers of "surfers" interacted with sun, sand, and surf. The popularity of the beach resulted in demands for facilities—changing rooms, refreshment rooms, shade—and accouterments (beach chairs, bathing costumes, towels, tanning lotions). At Bondi, the state built a seawall and promenade, removed sand dunes, and planted exotic species. In so doing, it radically altered the natural contours of the beach. The Pavilion, a beachfront pleasure palace, signified the coup de grâce of the natural beach at Bondi. Made from rendered brick and concrete with a glazed Cordoba-tile roof, the Pavilion introduced an exotic Spanish style to Sydney's most popular beach; it included changing facilities for 12,000 swimmers, a restaurant-cabaret, and a spacious veranda under which patrons enjoyed afternoon teas and sodas while watching bathers, and the surf, perform. Beachgoers made an estimated one million visits to the Pavilion in the four years after its opening in 1929 (Booth, 2012). But if the Pavilion is proof writ large of the power of human agency to reconstitute the natural environment, it is the Bondi surf lifesaver who valorizes corporeal matter. By promising to mediate the risks of biomatter and geomatter, the Bondi surf lifesaver gave voice to notions of a "reasonably predictable natural world" (Tuana, 2008, p. 208). The surf lifesaver "metaphorically transformed" Bondi "into a swimming pool by the flag, the cap, the line and the whistle" (Fiske, Hodge, & Turner, 1987, p. 65).

Descriptions of lifesavers as a higher form of corporeal matter began appearing soon after the formation of clubs (Brawley, 2006). Russell (1910) called them the "highest class" of Australian. Surf lifesavers, he wrote, are "the Samurais, the oligarchs, the elite—a gladiator caste, envied by all men, adored by all women" (p. 263). In an analysis of the representation of surf lifesavers during the interwar years, Kay Saunders (1998, p. 103) argues that they "literally and figuratively" superseded the soldier as a national hero. Titles such as *Vigilant and Victorious* (Brawley, 1995), *Gladiators of the Surf* (Galton, 1984), *Heroes of the Surf* (Harris, 1960), *Surf: Australians Against the Sea* (Maxwell, 1949) implied that the lifesaver could subordinate the surf and added to their status. Maxwell (1949) was explicit. "There could exist no conditions at all likely to intimidate these men," she declared (p. 87). The panegyrics of the governing association, Surf Lifesaving Australia (SLA), further reinforced this view. When South Curl Curl on Sydney's northern beaches opened its new clubhouse in 1937, *Surf in Australia*, the official voice of the association, proudly declared that it stood in "bold contrast against its surroundings," a "tower high above the wide beach" that "gives the impression of a mighty fortress with the beach and surf under its submission" ("New surf pavilion," 1937).

Events at Bondi on February 6, 1938, however, challenged these perceptions. On that day, subsequently known as Black Sunday, five bathers drowned during a mass rescue in an area that red-and-yellow flags designated a safe bathing zone. Orthodox accounts of Black Sunday refer to the deaths as a natural disaster, the result of freak waves and a collapsing sandbank. These representations not only absolve surf lifesavers of responsibility; they render them heroes for containing the number of fatalities when a ferocious flash rip swept some 200 bathers seaward. Critical analysis suggests that the drownings were more the result of an erroneous belief by lifesavers that they could control any surf conditions (Booth, 2015) and highlights the value of an interactionist ontology that takes seriously the dynamic relationships between nonhuman and human matter (Öhman, 2016).

In the language of new materialism, Black Sunday involved a "complex interaction" of matter that transcended "the ontological divide" between humans and their physical environment (Tuana, 2008, p. 193). Black Sunday constituted a concatenation of geomatter (sunny sky, low-pressure system, rough surf, sand bars, tides, rip currents), corporeal matter (surfbathers, surf lifesavers), and a specific social scene (Bondi, Sydney's most popular bathing destination). Within this framework, geomatter does not reside in relatively stable and predictable systems so familiar to social constructionism but is part of a dynamic system characterized by "differential periods of stability," "relative equilibrium," and "periods of real disequilibrium" (Connolly, 2013, p. 400). The 48 hours leading up to Black Sunday was a period of differential stability marked by moderate to rough seas, generated by two low-pressure systems in the Tasman Sea. On the Friday before Black Sunday, a local resident drowned in rough surf at Bondi, and on Saturday, Bondi lifesavers performed scores of rescues. The surf ran at unusual angles and there were strong rip currents on Sunday. In the afternoon, the tide dropped, exposing a sandbar offshore to those bathers who liked to keep their feet firmly planted in the sand. They simply waded through a shallow channel to the sandbar.

In the midafternoon, Bondi Beach experienced a period of substantial disequilibrium when a set of waves broke in deep water, washed over the sandbar, flooded the channels on either side, and raised the sea level. Some bathers lost their footing during the process that mimicked the sensation of a collapsing sandbar (Booth, 2015). The flooded channels had no capacity to return the water to the sea after it reached the shore. Thus the water flowed directly offshore and over the sandbar as one massive flash rip reaching speeds of 2 meters per second (i.e., a speed achieved only by world-champion swimmers and exerting a force at waist-deep capable of sweeping even the largest adults off their feet; Brander, 2015).

Surf lifesavers rushed to action from the beach and adjacent clubhouses, but they were unprepared for the level of disequilibrium. The reel-and-line was too slow for mass rescues, and the line tended to entangle bathers (Booth, 2015). In fact, one of the bathers who drowned that afternoon was hauled to shore with his wrist twisted in a line (Maxwell, 1949, p. 84). The most valuable rescue equipment during Black Sunday was a pleasure craft, an inflated rubber mat and popular wave-riding device known as a surf-o-plane. Many lifesavers used the mats to keep bathers afloat until their colleagues could return them to shore.

Why were Bondi surf lifesavers unprepared for the possibility of more intense disequilibrium in what was clearly already a period of relative instability? Why didn't they show greater precaution and close the beach or at least reduce the area available to bathers? The answer lies, first, in a culture in which the agency of the surf lifesaver (corporeal matter) transcends the agency of the surf (geomatter; Tuana, 2008, p. 197). It is a culture that predisposes people to poor risk assessment and poor advice. As noted above, from the beginning of organized surf rescues in Australia, lifesavers have advised bathers caught in rips to avoid panic. Not only does this advice totally ignore the physiological variations in response to life-threatening (and *ilinx*) situations (Gill & Beaven, 2007); it is offered to people who typically cannot "see," let alone "read," a rip. Many people feel confident venturing into the sea and even the surf—as long as they can stand on the seafloor. But "relax and don't panic" messages offer no comfort when those bathers are unexpectedly immersed in a moving current of water with a will of its own and a force over which they have no control (Brander, 2010, p. 112). A second reason for the lack of precautions prior to the events of Black Sunday concerned an unerring, and ultimately deadly, faith in socially constructed knowledge of the surf. Unfortunately, for those involved in Black Sunday and other comparable incidents, such faith belies the limited knowledge of rip geomorphology. Coastal geomorphologist Rob Brander (2015), a.k.a. Dr. Rip, maintains that there are "no generic rip current risk forecasting techniques" and that "despite substantial efforts . . . we still cannot accurately predict their occurrence or flow behaviour" (p. 359). Black Sunday is a potent reminder of the value of an interactionist ontology to the analysis of surfbathing as a physical culture. While one must be wary of adopting a monistic approach that flattens the social and the material worlds, care must be equally applied to avoid artificially separating the two, which, at least at the beach, are not so easily discernible (Dolphijn & Van der Tuin, 2012, pp. 91–92; see also Tillman, 2015).

CONCLUSION: A NEW ENVIRONMENTAL POLITICS AND ETHICS AT THE BEACH?

Surfbathing and surfing are good examples of physical cultures that have emerged from an "intermeshing" of human and nonhuman matter and that display evidence of geomatter inscribing itself on, and even in, corporeal matter (Alaimo, 2010, p. 2). Surfbathing and surfing illustrate the agency of nonhuman matter and are potent reminders that the interactions between humans and nonhumans are never predetermined. Again, the suntan, an early marker of surfbathing culture, is an apposite case. In recent decades, surfbathers have shied away from the dark tan that is as conspicuous on the sand today as pale skins were 60 years ago. Sunglasses, hats, and radiation-protection clothing are common; even lifesavers don long-sleeve tops and broad-rim hats.

Could an interactionist ontology that underscores the entanglement of corporeal and geomatter give new meaning to physical cultures such as surfbathing and surfing? Among disciples of new materialism, there is a sense of optimism that an ontology that directs attention to the essential and fragile interaction of human and nonhuman matters might cultivate a new ethic of care for the environment (e.g., Connolly, 2013). This optimism would be welcome at Bondi. Notwithstanding the popularity of Bondi, sewage from outfalls at adjacent North Bondi assailed beachgoers for much of the 20th century. Concoctions of corporeal and manufactured waste, including heavy metals and pesticides, and biomatter such as bacteria and viruses regularly flowed into Bondi Bay with northeasterly and easterly swells and winds (Beder, 1989). Activist-journalist John Pilger (1990, p. 10) once described Bondi as an "enduring environmental atrocity" and recalled hot summer winds serving a "cocktail of salt, sand and shit" to surfbathers. In the 1970s and '80s, Bondi was known as "Scum Valley." Stormwater complemented the sewage, flowing directly onto the sand from a double drain pipe—colloquially known as "the binoculars"—at the southern end of the beach. Surfers and surf lifesavers reported constant ear and eye infections and gastric problems. When the wind blew from the east, local boardrider Brad Johnson said, "you get a film or smell off the top of the water, it gets in your eyes, and make[s] them sting and water, and when you take a deep breath your chest seems to cloud" (Channon, 1974, p. 10; see also Beder, 1989, p. 325).

Large-scale public protests forced the state government to relocate the outfalls into a deeper marine environment (1,500 meters) offshore. While these protests would appear to articulate the optimism espoused by new materialism for a new ethic of care for the environment, local boardriders and lifesavers were conspicuously absent from the protests. "We've gotten used to [the pollution]," and "we're all a bit too complacent," admitted local boardrider Victor Ford (Channon, 1974, p. 10; see also Beder, 1989). At one point, SLA threatened to withdraw patrols and effectively close Sydney's beaches (Beder, 1989, p. 329). However, on the eve of the public protests in 1989, Clif Ryan, president of the Bondi Surf Bathers' Life Saving Club and a medical practitioner, declared that local lifesavers would not strike "in support of the anti-pollution cause," which he dismissed as a media campaign and deemed a threat to community well-being (Brawley, 2007, p. 271). Bondi surf lifesavers, Ryan insisted, were not "presenting symptoms associated with their contamination" and "were no more well or unwell than they had ever been" (p. 271). These sentiments articulate well the prevailing view that urban beaches are first and foremost human spaces and call into question any optimism that physical cultures might advance a new ethic of care for the environment.

The idea of sacrosanct human spaces, which Tuana (2008, pp. 197, 203) labels materialized denial and ignorance, still prevails at Bondi, as demonstrated by a recent drowning and the reconstruction of the North Bondi Surf Lifesaving clubhouse. The drowning occurred late in the afternoon on November 4, 2013, on a day described by a Bondi lifeguard as "eerie"; the weather was cold, the surf was mean, and the rip currents were as strong as they get (Bondi Rescue, 2014). Almost as soon as two young Japanese nationals began playing in the shore, a rip seized them. Lifeguards raced to the site and dragged one bather, who was already underwater, onto a rescue board. Scanning the water for the second bather, the lifeguards spotted him motionless a mere 10 meters away. But the surf would not let them reach the bather. As one lifeguard paddled to the drowning youth, a wave jacked up, lifted him, and threw his board into the air. A second lifeguard swam toward the patient, but before he arrived a wave hit the bather and pushed him under the surface. (The body was found several weeks later.) The surf triumphed in a period of intense disequilibrium that day. The two lifeguards displayed conspicuous grief and remorse but they never conceded defeat: Lifesaving culture demands that the surf is conquered and subjugated (Bondi Rescue, 2014).

A month prior to the drowning, the North Bondi Surf Life Saving Club opened new clubrooms. The structure was widely commended as a "sculptural masterpiece," "architecture at its best," and "perfectly suited to the surrounding environment" (De Brito, 2013). Irrespective of individual assessments of its aesthetic appeal, the clubhouse, like that at South Curl Curl referred to earlier, blatantly ignores the geoprocesses that operate at the beach, including cycles of sand movement and periods of disequilibrium in which waves erode the backshore. In contradistinction to "dissipative structures" favored by architects with an interactionist ontological bent and that incorporate biomatter and weigh lightly on geomatter (Armstrong, 2015), the North Bondi clubhouse has been "embedded" deep into the foreshore and breaks up the natural contours (Bevan, 2011). The clubhouse is a relic of an ontology that puts humans at the center of the universe for their own edification. The structure shares no continuity with the nonhuman matter that comprises and created Bondi, and it offers no acknowledgment to the vibrant interactions between human and nonhuman matter that underscore the physical cultures that make up contemporary Bondi.

NOTES

Sincere thanks to Juergen Martschukat for his helpful comments on an early draft of this chapter.

1. *Physalia utriculus*, a small relative of the larger and more venomous Portuguese Man-o'-war found in the Atlantic (*Physalia physalis*).

Environmental assault? Seawall, North Bondi Surf Life Saving clubhouse (left middle-ground), apartments, and sewage vent (center background) viewed from Bondi Bay (Photographer Franck Gazzola, published with permission).

2. Surf lifesavers provided a safety service through the auspices of volunteer clubs, two of which operated at Bondi: the Bondi Surf Bathers' Life Saving Club (the first formed lifesaving club in Australia), and the North Bondi Surf Life Saving Club.

REFERENCES

Alaimo, S. (2010). *Bodily natures: Science, environment, and the material self.* Bloomington: Indiana University Press.

Armstrong, R. (2015, January 22). Experimental architecture: Catalysing the transition from an industrial age to an ecological era. Lecture presented at Newcastle University, England. Retrieved from https://www.youtube.com/watch?v=PJGi4ro47Aw

Beder, S. (1989). *From pipe dreams to tunnel vision: Engineering decision-making and Sydney's sewerage system* (Unpublished doctoral dissertation). University of New South Wales. Retrieved from https://www.uow.edu.au/~sharonb/phd.html

Bevan, R. (2011, March 18). Architects carve their own space. *Australian.* Retrieved from http://www.theaustralian.com.au/arts/architects-carve-their-own-space/news-story/4ccb3eof8a1169284e1ddd8aa7ac1d55

Bondi Rescue. (2014, February 27). *Swimmer vanishes at Bondi Beach* [Video file]. Retrieved from https://www.youtube.com/watch?v=3HNpD4rWd8g

Booth, D. (2012). Bondi Park: Making, practicing, performing a museum. In M. Phillips (ed.), *Representing the sporting past in museums and halls of fame* (pp. 204–230). London, England: Routledge.

Booth, D. (2015). The myth of Bondi's Black Sunday. *Geographical Research, 53*(4), 370–378.

Booth, D. (2016). Origins in history and historiography: A case study of Bondi's first swimmer. *Journal of Sport History, 43*(1), 21–36.

Brander, R. (2010). *Dr Rip's essential beach book.* Sydney, Australia: UNSW Press.

Brander, R. (2015). Rip currents. In J. Ellis & D. Sherman (eds.), *Sea and ocean hazards, risks and disasters* (pp. 335–380). Amsterdam, Netherlands: Elsevier.

Brawley, S. (1995). *Vigilant and victorious: A community history of the Collaroy Surf Life Saving Club 1911–1995.* Sydney, Australia: Collaroy Surf Life Saving Club.

Brawley, S. (2006). Surf bathing and surf lifesaving: Origins and beginnings. In E. Jaggard (ed.), *Between the flags: One hundred summers of Australian surf lifesaving* (pp. 23–47). Sydney, Australia: UNSW Press.

Brawley, S. (2007). *The Bondi lifesaver: A history of an Australian icon.* Sydney, Australia: ABC Books.

Caillois, R. (2000). *Man, play and games.* Urbana: University of Illinois Press. (Original work published 1961)

Channon, B. (1974, June). Panache. *Surfing World,* pp. 8–15.

Connolly, W. (2013). The "new materialism" and the fragility of things. *Millennium: Journal of International Studies, 41*(3), 399–412.

Coole, D., & Frost, S. (eds.). (2010). *New materialisms: Ontology, agency, and politics.* Durham, NC: Duke University Press.

Corbin, A. (1994). *The lure of the sea.* Berkeley: University of California Press.

De Brito, S. (2013, October 3). It pays to sweat the small stuff. *Sydney Morning Herald.* Retrieved from http://www.smh.com.au/executive-style/culture/blogs/all-men-are-liars/it-pays-to-sweat-the-small-stuff-20130925-2ueas.html

Dolphijn, R., & Van der Tuin, I. (2012). *New materialisms: Interviews & cartographies.* Ann Arbor, MI: Open Humanities Press.

Fell, G. L., Robinson, K. C., Mao, J., Woolf, C. J., & Fisher, D. (2014). Skin β-endorphin mediates addiction to UV light. *Cell, 157*(7), 1527–1534.

Fiske, J. (1983). Surfalism and sandiotics: The beach in Oz culture. *Australian Journal of Cultural Studies, 1*(2), 120–149.

Fiske, J., Hodge, B., & Turner, G. (1987). *Myths of Oz: Readings in Australian popular culture*. Sydney, Australia: Allen & Unwin.

Flynn, P. J. (1987). Waves of semiosis: Surfing's iconic progression. *American Journal of Semiotics, 5*(3–4), 397–418.

Foran, F. (1928, November 11). Riding the board on a racing wave—here's how. *The Sun* (Sydney), p. 25.

Frost, S. (2011). The implications of the new materialisms for feminist epistemology. In H. E. Grasswick (ed.), *Feminist epistemology and philosophy of science power in knowledge* (pp. 69–83). Amsterdam, Netherlands: Springer.

Galton, B. (1984). *Gladiators of the surf*. Sydney, Australia: Reed.

Game, A., & Metcalfe, A. (2011). My corner of the world: Bachelard and Bondi Beach. *Emotion, Space and Society, 4*, 42–50.

Gershwin, L. (2015, October 15). The blue bottles are coming, but what exactly are these creatures? *The Conversation.* Retrieved from https://theconversation.com/the-blue -bottles-are-coming-but-what-exactly-are-these-creatures-48675

Gill, N., & Beaven, M. (2007). Physiology of risk. In D. Booth & H. Thorpe (eds.), *Berkshire encyclopedia of extreme sports* (pp. 236–240). Great Barrington, MA: Berkshire.

Harris, R. (1960). *Heroes of the surf: Fifty years' history of Manly Life Saving Club*. Sydney, Australia: Manly Surf Life Saving Club.

Hayward, N. K. (2003). Genetics of melanoma predisposition. *Oncogene, 22*(20), 3053–3062.

Jaggard, E. (2007). Bodysurfers and Australian beach culture. *Journal of Australian Studies, 31*(90), 89–98.

Law, J. (2009). Actor network theory and material semiotics. In B. Turner (ed.), *The new Blackwell companion to social theory* (pp. 141–158). Oxford, England: Blackwell.

Lenček, L., & Bosker, G. (1998). *The beach: The history of paradise on earth*. New York, NY: Viking.

Lowe, A. M. (1958). *Surfing, surf-shooting and surf-lifesaving pioneering*. Sydney, Australia: Lowe.

Maxwell, B. (1949). *Surf: Australians against the sea*. Sydney, Australia: Angus and Robertson.

McDermott, M. (2012). *Wet, wild and convivial: Past, present and future contributions of Australia's ocean pools to surf, beach, pool and body cultures and recreational coasts* (Unpublished doctoral dissertation). Edith Cowan University, Perth, Australia. Retrieved from http:// ro.ecu.edu.au/theses/517/

Meillassoux, Q. (2012). Interview. In R. Dolphijn & I. Van der Tuin (eds.), *New materialisms: Interviews & cartographies* (pp. 71–81). Ann Arbor: University of Michigan Press.

Mills, S. (1906a, March 7). Shooting the breakers. *Sydney Mail*, p. 606.

Mills, S. (1906b, March 7). The romance of the breakers. *Sydney Mail*, p. 634.

Neff, C. (2012). Australian beach safety and the politics of shark attacks. *Coastal Management, 40*(1), 88–106.

New surf pavilion at South Curl Curl. (1937, September 1). *Surf in Australia*, p. 11.

Nichols, W. (2014). *Blue mind*. London, England: Little, Brown.

Öhman, M. (2016). Embodied vulnerability in large-scale technical systems: Vulnerable dam bodies, water bodies, and human bodies. In F. Käll (ed.), *Bodies, boundaries and vulnerabilities* (pp. 47–79). Cham, Switzerland: Springer.

Pickering, A. (2005). Asian eels and global warming: A posthumanist perspective on society and the environment. *Ethics and the Environment, 10*(2), 29–43.

Pilger, J. (1990). *A secret country*. London, England: Vintage.

Potter, M. (2017, March 18). Commentary at Quiksilver Pro, Gold Coast, Australia, Round 4, Heat 4, 23:19, and Round 5, Heat 4, 18.34. Retrieved from http://www.worldsurfleague .com/events/2017/mct/1800/quiksilver-pro-gold-coast/heatanalyzer

Russell, E. (1910, January). Australia's amphibians. *Lone Hand*, pp. 252–265.

Saunders, K. (1998). "Specimens of superb manhood": The lifesaver as national icon. *Journal of Australian Studies, 22*(56), 96–105.

Schouwenburg, H. (2015). Back to the future? History, material culture and new materialisms. *International Journal for History, Culture and Modernity, 3*(1), 59–72.

Schnitzler, C., Button, C., Croft, J. L., & Seifert, L. (2015). A new qualitative typology to classify treading water movement patterns. *Journal of Sports Science and Medicine, 14*, 530–535.

Stewart, M. (1984). Beachstruck on Bondi. In R. Drew, J. Kingsmill, M. Stewart, M. Whitlam, C. McGregor, C. Holmes, & O. Strewe (eds.), *Bondi* (pp. 28–53). Sydney, Australia: James Fraser.

Storrie, A. (1905, February 25). Surf bathing. *Sydney Morning Herald*, p. 7.

Taylor, R. (2005). *Bondi: A biography*. Sydney, Australia: Historic Houses Trust.

Tillman, R. (2015). Toward a new materialism: Matter as dynamic. *Minding Nature, 8*(1), 30–35.

Tuana, N. (2008). Viscous porosity: Witnessing Katrina. In S. Alaimo & S. Hekman (eds.), *Material feminisms* (pp. 188–213). Bloomington: Indiana University Press.

Whatmore, S., & Hinchliffe, S. (2010). Ecological landscapes. In D. Hicks & M. Beaudry (eds.), *The Oxford handbook of material culture studies* (pp. 440–458). Oxford, England: Oxford University Press.

Williams, F. (1907, January 16). The call of the surf. *The Sydney Mail*, pp. 152, 154.

12 · BODIES OF WATER

Intra-actions among Water, Sport, and the Body Politic

KYLE S. BUNDS AND MICHAEL D. GIARDINA

I (Kyle S. Bunds) first learned about water disparity while studying medieval European history—examining how those water systems worked. I learned of the social disparities and who had access to water (Magnusson, 2001). This had a profound impact on me and made me realize how precious water is for people to live. I entered into the water charity phenomenon through an introduction to the Royal Bank of Canada's partnership with One Drop and the All in for One Drop poker tournament (One Drop became the official charity partner of the World Series of Poker in 2012). I began to think about water charity more explicitly and the connections between sport and raising money for charity. Through this realization that sport can impact water charity, I studied the phenomenon (and the phenomenon made me anew).

Charity: Water emerged as the first mediated product I watched. There, I saw an active push to have people be physically active in order to raise money for building water systems. The organization shows how they have produced successful water systems for people and therefore deserve your donation. Examining these connections, I got involved with an organization we will call Sport for Water. Water charity as a phenomenon through Sport for Water occurs thus, as explained through the witnessed events of Kyle Bunds.

* * *

In the mountainous village of 18,000 in the Southern region of Ethiopia, a woman washes her body in an open water source. The water flows downstream two kilometers through the village. An animal farther downstream drinks the exposed water, taking in the microbial agents (Rose et al., 2011) passed along by the woman cleansed via the dirty water. Defecation from a goat rests slightly downstream from the drinking animal. A boy collects water in a yellow jug for his family to use for drinking, cooking, and cleaning, taking in the flaked skin, the defecation, and the dirt.

A villager witnesses Sport for Water partnered with the International Development Agency for Change (IDAC) solving a clean drinking water problem in another village and reaches out to IDAC to discuss the need for clean water in her village. IDAC witnesses the need and interacts with the local people. Through interactions with the villagers, IDAC is moved to act, and through its ability to act as an organization, a partnership is brokered. In this instance, IDAC and Sport for Water choose to utilize a successful regional water system implementation supported by monies raised by a race to showcase how fundraisers and donors have benefitted the village impacted by the water system. Sport for Water decides to dedicate proceeds from an upcoming running event to the village in the Southern region.

In order to promote the race and raise funds for the new village project, Sport for Water creates videos, images, website text, pamphlets, and other publically available material telling the story of how fundraisers make the water systems possible. Telling the story from the perspective of the organization to the runners and potential donors, Sport for Water changes (necessarily due to language barriers) the tale of dirty water and need to fit the narrative it believes will make new water systems possible. The villager is present in representation as a victim, with the fundraiser as a savior (Chouliaraki, 2010), and is only known to fundraisers as a representation—the body is absent.

A fundraiser and runner consciously decides to take up this cause, to run in a race specific to helping a particular village in the Southern region of Ethiopia receive water. The runner/fundraiser is changed and changes insofar as she now runs for something else but nevertheless is impacted socially and physically by her decision to participate in embodied philanthropy (Bunds, Brandon-Lai, & Armstrong, 2016). The act becomes money, and the money changes the organization's ability to implement solutions. The solutions are conceptualized and then realized by the villagers' laboring, sweating, and building. The laboring, sweating, and building change the landscape, the water flow, and the access. The clean water enters the villager's body, having gone from dirty, to brokered, to represented, to action, to money, to labor, to clean drinking water, connecting and permanently changing each along the way. The villager is now and will be different in body, regardless of how long the water system is in place. The fundraiser may or may not run again but is forever present in the body of the villager. The organization may or may not build another water system but is forever changed by the decision to build a water system in the Southern region of Ethiopia.

*   *   *

Karen Barad (2003) famously begins a rather scintillating discussion on how matter comes to matter by proclaiming, "Language has been given too much power. . . . At every turn lately every 'thing'—even materiality—is turned into a matter of language or some other form of cultural representation" (p. 801). Water charities, which exist for the sole purpose of raising funds in the Global North to provide clean drinking water to those in the Global South (Bunds, 2016; 2017), actively represent both clean drinking water and the people in need of clean drinking water. As introduced in the proem, in this chapter, we focus specifically on water charities that utilize sport in the Global North (primarily running events) as a way to raise money for water systems in the Global South. Of interest is the utilization of the mobile, physical, sporting body as a means to "aid" bodies that are presented as immobile—physically, socially, or economically.

How a water charity represents water and people in need of it is important because, as Bostrom and Uggla (2016) note, "the environment cannot plead its own case but must be represented" (p. 355). In this way, it might seem matter-of-fact to view dirty water—as the omnipresent picture of oppression in media and promotional accounts used by water charities might have us do—as being represented as "inert, stable, concrete, resistant to sociohistorical change" (Pellizzoni, 2016, p. 312).

Yet if we allow that materiality in fact "exhibits agency, inventive capacities, generative powers" (Pellizzoni, 2016, p. 312), matter can become the foundational character (Clark, 2011). Matter as foundational allows a focus on the intra-action rather than the interaction. Within a relationship predicated on interaction, entities maintain a sense of individuality *a priori* to the relationship. However, within intra-action, the ability to act comes from and emerges within the relationship itself. Water charity is thus about not just a lack of clean water itself but the intra-action of water with human and nonhuman entities, including human bodies, discourses on the Global North and Global South, local and global politics, and changing weather patterns, to name a few. Put differently, water charity is not a matter of simply clean or polluted water but a phenomenon that is made and unmade through intra-actions between nature, culture, science, and technology involving people, animals, media, nature, and the environment. Through these intra-actions, we are all brought together into the water charity phenomenon, through which we are made to or become the afflicted or nonafflicted, the thirsty or the quenched, the problem or the solution.

To borrow from Barad (2003), then, the central concern at the heart of this chapter is not about examining the interactions among the various agents involved within water charity but about understanding the intra-actions in the phenomenon of water charity (the "inseparability of observed object and agencies of observation" [p. 814]). By studying these intra-actions, we can see how, for example, differences come to be made and unmade. It is unlikely most readers of this book in the Global North will interact with water crisis or charity, but we all intra-act with the water crisis *phenomenon* and are therefore responsible to and for the matter produced therein: the discourses, the material, and the subject positions. Importantly, as Barad (2003) suggests, in interaction, responsibility can be skirted; through intra-action, responsibility is distributed evenly among constitutive entities. Intra-action thus erases cause and effect, individual agency, and subject-object distinctions (Barad, 2003) in favor of simultaneity and the always-changing (rather than always already) artificiality of the phenomenon in question, or what Barad refers to as a "relational ontology." And in the case of nature, of which water is part and parcel, Barad (2007) has argued that "nature is neither a passive surface awaiting the mark of culture nor the product of cultural performances" (p. 145).

In this chapter, then, we consider water charity as made and unmade through intra-actions among nature, culture, science, and technology. Importantly, we study the phenomenon and the intra-action present therein, not presuming "the prior existence of independent entities/relata" (Barad, 2003, p. 815).[1] More specifically, we consider the intra-action(s) present in international water charity. To do so, we lean largely on six years of in-depth ethnographic research, working in and with water charities in four countries on three continents. Rather than invoking some form of authorial (self-)reflexivity, we read the landscape of water crisis / water charity through Barad's (2003) method of diffraction—that is, through a diffractive reading of the intra-actions. This Baradian approach, which is informed in large but not exclusive measure by Donna Haraway's exposition of diffraction as outlined in "The Promise of Monsters," reorients our vision, or how we "see" water crisis come into being. As Haraway (1992) writes,

> Diffraction does not produce "the same" displaced, as reflection and refraction do. Diffraction is a mapping of interference, not of replication, reflection, or reproduction. A diffraction pattern does not map where differences appear, but rather maps where the effects of difference appear. (p. 70)

We begin this chapter by presenting the various *objects* that emerge in and form the water charity phenomenon before discussing intra-actions as present in the water charity phenomenon and why matter *matters* in water charity.

THE WATER CHARITY PHENOMENON

In this section, we focus on the intra-acting components in the water charity phenomenon. Barad (2003) informs,

> Phenomena are ontologically primitive relations—relations without preexisting relata. The notion of intra-action (in contrast to the usual "interaction," which presumes the prior existence of independent entities/relata) represents a profound conceptual shift. It is through specific intra-actions that the boundaries and properties of the "components" of phenomena become determinate and that particular embodied concepts become meaningful. (p. 815)

Specifically, we are concerned with the relata/entities not as they preexist relations but rather as they emerge through specific intra-actions in the water charity phenomenon. In what follows, we break down the proem at the beginning of the chapter and break down the phenomenon by each intra-acting component. We lean heavily on the fieldwork experiences of Bunds during his time working with water charities in Miami, Vancouver, London, and Ethiopia as a source for the identified intra-acting components.

Water

Although "dirty" water is something we may not always think about, water is often contaminated, and not only due to industrial pollutants or lax environmental regulations. Perhaps in a world without human and animal interference, we could consider situations in which water would remain "clean," but the fact is that pollutants emerge in various forms, which we will refer to as contaminants. One of the main sources of global water pollution that results in long-lasting and sometimes permanent (if not corrected) damage of water systems is referred to as *organic pollution* (Sen, Sonmez, Canpolat, Alp, & Kocer, 2013). These organic pollutants "originate from domestic sewage (raw or treated), or urban run-off, industrial effluents and farm water" (Sen et al., 2013, p. 335). Organic pollutants can be sources of pathogenic organisms. The pollutants can impact water quality and must be removed to combat the spread of disease. But pollutants can also be introduced into the system by changes in the physical environment itself, such as the natural blockage of water reserves or the (over)abundance of wildlife waste in a given water source. Yet water only emerges as "dirty" or "polluted" once it enters the organic body of the person consuming it; it is made dirty and unmade clean in the moment of its human consumption.

The Villager

People in villages and cities who do not have clean or made-clean water are obviously susceptible to water-borne diseases. In fact, one Ethiopian village of 18,000 that Bunds visited kept a tally of the top 10 causes of death for villagers. Nine of the top 10 reasons for death could be traced back to water and sanitation issues, with general "trauma"—consisting of natural causes, injury, and so on—being the only cause of death not directly related to water. Being made aware of "dirty" water (and thus in some manner of fashion calling it into being), witnessing death brought about by dirty water, and managing capital shortage issues that get in the way of obtaining clean water all contribute to the dirty water narrative and fashion the water charity as its im/material savior.

For example, a village Bunds visited in southern Ethiopia did not have a clean water project under way and was 2.5 kilometers down the mountain from a clean water spring. Villagers attended school in a village down the mountain until they were needed to work in the household or on the land. There was little hope and, as far as Bunds could tell, little desire to leave the village. Villagers died of water-borne diseases at a greater rate than any other source of morbidity in this village. Problematically, nobody in the village had the knowledge, equipment, or skill to deliver fresh water from the spring to the villagers.

In this example, as well as others, villagers lacked multiple forms of capital (in the Bourdieusian sense). There could be a lack of cultural capital in the knowledge formation of engineering water systems; there is often a lack of economic capital, resulting in the inability to create the necessary capital for the development of a water system; and there is often a lack of social capital, wherein those who are impoverished are unable to influence policy makers to help them build water systems.[2] "Dirty" water enters into the body and makes the body sick. The body becomes that which it consumes; the body is altered.

Water Charity Representative

A water charity is involved in "cleaning" dirty water or "finding/accessing" clean drinking water in several different ways. One common way includes villages reaching out to established water charities. This occurred often for the water charity Bunds worked with in Vancouver, here called Sport for Water. This water charity employed full-time staff from Ethiopia to oversee water projects in that country. Repeatedly, individuals from communities or other organizations that know of water issues in a particular region will contact Sport for Water, and Sport for Water will conduct a site assessment, talk to the local people, learn what water system solutions might be applicable, and then agree to undertake the project.

Another way a water charity can become involved is through an entirely new water charity. As in the examples of Charity: Water founder Scott Harrison,[3] or

Sport for Water founder Brian, an individual "comes across" a village in need of water or is told the story of need for water. In this scenario, the individual, before ever partaking in "water charity," *sees* the dirty water, *talks* to the villagers, is moved to an *emotion*, and is compelled to *act*. In this instance, the charity is changed—from a call to action, from seeing the dirty water (or its effects), and/or from talking with villagers directly affected by water crisis.

A third way the water charity representative becomes involved in a water project is through connections with other charities. For example, in the proem, Sport for Water representatives partnered with IDAC to work on a water project. It was the Sport for Water representative who reached out to the IDAC representative to obtain help on the water project.

Fundraising

The water charity, and by extension its philanthropic activity in afflicted areas, cannot exist without the raising of funds (i.e., donations). As Bunds learned in London with London Water Charity, there is a definitive need for fundraising on two fronts. First, fundraising allows the water charity itself to exist. This is a primary focus of water charities and any nonprofit in general. If the organization is unable to secure funding, it cannot operate. Thus many water charities operate to in part perpetuate their own operational existence. Second, fundraising permits the development and implementation of the actual water system. Without such funding, clearly no water systems could be built. To this end, the water charity dedicates resources to create promotional materials meant to (re)present the water crisis in a particular village (a presentation that may or may not be divorced from the reality of the situation on the ground).[4]

Message Creation

Message creation, or the creation of a promotional narrative through which to solicit external philanthropic donations or other forms of funding (e.g., government or foundation grants), plays a key role in leveraging and monetizing those in need. Often these promotional videos or website renderings (images, charts, infographics, etc.) include a disarticulated individual in need, caricatured storytelling, or spliced videos from various locations that activate or deploy what Hannah Arendt (1963) might call a "politics of pity" toward those who are suffering in a far-off place and for whom a collective "we" should feel pity. Drawing on Boltanski's (1999) work on distant suffering, such a promotional narrative can be understood not as the leveraging of activist notions of social justice but as the amplification of passive, if not distant, observation: "observation of the unfortunate by those who do not share their suffering, who do not experience it directly, and who, as such, may be regarded as fortunate or lucky people" (p. 3).

Historically, one way the self–distant stranger relationship has formed in humanitarian communication is through the use of what Chouliaraki (2010) describes as "shock effect" and "positive image." Chouliaraki (2010) in particular charts a long history of "shock effect appeals," with the intent behind the appeal being to "represent suffering in its plain reality" (p. 5). Most notably, she utilized examples from Oxfam in 1956 and the Red Cross in 1961 to indicate how campaigns depicted bodies clearly in severe need of nutrition or bodies that were half-naked. Chouliaraki also found that the victim was the focus of the campaigns, and the films were designed to juxtapose the viewer with the sufferer, thus moving the "Westerner" to act on account of the guilt they felt. Again, the key is the distance between the sufferer and the viewer.

However, as Bruna Seu (2010) acknowledges, this type of shock-inducing messaging leads to donor fatigue; people in general do not like the "guilt-tripping message of the 'shock effect' campaign" (cited in Chouliaraki, 2010, p. 7). Therefore, according to Chouliaraki, humanitarian agencies moved toward "positive image appeals" (p. 7), which focused not on the plight of the down-trodden but on the agency of the person in need. In this way, the donor is a savior, not the persecutor.

Related to the utilization of sport within this creative messaging, Thorpe and Rinehart (2013) suggest that "sport has become a particularly effective vehicle for exploiting affect" (p. 121). Humanitarian campaigns work in part based on the extratextual knowledge an individual donor brings to his or her reading of a given campaign: preconceived notions of poverty in the Global South, feelings of sympathy toward the less-privileged, shame at living a privileged lifestyle, and a general need to feel a part of something bigger than oneself (see Bunds, Brandon-Lai, & Armstrong, 2016; Giardina, 2010). These promotional projects are thus intentionally created for consumption by targeted audiences in the United States and the rest of the Global North (Boltanski, 1999; Bunds, 2017; Chouliaraki, 2010). Particularly in these types of campaigns, the viewing public is asked to donate money, act in some way (such as volunteering or taking part in a charitable "thon"), learn about the problem, and get others to donate as well by telling the charity's story to others. Ideally, the consumer becomes a donor and fundraiser him- or herself by engaging in some form of fundraising event.

By way of example, Charity: Water operates a website called mycharitywater .org, which also acts as a personal website through which the now active participant can solicit donations on his or her behalf as he or she engages in an activity such as running or biking or ask for donations for his or her birthday. Other charities may opt to utilize organized runs (i.e., 5K or 10K road races, similar to the Komen Foundation's "Race for the Cure" 5Ks). Organizations such as Team World Vision use already established runs like the Miami Marathon, whereas charities such as Sport for Water stage their own events. From the standpoint of the organization, in the case of staging its own run, Sport for Water must adapt its message to the people based on the targeted project and village and design activities that fit with the particular cause, thus facilitating the fundraiser. In the case of those running, many donate money themselves and use social networks to obtain donations from friends, family, business partners, and so on and spread knowledge about why they are raising money.

Perhaps most important is the function of being able to run at all. During our research, we have been told that sport is a good way to raise money because "anyone can do it," but there seem to be multiple types of cultural and economic capital that can keep some from participating but allow others to do so. Running to raise money involves all types of capital, including having the time to engage in running, safe areas in which to run, knowledge of training techniques, the ability to get people to donate, and the physical capacity to participate. Often the runners join groups and build their social networks with other like-minded philanthropists. In this stage, the fundraiser as philanthropist first engages with the building of all forms of capital. If a person has the time, the ability to fundraise, the interest in fundraising, and an interest in running; is moved to emotion by the message; and decides to act, then the person changes his or her routine to become a part of the organization and water charity. The end goal, of course, is still one of capital: In this case, money goes from the fundraiser to the charity, which begins putting plans into place that will help build water systems through the donated money. At this point, through a GoFundMe page, a my.charity.org page, or a number of other sources, individual donors to a fundraiser's cause are made aware of how much money the fundraiser has acquired and donated to the charity. The individual raising money is free to share this information with family, friends, and strangers in order to be viewed and acknowledged as an individual providing money to those in need.

Understanding Water Charity

In her book *Meeting the Universe Halfway: Quantum Physics and the Entanglement of Matter and Meaning*, Barad (2007) details how diffraction allows creative insights and the ability to read for differences. She follows up on that idea in her interview with Dolphijn and Van der Tuin (2012): "What I propose is the practice of diffraction, of reading diffractively for patterns of differences that make a difference" (p. 49). Barad notes that understanding things diffractively is about understanding the entanglement.

By reading the phenomenon of water charity by examining patterns of differences that make a difference, we have attempted to present the intra-action of objects in the water charity phenomenon. In the interview with Dolphijn and Van der Tuin (2012), Barad suggests that through this intra-action, we do not understand these things separately as existing entities that interact but think of them in intra-action. In this way, we begin to shift understanding from inter-action to intra-action, "where there are interactions through which subject and object emerge, but actually as a new understanding of causality itself" (Dol-phijn & Van der Tuin, 2012, p. 55).

Through a Baradian lens, we could surmise that within the phenomenon of water charity, what we really see are different experiments wherein the apparatus and the ontology change. Barad (Dolphijn & Van der Tuin, 2012) suggests that when an adjustment is made to the apparatus or the objects, the ontology, the apparatus, and the phenomenon all change. In this way, each water charity proj-ect is its own phenomenon to be understood differently from every other water charity project. However, what one can glean from examining the phenomena of water charity projects as separate entities is a causal relationship through the intra-action of objects. We can learn what works, and when something does not work, we can understand it through the differences that make a difference.

To illustrate, let us encapsulate how water charity in toto *could* work. Mate-rially changing the landscape through the intrarelated actions of numerous objects, "clean" water is thus produced and can be consumed by the aid receiv-ers. This quite literally changes the composition of the body, wherein there is a greater chance for health. Ideally, this health leads to additional resources for gaining capacities for those in need. For example, less time spent going to get water can allow children (especially girls, at least in the villages we are directly familiar with) the ability to accumulate knowledge; less time spent fetching and searching for water, more time learning, and more resources put toward devel-opment in other areas, including agriculture, can lead to economic capital development; healthy bodies receiving clean water can lead to positive healthy investments in physical capital; and time gained can be spent on building social networks with other local "marketplaces." This was a key and consistent theme emerging from all the water charity executives with whom Bunds worked: the ultimate goal being to allow villagers financial and life freedoms similar to those in the Western world (the problematic imposition of Western cultural and economic ideals notwithstanding).[5]

Having successfully implemented a new water system, the charity is able to leverage the clean water produced therein and begin to focus on another area with "dirty" water. An individual receiving aid is no longer conceived of in the phenomenon of water crisis as a person but rather as an *image*, though one which in many ways remains permanently affected by the phenomenon. A vil-lager could also reenter the phenomenon in totality when things go wrong with implemented water systems. In either case, the charity produces a narrative that encapsulates *its* good work in the service of *its* continued existence.

Barad (2012) once said, "Diffractive readings bring inventive provocations." In that spirit, we ask the following: If water charities didn't exist, would dirty water exist? In some ways, both are the outcome of relations between each other and other material and nonmaterial objects (as Latour might say). What, for example, is the intra-action between "water charity" and dirty water? This schema presupposes that both water charity and dirty water (a human and nonhuman intra-action) consist of entangled agency (i.e., having the ability to act), which emerges from within the relationship, not outside of it (see Barad, 2003, 2007). At the same time, what is the intra-action between water systems and public health? Between healthy bodies and diseased bodies?

DISCUSSION

Conceptualizing the water crisis/water charity phenomenon as a series of intra-actions among actors who emerge out of the phenomenon rather than as a linear cycle of interactions among separate stakeholders allows us to better understand how the effects of these intra-actions come into being. Put another way, the water charity industry has primarily been considered as a progression of separate interests, industries, and sectors: A water source is either polluted or otherwise inaccessible, which causes individuals in a village to become ill or otherwise encounter some form of water-based hardship; a water charity becomes active in that space, utilizing the tools of fundraising and marketing/promotional culture to raise money and/or awareness of an issue to volunteers and donors, who decide to become involved in a particular cause or effort; and the money that is raised is then used to build water systems, of which the clean water produced is consumed by those in the village. In each of these points of articulation, there is an implicit causality: Polluted water makes someone sick; I witness people in need, so I act; water systems are built, and people have clean water; and so on. But if we were to read the above water charity *phenomenon* diffractively for "patterns of difference that make a difference" rather than offering critiques of the various interactions that occur (Barad, quoted in Dolphijn & Van der Tuin, 2012), we might be better positioned to understand how the water charity phenomenon comes to matter through specific intra-actions.

Returning to the previous section, we might thus reconfigure the various actors in and generally accepted arc of the water charity industry as a series of intra-active events that give rise to the water charity phenomenon as it is presently understood. To wit, we might understand—or perhaps better put, come to view the emergence of—the water charity phenomenon in the following series of intra-active events:

- the presence of contaminated or polluted water and/or the correlative absence of clean drinking water bringing a public health crisis into being
- discourses on the suffering of those in the Global South

- discourses on the culpability of those bodies of privilege in the Global North
- discourses on the poverty of capitalism, the exploitation of the Third World, and the privileging of economic growth over public health
- discourses on water purification, water delivery systems, and hydrological science
- discourses of charity and philanthropy
- the role of politics and governments
- economic development
- disease and death
- water charity as material industry

We might represent this water charity phenomenon as shown in figure 11.1.

Although we all become implicated in the phenomenon, Barad (2012) maintains that it is through these intra-actions that we become separated into new and different subject positions: Global North/Global South, healthy/unhealthy, advantaged/disadvantaged; quenched/thirsty, and so on.

Let's look at one of these intra-actions more closely. To do this, we can look to the intra-actions between discourses of privilege and the material consequence

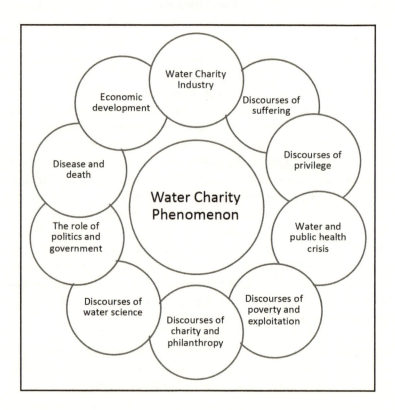

of dirty water. Throughout the vast majority of water charity promotional efforts and educational endeavors, the effects of dirty water—and the concomitant deaths and diseases often situated alongside it—on consumers in the Global North are positioned as something out of reach, out there, affecting "the other" in some far-off land (usually an African nation). Feelings of pity or shame are operationalized through images depicting abject poverty, glasses of brown water, and unhealthy bodies. Increasingly, the consumer is also hailed through *economic* differentiation. The Water Project, for example, actively equates "poverty" with "lack of clean water" in its most prominent infographic (see The Water Project, 2019). Rather than appeal to diseased suffering, the consumer is presented with the material conditions of economic poverty, such as work hours lost collecting water, the inability to grow food, the exchange between investment and economic return ("For every $1 invested in water and sanitation, there is an economic return of between $3 and $4!"), and so forth. The solution presented, in this particular case, is to "Help Unlock Potential"—an outcome in keeping with the economic appeal of helping those in need (usually through some form of fundraising outlined previously in this chapter). Through this intra-action, bodies of privilege come into being through the water charity phenomenon and are ordered and made meaningful over and against those lacking such economic power.

Consider as well how "healthy" and "unhealthy" bodies come to be made and remade in the service of the water charity phenomenon. Running events, charitable "thons," and other sport- and physical activity-themed fundraising events amplify the extent to which those who participate in such events are "healthy" or otherwise "fit" to participate in physical activity. But this form of health is not simply a matter of being "strong," displaying good cardiovascular health, or being able to walk or run midrange (i.e., 5K) distances. Rather, it is a holistic kind of health exemplified by voluntary participation in physical activity: Men and women, often in the latest sportswear and apparel, voluntarily pay an entrance fee to run a race, putting their bodies to use to raise money. Yet this health is not restricted simply to biological health and well-being; it is an artifact of having social and economic capital. The "unhealthy" body is similarly made and remade in this intra-action between the capacity to *voluntarily* expend physical energy in a leisure pursuit and the requisite *necessity* of a villager traversing great distances carrying a 40-pound water jug to *stay alive*. In this way, unhealthiness can be thought of as more than being afflicted with a water-borne illness; it can also be thought of in nonmaterial terms such as quality of life.

We offer one last example, this time in the intra-action between the role of politics and governments and discourses of charity and philanthropy. In the 20th century, there was a large shift, predominately in the United States and the United Kingdom, in popular and political attitudes toward the free market as a result of what Hayek (2007) deemed the failure of the state. As Samantha King (2008) has detailed at length, it was largely because of the profit-driven nature of corporations and the lack of governmental funding for social welfare programs that corporate philanthropy, cause-related marketing, and social marketing began to saturate the social landscape in the first place. There are thus neoliberal undertones that currently orient the volunteer market as one that is mobilized to reach particular ends by organizations that in some instances see "Northern private interest . . . clash with local/Southern development priorities by usurping either domestic suppliers or non-profit organizations" (Levermore, 2010, p. 253, as cited in Hayhurst, 2011, p. 4). The water charity *industry* that emerges from this tension and political-economic condition is both precarious (in the material sense of it needing to perpetuate itself financially in order to maintain the ability to provide water systems) and normatively recognizable as a legitimate charitable cause (e.g., in the way that the Susan G. Komen Foundation's "Race for the Cure" has become an indelible part of North American popular culture in the last 20 years).

What we see, then, is that the water charity phenomenon is quite clearly more than a series of interactions existing separate from and prior to their coming into contact with one another. Rather, in viewing the phenomenon intra-actively, we are able to not only describe the circuit of water charity production and consumption but come to understand how it comes into being, what subject positions are made and remade through such intra-actions, and how best to move forward toward tackling the original problem at hand: that of a lack of clean water.

CODA

Water charity and the intra-actions within are an interesting phenomenon to study. If one were to look at impacts, one might find that millions of people have been provided with clean drinking water, and one will also find that many water charities have failed to deliver on their promises. However, there is much more to the phenomenon than simply understanding whether there is a return on investment of sorts. When taking this up, it becomes apparent that the privileged or Global Northern body is remade yet remains the decision-making body—as Charity: Water has said in their public service announcements seeking funding, communities in the Global North struggle with dirty water "until they get a little help." This relationship of savior and sufferer is reiterated consistently throughout representations of water charity and (re)enforces this North/South divide. Using Barad, we see causality in the relationship between the objects and the apparatus in the water charity phenomenon. While we might argue that each project perhaps should be looked at as its own phenomenon given Barad's belief that changing the object or apparatuses involved in the "experiment" would change the phenomenon, taking these "experiments" together, we can see how the individual objects in the phenomenon intra-act, causing the water charity project to either work or not work. However, these cannot be assumed prior to the intra-action whereby each object interacts as preexisting entities but rather as they emerge from the water project. Then we can see the intra-action between water systems and public health and between healthy bodies and diseased bodies.

NOTES

1. Barad (2003) defines *relata* as "would-be antecedent components of relations. According to metaphysical atomism, individual relate always preexist any relations that may hold between them" (p. 812).
2. This was the case in Malawi, where the water charity the first author worked with in London witnessed government officials unwilling to help communities that did not first help themselves; for more, see Bunds (2016).
3. Scott Harrison is the real name of the founder of Charity: Water. Neither author worked with Scott Harrison or Charity: Water, but his organization was utilized as an exemplar water charity by water charity organizers with whom we worked.
4. If, for example, the water charity suggests it will provide funds for village *x*, then it is beholden by law to utilize any money an individual provides in the name of village *x* for building a water system in village *x*. This was a mistake made by London Water Charity once. London Water Charity had leftover funds from a grant it received that enabled London Water Charity to focus on a couple particular villages. However, because the water charity has a board that measures return on investment and were beholden to the funding agency, they could not then use that money for another project that needed the money. Therefore, it is important that water charities utilize a diverse set of campaigns and money sources to provide money for water system projects on the ground and for funding the jobs that water charity employees inhabit.
5. This has also been a consistent finding in the sport for development research conducted by Darnell (2007), Hayhurst (2011), and Thorpe and Rinehart (2012), among others.

REFERENCES

Arendt, H. (1963). *On revolution.* New York, NY: Penguin.
Barad, K. (2003). Posthumanist performativity: Toward an understanding of how matter comes to matter. *Signs: Journal of Women in Culture and Society, 28*(3), 801–831.
Barad, K. (2007). *Meeting the universe halfway: Quantum physics and the entanglement of matter and meaning.* Durham, NC: Duke University Press.
Barad, K. (2012). Matter feels, converses, suffers, desires, yearns, and remembers. In R. Dolphijn & I. Van der Tuin (eds.), *New materialism: Interviews and cartographies.* Retrieved from http://quod.lib.umich.edu/o/ohp/11515701.0001.001/1:4.3/-new-materialism -interviews-cartographies?rgn=div2;view=fulltext

Boltanski, L. (1999). *Distant suffering*. Cambridge, England: Cambridge University Press.

Boström, M., & Uggla, Y. (2016). A sociology of environmental representation. *Environmental Sociology*, 2(4), 355–364.

Bruna Seu, I. (2010) "Doing denial": Audience reaction to human rights appeals. *Discourse Society*, 21(4): 438–457.

Bunds, K. S. (2016). On the messiness of activism from the inside: Global water charities, organizational ethnography, and the politics of change. *Review of Education, Pedagogy, and Cultural Studies*, 38(3), 236–259.

Bunds, K. S. (2017). *Sport, politics, and the charity industry: Running for water*. New York, NY: Routledge.

Bunds, K. S., Brandon-Lai, S., & Armstrong, C. (2016). An inductive investigation of participants' attachment to charity sports events: The case of team water charity. *European Sport Management Quarterly*, 16(3), 364–383.

Chouliaraki, L. (2010). Post-humanitarianism: Humanitarian communication beyond a politics of pity. *International Journal of Cultural Studies*, 13(2), 107–126.

Clark, N. (2011). *Inhuman nature: Sociable life on a dynamic planet*. London, England: Sage.

Darnell, S. C. (2007). Playing with race: Right to play and the production of whiteness in "development through sport." *Sport in Society: Cultures, Commerce, Media, Politics*, 10(4), 560–579.

Dolphijn, R., & Van der Tuin, I. (2012). *New materialism: Interviews & cartographies*. Ann Arbor, MI: Open Humanities Press.

Giardina, M. D. (2010). One day one goal? PUMA, corporate philanthropy and the cultural politics of "brand" Africa. *Sport in Society: Cultures, Commerce, Media, Politics*, 13(1), 130–142.

Haraway, D. (1992). The promises of monsters: A regenerative politics for inappropriate/d others. In L. Grossberg, C. Nelson, & P. Treichler (eds.), *Cultural Studies* (pp. 295–337). New York, NY: Routledge.

Hayek, F. A. (2007). *The road to serfdom: Text and documents the definitive edition*. B. Caldwell (ed.). Chicago, IL: University of Chicago Press.

Hayhurst, L. M. C. (2011). "Governing" the "girl effect" through sport, gender and development? *Postcolonial girlhoods, constellations of aid and global corporate social engagement*. Ann Arbor, MI: Proquest.

King, S. J. (2008). *Pink ribbons, inc.: Breast cancer and the politics of philanthropy*. Minneapolis: University of Minnesota Press.

Magnusson, R. (2001). *Water technology in the middle ages*. Baltimore, MD: Johns Hopkins University Press.

Pellizzoni, L. (2016). Catching up with things? Environmental sociology and the material turn in social theory. *Environmental Sociology*, 2(4), 312–321.

Rose, J. B., Epstein, P. R., Lipp, E. K., Sherman, B. H., Bernard, S. M., & Patz, J. A. (2001). Climate variability and change in the United States: Potential impacts on water-and foodborne diseases caused by microbiologic agents. *Environmental health perspectives*, 109 (Suppl 2), 211.

Sen, B., Alp, M. T., Sonmez, F., Kocer, M. A. T., & Canpolat, O. (2013). Relationship of algae to water pollution and waste water treatment. In W. Elshorbagy & R. Chowdhury (eds.), *Water treatment*, pp. 335–354. Retrieved from https://www.intechopen.com/books/water-treatment

The Water Project. (2019). Poverty in Africa begins with a lack of clean water. Retrieved from https://thewaterproject.org/infographic-poverty-in-africa

Thorpe, H., & Rinehart, R. (2013). Action sport NGOs in a neo-liberal context: The cases of Skateistan and Surf Aid International. *Journal of Sport & Social Issues*, 37(2), 115–141.

13 · FEMINIST NEW MATERIALISMS AND THE TROUBLING WATERS OF THE 2016 RIO DE JANEIRO OLYMPIC AND PARALYMPIC GAMES

MARY G. MCDONALD AND JENNIFER J. STERLING

The 2016 Olympic and Paralympic Games took place within the striking topography that encompasses Rio de Janeiro, Brazil (hereafter referred to as Rio). The city of Rio has long been influenced by the mountains, bays, coves, lagoons, and ocean that encompass it and their interaction with humans and non-humans, including urban infrastructures. In 2012, UNESCO designated portions of Rio's Caricoa landscapes as a World Heritage site. Olympic visitors to the city encountered sites of its "naturally occurring" and "man-made"[1] beauty, including Sugarloaf Mountain, the granite peak rising atop a peninsula at the entrance of Guanabara Bay; the celebrated reinforced-concrete and soapstone "Christ the Redeemer" statue that tops Corcovado Mountain over Guanabara Bay; the unique biome of Tijuca Forest, among the world's largest urban tropical rainforests; and the famed Copacabana Beach, 2.5 miles of white sandy shoreline bordering the Atlantic Ocean and home to a vibrant ecosystem and beach culture.

Several Olympic and Paralympic outdoor water sports took place in or adjacent to the Atlantic Ocean, where these iconic landscapes served as a backdrop to competitions such as the triathlon, marathon swimming, canoe sprint, rowing, kayaking, and sailing. Both the Olympic and Paralympic triathlon held their swimming events off Copacabana Beach, as did marathon swimming. Canoe sprint, rowing, and kayaking were held in Rodrigo de Freitas, an attractive lagoon in the South Zone of Rio at the base of the Corcovado Mountain. And Olympic sailing's regattas took place in Marina da Gloria in Guanabara Bay near Flamingo Park, a sprawling public recreation area.

Held in picturesque venues in a stunning city by the sea, the Olympic and Paralympic megaevents garnered worldwide attention, with an estimated 2.5 billion people following the competitions via the global media (Drehs & Lajolo, 2017). Perhaps befitting Rio's hosting of two United Nations Conferences on Development and the Environment (1992's "Earth Summit" and Rio + 20 in 2002), Rio 2016 organizers promised to provide "Green Games for a Blue Planet" in their Olympic bid documents. In line with the International Olympic Committee's (IOC's) Olympic Agenda 2020, more than half of the 25 "legacy goals" for the Rio Games focused on the environment. This theme also played out in the opening ceremonies, in which

> visuals showed the impact of rising sea levels if climate change isn't stopped. The Olympic rings were represented in a formation of trees, and every athlete was invited to plant a tree as part of the games' bid to offset its 3.6 million-ton carbon footprint. (Going for the green, 2016, para. 5)

Despite the construction of a new metro transportation line, a tramway, electric low-noise city vehicles, and a renovated port area (Boykoff & Mascarenhas, 2016), many of the proposed infrastructural and environmental legacies, including remediation of Rio's polluted waterways, never fully materialized as anticipated (Going for the green, 2016). Regardless of the promises of progressive development, the social and economic legacies of Rio 2016 proved to be limited and entrenched within the political character of the Games. As with previous Games, economic and socially disenfranchised local residents were displaced to build Olympic and Paralympic venues and housing. The residents' removal were justified under the guise that these acts of gentrification would actually provide for urban renewal and economic prosperity instead of an upward transfer of wealth (Boykoff & Mascarenhas, 2016).

Numerous media accounts pointed to a crippling recession, cuts to public spending, corruption, and social protest, as well as a scandal surrounding President Dilma Rousseff's impeachment, as limiting broader plans for the city while saddling government entities with additional debt. These shortcomings extended to unfulfilled promises to "regenerate Rio's magnificent waterways," including that of the polluted Guanabara Bay via sanitation improvements (Balch, 2016, para. 2). While the specifics of the Rio case are unique, and its failures ostensibly were due to a lack of state funds and an unrealistic timeline, similar promises of widespread prosperity and infrastructural improvements have also largely gone unfulfilled within other Olympic host cities in recent times (Boykoff & Mascarenhas, 2016).

Befitting this mediated megaevent spectacle, worldwide attention during the Games centered on both the venues' apparent beauty and the athletes competing in them. In 2016, global attention was additionally turned toward Rio's troubled waters during and in the lead-up to the Games. One such highly mediated event occurred during the Olympic Games competition, as an indoor pool venue devoted to water polo had to be drained and refilled after it turned from aqua blue to green.[2] While this case drew a great deal of coverage, most of Rio's water troubles, and the ones that we will focus on in this chapter, center on outdoor venues, particularly beaches, lagoons, and bays. Despite the aforementioned assurances of remediation prior to the Games inferring a commitment to a healthy (blue) planet, these spaces, especially waterways, continued to be discussed in relationship to their polluted status as teeming with raw sewage, discarded plastics, bacteria, and viruses. Metals and harmful chemicals—the byproducts of industrial waste—also infiltrated Rio's watercourses, including those doubling as venues during the 2016 Olympic and Paralympic Games (Barchfield, 2016).

In what follows, we investigate what Rio's waters might offer regarding the materiality and intrarelationships (Barad, 2007) between bodies and bodies of water within and beyond sport. To do so, this chapter draws on selected insights from a knowledge project infrequently engaged within sports studies scholarship—feminist (new) materialisms. Bennett (2010) offers a provisional starting point toward defining elements of this project, particularly the focus on water, arguing for the "vitality" of "thing-power" as the understanding that "action" is performed not only by humans but by nonhuman actors, including animals, objects, and ecologies that constitute complex assemblages. As Neimanis (2014) further explains, this sensibility seeks "to decenter human privilege, authority and agency, while still acknowledging the specific role and responsibilities of human bodies and calling them to account for the ways in which they affect other bodies of all kinds" (p. 14).

Feminist (new) materialisms—as articulated by Barad (2007), Alaimo (2008), Haraway (2007), Bennett (2010), and Neimanis (2014), to mention but a few—are instructive in helping scholars engage alternative ontological and epistemological approaches. These approaches, in turn, offer new ways to theorize evolving issues and ethical concerns around sports while maintaining a focus on the inequalities inherent in the binary hierarchies they trouble. Applied to the case of Rio's polluted waterways as competitive sports venues, a feminist new materialist insight demonstrates (sporting) bodies as not simply material but also ensconced within the material world. This perspective is helpful in providing an ontological reorientation encouraging critical sport scholars to bridge the nature-cultural divide—a binary critiqued by many feminists in the field but still frequently operating within scholarship about sport.

These insights—as well as similar ideas found within Indigenous thought, feminist ecocriticism, feminist posthumanism, and animal studies, in addition to science and technology studies scholarship—demand a reworking of notions of "agency" beyond dominant understandings of rational human exceptionalism and intentionality. Working instead toward complex understandings of the interactivity of the "environment," humans, and "more-than-humans" (Alaimo, 2010) allows us to shift our analytic gaze to take Rio's waters seriously, opening "the possibility of a politics and poetics of nature through a broader perception of reality that includes ethics, genetics, chemistry, politics, and biotechnological advances" (Oppermann, 2013, p. 81).

Organizationally, the remainder of the chapter consists of two principal and contrasting parts. First, we discuss dominant discourses of Rio's Olympics and water venues. We use this brief narrative analysis heuristically in order to contrast this example with that of a different, ontologically orientated sensibility—specifically, the first section documents the narratives that surround the Games as emblematic of human exceptionalism via a recounting of human achievement and progress. Situated within these sensibilities, dominant understandings of Rio's troubled waters are also worrisome, as too frequently users conceive of water as an object: a fruitful resource that serves humanity, or endangers it, despite of or due to "man's" often-dubious stewardship. In sum, this first section acknowledges the ways in which technological solutions that promise remediation of Rio's water dilemmas, while never fully realized, still problematically appeal to human progress and reason. These techno- and anthropocentric notions are also hallmarks of modern sport more generally and the Olympic movement in particular.

The next section offers an alternative approach that draws from feminist posthumanism and feminist new materialism writings about water: writings that disrupt and displace human actors and actions away from the center of their analysis. This later discussion notes how a focus on water itself can queer dominant Western binary understandings, including active-passive, material-discursive, masculine-feminine, nature-culture, subject-object, and human-nonhuman. Both feminist posthumanism's and new materialism's conceptual horizons rest "on the assumption that the gendered dichotomies hidden in the phallogocentric Western thought can be surmounted by adopting a relational ontology" (Oppermann, 2013, p. 75). Stated differently and more succinctly, this latter section reorients our conceptual focus on the troubling waters of the 2016 Rio Olympic and

Paralympic Games to illuminate important intrarelationships that feminist new materialists continue to direct attention toward (Barad, 2007).

In this way, our argument is strategic, as we center Rio's water in relationship to the 2016 Games and then engage sensibilities from feminist new materialisms to make visible—and then to trouble—standard ontological and epistemological assumptions that ground much sports studies scholarship. Water provides a material flow that moves within and through both human and nonhuman bodies, providing a compelling posthumanist analytic vantage point. Against the commonsense narratives of human exceptionalism generated through Olympic competitions and conservation management, the concluding portions of this chapter emphasize the interconnected relationships between human and more-than-human bodies and actions as starting points to think differently and ethically about water. As such, the contents of this chapter are offered in the hopes of nurturing alternative "imaginings of water" as agentic and productive of life (Neimanis, 2014, p. 12).

HUMAN EXCEPTIONALISM, THE OLYMPICS, AND TROUBLED WATERS

Organizers of the 2016 Olympics in Rio adopted an official slogan for the mega-event: "A New World." Their creation of a slogan is in line with previous Olympics, which have crafted their own mottos. For example, London organizers in 2012 coined the phrase "Inspire a Generation" (Belen, 2016). The International Olympic Committee (IOC) and the International Paralympic Committee (IPC) themselves have also long embraced mottos that have reflected their humanist ideals: the IPC through its official logo and motto embodying "Spirit in Motion" and the IOC in the form of Olympism, partially encapsulated by the motto "Citius, Altius, Fortius," which is Latin for "Faster, Higher, Stronger." Pierre de Coubertin, who helped initiate the modern Olympics in 1896, borrowed the hendiatris from his friend Henri Didon, a Dominican priest (Schantz, 2015).

The motto "Faster, Higher, Stronger" encapsulates a quest for "the idea of human perfection and progress, originally one of the core concepts of Coubertin's humanism" (Schantz, 2015, p. 841). As feminist sport scholars have long argued, much as with humanism itself, modern sport is similarly saddled with early preoccupations about the alleged exceptionalism of the white, male, middle-classed body. While "Coubertin's Olympism has been modified, reinterpreted or simply ignored by Olympic stakeholders over the years," tenets remain (Schantz, 2015, p. 841). Technological enhancements such as improved equipment and facilities, as well as the rise of "the so-called NBIC technologies and sciences (nanotechnology, biotechnology, information technology and cognitive science)," have merged with a continuous quest for the pursuit of human development and athletic mastery within the Olympic and Paralympic Games, seemingly for the good of global societies (Schantz, 2015, p. 841).

In Rio, the pursuit of athletic mastery was differently on display within different sporting disciplines—where athletes competed variously against themselves, opponents from other nations, and at times, with assistive technologies and against the "natural elements." For example, the sailing competitions provided unique challenges, as conditions within Guanabara Bay included "relatively flat water, with a lot of currents and shifting winds," while other sailing-class courses provided "for the surf and swell of the open ocean" (Clarke, 2015, para. 11). Prior to the start of the sailing events, Josh Adams, the managing director of U.S. Olympic Sailing observed that "ultimately, what these Olympic Games will test is who is the most complete sailor" (Clarey, 2016, para. 10).

This hegemonic construction of the Olympics is not simply about individualistic masculine mastery and human development. According to former IOC president Jacque Rogge, the enactment of official Olympic legacy programs is similarly designed to acknowledge that "every city that hosts the Olympic Games becomes a temporary steward of the Olympic Movement" (cited in International Olympic Committee, 2013, p. 3). According to Rogge, this means that each city

> has a once-in-a-lifetime chance to showcase the celebration of the human spirit. And each creates a unique set of environmental, social and economic legacies that can change a community, a region, and a nation forever. (cited in International Olympic Committee, 2013, p. 3)

The Rio 2016 slogan "A New World" evokes a related vision of hope and (human) progress. In announcing the 2016 motto, IOC chair Thomas Bach noted that the phrase "is what the Olympic movement is all about . . . changing the world for the better" (Belen, 2016, para. 4). Reflecting the increased commodification of the Games over the past half-century, 2016 Olympic brand director Beth Lula additionally observed, "The slogan is the brand's promise to society" (Belen, 2016, para. 4). The motto's multiple meanings perhaps first harken to the alleged 16th-century "discovery" of the Americas (including portions of the host continent of "South America"), named by Europeans as the "New World." The phrase also evokes a series of technological transformations, including new modes of human interactivity via social media as well as cross-cultural exchanges that have seemingly helped connect humanity in a "borderless" world. Appeals to this "New World" apparently also encompass the Olympic

Agenda 2020's stated commitment to sustainability and broader missions to find necessary solutions to combat ominous problems such as the multiple threats to people presented by global climate change (Keith, 2016).

Most of the pre-event media attention focused not on broader issues of climate change but rather on the failure of Rio's government officials to live up to the promises made during Rio's Olympic bid in 2009. Particularly criticized were the shortcomings of Rio's commitment to regenerate its waters, with over $1 billion (USD) promised in infrastructural improvement and programs to cleanse the polluted bays and beaches, thus "setting a new standard of water quality preservation for the next generation" (cited in Khazan, 2016, para. 5). This commitment to providing the "man-made" technology and science necessary to remediate these water problems ahead of the Games was never completely fulfilled—nor could it likely have been. Prior to the Olympic and Paralympic Games, Rio governor Luiz Fernando Pezão suggested Guanabara Bay's promised 2016 cleanup date would most likely be accomplished in 2035 (Boykoff & Mascarenhas, 2016). Such broken cleanup promises reinforce an enduring trajectory in Rio, where "water pollution has long plagued Brazil's urban areas, . . . most sewage isn't collected, let alone treated. . . . [and] much of the waste runs through open-air ditches to fetid streams and rivers that feed the Olympic water sites" (Meech, 2015, para. 8).

While often acknowledging water's vulnerability (Neimanis, 2014), dominant media narratives surrounding the Rio Games are important to further analyze the assumptions they reveal. Too frequently articulating water as a controllable resource necessary for human consumption and economic viability, these dominant framings obscure other ways of thinking through water. In particular, dominant human exceptionalist narratives too frequently conceal new materialist understandings of water as generative of "life itself" (Chandler & Neimanis, 2013, p. 61). In the following sections, we highlight two of these problematic and intertwined constructions: concern for athlete and tourist health and the limits of technological solutions.

Concern for Athlete and Tourist Health

Prior to the megaevent, cautionary tales were directed toward Olympic-inspired tourists who descended on the picturesque beaches of Rio, such as Copacabana. Concern was especially expressed for the nearly 1,400 athletes competing in outdoor water events such as the triathlon, canoe sprint, marathon swimming, kayaking, sailing, and rowing. Dr. Daniel Becker, a local pediatrician who treats Rio's poor, summarized the situation succinctly (if not crassly): "Foreign athletes will literally be swimming in human crap, and they risk getting sick from all those microorganisms. . . . It's sad but also worrisome" (cited in Jacobs, 2016, para. 6).

In a similar vein, the *Atlantic* noted, "Athletes might be sailing and swimming through raw sewage, ruining matches and landing in hospitals as a result" (Khazan, 2016, para. 4).

Prior to the start of the Olympics in July 2016, the World Health Organization (WHO, 2016) offered the following advice in regards to Rio's polluted waterways:

> The degree of exposure experienced by athletes (splashing, water ingestion, full body contact) will depend upon the sport, and the options available for minimizing the risks will depend on the degree of inherent exposure. It is suggested that all athletes should cover cuts and grazes with waterproof plasters prior to exposure, try to avoid swallowing the water, wash/shower as soon as possible after exposure and, as far as possible, minimize their time in the water and avoid going in the water after heavy rainfall if possible. Sports where there is little contact with the water will be taking place on sites not recommended for full body contact with the water, so it is recommended that people don't jump in the water or throw people in. (p. 3)

The report additionally posited that given their emergence in water, swimmers in outdoor venues would be at the highest risk for illnesses such as respiratory tract infections and mild gastroenteritis. While the WHO discussed the latter as "usually short-lived and self-limiting in healthy people" (WHO, 2016, p. 2), some media reports noted that this illness could cause diarrhea—and less commonly, inflammation of the brain or heart (Barchfield, 2016). Still other coverage argued that both tourists and athletes would be at risk for contracting hepatitis A through water exposure (Dockray, 2016).

Much was made of Belgian sailor Evi Van Acker's severe intestinal infection that she allegedly acquired while training in Guanabara Bay one month prior to the Olympics, an infection that continued to "sap" her energy weeks later (Levy, 2016). Several sailing and outdoor water sport national teams arrived in Rio early to train, partially to acquire immunity resistance that many, although not all, Rio residents have seemingly acquired against some water-borne infections. However, this quality does not extend to all Rio residents equally: Those with weakened immune systems, babies, and older adults fall outside the parameters of WHO's aforementioned "healthy people" and are the most vulnerable.

This state of affairs clearly points toward additional political observations that can be made about Rio's water troubles. Dominant English-language media narratives frequently centered on questions about the cleanliness levels of Rio's waterways and their subsequent impact on the health and safety of those interacting with them, playing into lingering colonial constructions of Brazil as a "less developed," even "uncivilized," culture. These narratives are perhaps best captured by a story in the *Atlantic,* which proclaimed in its byline, "Raw sewage flows into many of Rio's Olympic venues every day. As the prospect of a full cleanup before the Games dims, the world is left wondering, who will get sick, and how?" (Khazan, 2016, para. 1). Initially prioritizing concerns about the health of athletes and tourists, the article goes on to confirm the idea of the city as contagion by outlining the enormity of Rio's water troubles and acknowledging the impact on the health and vitality of local residents.

Issues of inequality within water troubles in Rio are further evident, albeit less reported. While even wealthy residents do not have adequate sanitation, Rio's poor are disproportionally affected by water pollution, poor sanitation, and water-borne illnesses (Konkel, 2016). Those who practice subsistence fishing in Guanabara Bay are exposed to additional health issues as they capture and eat "seafood contaminated with heavy metals [and] industrial chemicals including PCBs and hydrocarbons from petroleum products" (Konkel, 2016, para. 15). And there are additional issues of power and economic control at play in the distribution of these concerns. As reported in the *Guardian* in its discussion of Rio 2016, "the worst affected areas are further to the north of the city, where low-income favela communities are concentrated and government investment in water systems and sewage treatment is inadequate" (Watts & Vidal, 2016, para. 18). Racial segregation patterns reveal that the premier property in the desirable South Beach Zone is predominately inhabited by international visitors and Brazilians with privilege. In this way, access to clean water and sanitation is an unequally distributed commodity with important and vastly different use and exchange values within Rio (Ioris, 2015).

The absence of these critical perspectives that recognize the politics of water and space reflect the constitution of what Linton (2014) has termed "modern water." Modern water is meant to signify a normative view of water as a simple, natural substance presumably existing outside the influence of place, culture, politics, and history. Beginning in the 17th century, scientists helped usher in this new paradigm. Where once local people thought of water "with the particularities of culture and place," increasingly water gained the status of a "common chemical compound" (Linton, 2014, p. 113).

One consequence of this shift has been to "render water ahistorical, to reduce it to an abstract quality and make it amenable to the application of instrumental reason" (Linton, 2014, p. 113). The 20th century saw the emergence of a new relationship between water and the state, including a "view of water as a 'resource' to be 'developed' and 'managed'" (p. 113). This process helps constitute a dominant view that separates "water from ecology and from human society," which in turn fuels the fantasy that water can be manipulated "without profound social consequences" (p. 113). This abstracted conception of water has further enabled water to serve as a repository of human waste, chemicals, and various discarded remnants of consumer capitalism (Bennett, 2010). While these abstracted notions have been challenged and resisted, underlying assumptions of the Games as a global spectacle continue to celebrate similar sensibilities of corporeal mastery and influence, which in turn appeal to human exceptionalism and Western liberal individualism. Such constructions hail anthropocentric sensibilities that promote binary thinking, falsely separating "humans" as apart from—and in control of—"nature" (and in the particular case of Rio, as apart from—and in control of—water).

To be clear, the concern for socially just human welfare and well-being for all people regardless of status is both necessary and important (Neimanis, 2014). However, as the preceding discussion about the constitution of modern water makes clear, there is much more at play here that requires deeper analysis. While gender, class, and racial relations are often obscured in debates about the impact of Rio's water, very little attention is devoted to the potentially harmful effects of pollution on the more-than-human life and complex ecosystems sustained through Rio's waterways. These absences reify hierarchies of inequality while positioning science and technology as the best hope to address "environmental" problems.

The Limits of Technological Solutions

The previous section noted that dominant narratives of Rio's troubled waters do very little to disrupt human-centered ways of knowing and being. As pointed out within feminist new materialist accounts, similarly centered analyses, hopeful solutions, and movements toward "sustainability" too frequently ignore or minimize nonhuman ecological impacts—that is, plans to create sustainable development frequently are "articulated too firmly to a technocratic, anthropocentric perspective" (Alaimo, 2012, p. 563). In contrast, Alaimo (2012) advises that the material self cannot be abstracted from simultaneous economic, material, political, and scientific land and waterscapes and thus advocates for the necessity of recognizing human and more-than-human entanglements as "trans-corporeal epistemologies" that "are uncertain, experimental, amateurish, contingent and engaged" (p. 561).

Alaimo offers a considerably different critical path forward than that offered through the narrowly defined preoccupation with human health and well-being described above. This is also a different approach than the interventions once proposed to remediate Rio's polluted waters within Rio's Olympic bid document and the 2013 "Sustainability Management Plan: Rio 2016." Rather, read from Alaimo's (2012) perspective, Rio's plans were doomed to fail from the start, as they largely reproduced the problem of technocratic notions of sustainability—in other words, too frequently calls for environmental redress were "linked to systems management and technological fixes," which falsely "[present] a rather comforting, conventional sense that the problem is out there, distinct from one's self" (p. 561). Given its abstract human-centeredness, Rio's "Green Games for a Blue Planet" plan does not provide an explanation of the social, economic, environmental, and technological politics and histories of how waterscapes such as Guanabara Bay came to be polluted in the first place. This process of downplaying important contexts is not simply limited to accounts surrounding the Rio Olympic Games but is again representative of the larger onto-epistemological project that constitutes modern water.

Such a construction further avoids a deeper engagement with what potential impacts of this state of affairs represent for diverse groups of residents—human and nonhumans, including fish, wildlife, microbes and the climate, to mention a few. Instead, Rio's plan offers several technological-driven promises, including

> expansion of secondary treatment (equipment purchase and installation) at the Alegria Sewage Treatment Plant. . . . Building a sewer collector trunk. . . . [and] Replacement of sewer networks. . . . Erecting and reforming eco-barriers (physical barriers to contain garbage floating in bodies of water) and eco-points (points for receiving recyclable materials removed from the eco-barriers and/or donated by local residents) in rivers that flow into Guanabara Bay. (Rio 2016 Organizing Committee, 2013, p. 41)

Even with full economic support, this appeal to the eminence of water management and technological solutions still "obscures power differentials, political differences, cultural values, and everyday human practices" (Alaimo, 2016, p. 172). While technological perspectives are helpful in many cases, an uncritical embrace such as that offered via discourses and practices of sustainable development actually "proceed[s] with the presumption that human agency, technology and master plans will get things under control" (p. 173). This viewpoint ignores questions of scale as well as the often "unintended effects of human activity, and its interactions with other forces" (p. 173).

The proposed technocratic solutions offered by planners and organizers in Rio and beyond are ontologically and epistemologically aligned with Enlightenment fantasies that privilege a particular form of human need that creates deleterious effects and consequences. Instead, again citing Alaimo (2008), a more useful way forward is to think through

> trans-corporeality—the time-space where human corporeality, in all its material fleshiness, is inseparable from "nature" and "environments." Trans-corporeality as

a theoretical site, is a place where corporeal theories and environmental theories meet and mingle in productive ways. (p. 238)

As these insights demonstrate, there is another starting point for understanding Rio 2016, and that is to eschew modernist conceptualizations that overwhelmingly center on humans' experiences with water and waste, as well as on remediation—however technologically sophisticated and scientifically informed. Instead, it is important to look below the surface of the trash- and toxin-filled Rio waterways to search for other possibilities and ways of thinking and knowing that recognize complex human and more-than-human entanglements.

ONTOLOGICAL SHIFTS AND TROUBLING WATERS

Scholars and activists connected to Indigenous and feminist traditions have argued persuasively about the promises of beginning with water as a philosophical starting point (Neimanis, 2012; Chandler & Neimanis, 2013). After all, the very prospect of life is water-contingent. The human body is between 60% and 90% water, and the ocean is a "main life support system controlling the planet's temperature, climate, and key chemical cycles" not just for humans but for all forms of life (Mitchell, 2009, p. 22).

This perspective is aligned with transcorporeal sensibilities in which human bodies do not stand apart from, and thus are not in control of, the "natural" world—a central myth of humanism and human exceptionalism. Rather, transcorporeality embraces the co-constitution of nature and culture, or what Haraway (2007) refers to as "naturecultures." Transcorporeal sensibilities posit that bodies of water and an array of other more-than-human and human bodies are intimately entangled in ways that matter. This conceptualization offers a relational ontology suggesting that different bodies are not separate but instead "intra-act" in dynamic articulations of becoming (Barad, 2007).

Read from this perspective, water (intra)acts "as a gestational milieu" with generative possibilities (Chandler & Neimanis, 2013, p. 65). A relational ontology of watery bodies, then, highlights their gestational possibilities: "Mammal, reptile, or fish; sapling or seed, river delta or backyard pond—all of these bodies are necessarily brought into being by other bodies of water" (Neimanis, 2012, p. 98). Chandler and Neimanis (2013) further propose that engaging with this gestational milieu offers an ethical way of being, as "water calls upon us, as human sovereign subjects, to amplify our own facilitative capacity as watery milieu in our interactions with all bodies of water" (p. 65). This articulation reflects how Barad's (2007) intra-actions and Chandler and Neimanis's (2013) gestational milieu meet in Haraway's (2007) naturecultures, providing a useful approach for unpacking—and also troubling—dominant understandings of Rio's troubled waters.

The multiplicity of intra-actions among different types of watery bodies and bodies of water are also evident when we consider the fragility of the gestational milieu, for a variety of lives are threatened by contaminated water, including those both within and beyond Rio's polluted waterways and ocean surf. As Neimanis (2012) makes clear, worldwide aquatic species face a greater risk of extinction than the risks that now confront mammals and birds. Many of the threats to waters are due to human-produced and human-consumed hazards, such as "automotive fluids, household solvents, pesticides, mercury and other toxins that make their way from human homes to culver to sea" (Neimanis, 2012, p. 106). Along their routes and throughout their circuitous destinations, among those profoundly affected are "those animal bodies that dwell at or near the bottom of an aquatic habitat—such as fish eggs and filter feeders—where pollutants tend to settle" (p. 106), disrupting the gestational possibilities of species both directly and indirectly.

The multiple lives affected by the pollutants and waste that line Rio's waterways include animal, and particularly marine, life that inhabit these spaces—not just Olympic athletes, tourists, and local residents. Not surprisingly, the large population and variation of fish that both constitute and dwell in Rio's hydrography are illustrative of these impacts. This is seen most clearly in Guanabara Bay, Brazil's second largest costal bay and the site of Olympic sailing events. This body of water receives both the "fresh" water of more than 50 tributaries as well as raw sewage, hospital waste, and debris collected from nearby neighborhoods (Fistarol et al., 2015). In particular, excessive nutrients from sewage runoffs often result in "excessive plant and algae growth in parts of the bay," which utilizes excessive oxygen and restrict its availability to native aquatic species such as fish (Konkel, 2016, para. 19). While Fistarol and colleagues (2015) explain that the eutrophication, or nutrient enrichment, of coastal waters is a global problem, the intensity of Guanabara Bay's pollution—its toxic algal blooms and several oil spills—has resulted in the destruction of aquatic habitats and a number of "fish kills," or large-scale localized deaths, in recent years (Fistarol et al., 2015; Konkel, 2016).

These deadly effects of Rio's troubled waters were also apparent during the lead-up to the Summer Games in January 2016, when "thousands of dead fish washed up on the shores of Rio's Guanabara Bay" close to Olympic and Paralympic venue sites (Dead Fish, 2016, para. 1). Furthermore, prior to the Games, scientists traced the negative impact of pollutants on species throughout the trophic, or food-chain, levels of the Guanabara Bay ecosystem, from the presence of antibiotic resistance genes within planktonic microorganisms (Fistarol et al.,

2015) to disappearing tucuxi, or freshwater dolphins. According to researchers at Rio de Janeiro State University, the presence of the tucuxi has been greatly reduced from around 800 in the bay in 1995 to about 35 in 2016 (Young, 2016). In sum, rather than impacting aquatic species individually, pollutants (intra)act across gestational milieu, including humans, highlighting watery fragilities.

Yet much as with the humans and nonhumans residing near or in the sea, water's relationship to toxins is far from simple. For instance, water moving from the Atlantic Ocean and back out of Guanabara Bay also demonstrates that the impact of pollutants varies by force of currents, location in the bay, and tidal stages. The strongest effect, as with other coastal bays, is near the bay's mouth, with contaminants settling closer to the inner body. High tide offers a diluted effect, which enhances the water quality (Fistarol et al., 2015). This attention to divergent flows, circulation of both life-forces and contaminants, and disparate effects further reveals ways in which water matters (Chen, MacLeod, & Neimanis, 2013).

As these examples illustrate, centering Rio's troubled waters helps blur and trouble binaries—binaries that feminists and other critics argue constitute hierarchical power relations—to instead acknowledge the multiple ways in which humans and more than humans are intertwined and affected. In doing so, this lens "draws human attention sideways, away from the ontologically ranked Great Chain of Being and toward a greater appreciation of the complex entanglements of humans and non humans" (Bennett, 2010, p. 112).

The recognition of a diversely embodied and inhabited watery milieu offers a much different sensibility than understandings of water as a controllable resource. This is not to imply that "human and non-human natures are the same or undifferentiated" (Neimanis, Asberg, & Hayes, 2015, p. 486). Rather, it is to highlight that a watery milieu offers an "index of belonging rather than a reduction to sameness" (Neimanis et al., 2015, p. 486). Furthermore, Neimanis and colleagues (2015) point out that "human vulnerabilities to and responsibilities for" water troubles in Rio and beyond

> are not equally distributed across human populations. . . . A history of colonization, global capitalism, and regional geophysical determinants direct flows of power and maintain important differences between communities while differences of age, gender, race, and ability/mobility are similarly salient. (p. 486)

Centering water does not mean seeking mastery of water, though human mastery is often the dominant masculine narrative of sports held in aquatic venues such as those offered in Rio. More broadly, the same will to human mastery also underlies conceptualizations of (modern) water as apart from other lives, as a utilitarian commodity and resource. Instead, centering water involves embracing an ongoing, interconnected, ethical, and onto-epistemological process suggesting that human bodies are entangled with multiple non- and more-than-human bodies. Starting with water requires coming to "know *with* [emphasis added] water, seeds, fish, pollution, sediment, and politics" (Neimanis et al., 2015, p. 488).

This moves us toward what Barad (2007) calls an "ethics of mattering." Such an ethics "is not about right or correct responsibility to a radically exteriorized other, but about responsibility and accountability for the lively relationalities of becoming of which we are a part" (p. 393). Haraway (2016) contends that humans must embrace a similar ethical position—that is, to become "response-able" in new ways, thus helping shape "conditions for multispecies flourishing" (p. 29). Again, this does not mean that every body, human or more than human, is "'response-able' in the same ways. The differences matter—in ecologies, economies, species, lives" (p. 29).

Such a feminist new materialist perspective compels us to embrace multiple and divergent ways of being and knowing. However, it is important to acknowledge that interconnected relationships and the processes of becoming are uncertain and without guarantees as to their effects. As Haraway (2007) also reminds us that there is no assured outcome in reversing the status of important entanglements, including those of Rio's polluted waterways: There is "no assured happy or unhappy ending, sociologically, ecologically, or scientifically. There is only the chance of getting on together with some grace" (p. 15)

CONCLUDING THOUGHTS

This analysis has offered diverse ways to think about material relationships and entanglements thanks to the vantage points offered via the troubling waters of Rio. Among inequalities, corruption scandals, and failed promises, dominant narratives promoted through the Olympic movement articulate particular notions of humanism and human exceptionalism—despite evidence of human-fueled "environmental" devastation as well as the important human and more-than-human connections exemplified by Rio's watery milieus.

That humanist narratives would find a home within the contemporary Olympics is not surprising given that the Games—and indeed modern sport itself—emerged from 19th-century Enlightenment ideals as anthropocentric, "institutionalized, standardized, employing systematic training and preparation in search of records and aiming at the betterment of the human being and of humanity" (Schantz, 2015, p. 841). Olympic legacy projects articulate similar ideologies as opportunities to assist with host cities' development. Clearly the centrality of the human and, by extension, commitments to human reason and human progress remain commonsense norms within the Olympic movement.

In contrast, as we have argued, troubling Rio's waters offers a productive site to join those feminists seeking to queer powerful binary assumptions that ground normative ontologies, thoughts, and actions, including binaries of active-passive, material-discursive, masculine-feminine, nature-culture, subject-object and human-nonhuman. This chapter is not meant to serve as a comprehensive account of how to think with water but rather to point toward the necessity of emphasizing processes of doing and becoming (Barad, 2007; Butler, 1990). In that spirit, it offers an alternative way to conceptualize the material world. Interrogating a complex set of intrarelationships instead moves beyond normative understandings of water—understandings locked within commonsense appeals to the substance as a resource for athletes, tourists, and city residents. As such, this chapter both suggests and points toward alternative accounts that reaffirm the materiality and vitality of water and life itself.

NOTES

1. We consciously use "man-made" to signify a longer history in which engineering and building were over determined with a presence of men and masculine norms. Consistent with postcolonial observations, we also readily acknowledge that notions such as "man" and "human" are embedded in hierarchical relations structured by gender, race, class, and nation. We also place such phrases as "man made," "naturally occurring," "nature," and "environment" in quotation marks in several places throughout the chapter to acknowledge that these terms exist within commonsense phrasing. We also do so to acknowledge that these constructs have histories—which frequently serve to bifurcate naturecultures.

2. Reports indicate that pool managers placed too much hydrogen peroxide in the water, causing a chemical interaction with the existing chlorine to create the green hue. The pool was subsequently drained and refilled in order to provide the more familiar aqua-blue hue for synchronized swimming.

REFERENCES

Alaimo, S. (2008). Trans-corporeal feminisms and the ethical space of nature. In S. Alaimo & S. Hekman (eds.), *Material feminisms* (pp. 237–264). Bloomington: Indiana University Press.

Alaimo, S. (2010). *Bodily natures: Science, environment, and the material self.* Bloomington: Indiana University Press.

Alaimo, S. (2012). Sustainable this, sustainable that: New materialisms, posthumanism, and unknown futures. *PMLA, 121*(3), 558–564.

Alaimo, S. (2016). *Exposed: Environmental politics and pleasures in posthuman times.* Minneapolis: University of Minnesota Press.

Balch, O. (2016, February 1). Funding problems hit plan to clean Rio's polluted waterways ahead of Olympics. *Guardian.* Retrieved August 25, 2016, from https://www.theguardian.com/sustainable-business/2016/feb/01/funding-problems-hit-plan-clean-rios-polluted-waterways-olympics

Barad, K. (2007). *Meeting the universe halfway: Quantum physics and the entanglement of matter and meaning*. Durham, NC: Duke University Press.

Barchfield, J. (2016, August 1). Rio 2016: Swimmers need to ingest only three teaspoons of water to be almost certain of contacting a virus. *Independent*. Retrieved August 25, 2016, from http://www.independent.co.uk/sport/olympics/rio-2016-water-pollution-virus-risk-danger-swimming-sailing-rowing-chance-of-infection-almost-a7165866.html

Belen, N. (2016, June 16). Rio 2016 Games unveils Olympic slogan: A new world. *Rio Times*. Retrieved August 26, 2018, from http://riotimesonline.com/brazil-news/rio-sports/rio-2016-games-unveils-olympic-slogan-a-new-world/

Bennett, J. (2010). *Vibrant matter: A political ecology of things*. Durham, NC: Duke University Press.

Boykoff, J., & Mascarenhas, G. (2016). Rio 2016: Urban politics and environmental impacts. *IdeAs*, 7, 1–5. doi:10.4000/ideas.1398.

Butler, J. (1990). *Gender trouble*. London, England: Routledge.

Chandler, M., & Neimanis, A. (2013). Water and gestationality: What flows beneath ethics. In C. Chen, J. MacLeod, & A. Neimanis (eds.), *Thinking with water* (pp. 61–83). Toronto, Canada: McGill-Queens University Press.

Chen, C., MacLeod, J., & Neimanis, A. (2013). Introduction: Toward a hydrological turn. In C. Chen, J. MacLeod, & A. Neimanis (eds.), *Thinking with water* (pp. 3–22). Toronto, Canada: McGill-Queens University Press.

Clarey, C. (2016, August 10). Sailing in Rio? Compared with nature, the pollution is a breeze. *New York Times*. Retrieved September 1, 2017, from https://www.nytimes.com/2016/08/11/sports/olympics/sailing-rio-games.html?_r=0

Clarke, J. (2015, August 23). Sailing through the trash and sewage of Guanabara Bay. *New Yorker*. Retrieved August 25, 2016, from https://www.newyorker.com/sports/sporting-scene/sailing-through-the-trash-and-sewage-of-guanabara-bay

Dead fish wash up on the shores of Rio bay near Olympic venue. (2016, January 16). *Reuters*. Retrieved August 25, 2016, from http://www.reuters.com/article/us-olympics-sailing-pollution/dead-fish-wash-up-on-shores-of-rio-bay-near-olympic-venue-idUSKCN0UR2H720160113

Dockray, H. (2016, August 5). 11 absolutely terrifying illnesses you can get at the Rio Olympics. *Mashable*. Retrieved August 25, 2016, from https://mashable.com/2016/08/05/13-diseases-get-rio-olympics/

Drehs, W., & Lajolo, M. (2017, August 10). After the flame. *ESPN*. Retrieved August 19, 2017, from http://www.espn.com/espn/feature/story/_/id/20292414/the-reality-post-olympic-rio

Fistarol, G. O., Coutinho, F. H., Moreira, A. P., Venas, T., Cánovas, A., de Paula Jr., S. E. M. . . . Thompson, F. L. (2015). Environmental and sanitary conditions of Guanabara Bay, Rio de Janeiro. *Frontiers in Microbiology*. doi:10.3389/fmicb.2015.01232.

Gaard, G. (2010). New directions for ecofeminism: Towards a more feminist ecocriticism. *ISLE*, 17(4), 643–665.

Going for the green: No medals for the Rio Olympics' environmental legacy. (2016, August 23). *DW*. Retrieved August 27, 2016, from http://www.dw.com/en/going-for-green-no-medals-for-the-rio-olympics-environmental-legacy/a-19495318

Haraway, D. (2007). *When species meet*. Minneapolis: University of Minnesota Press.

Haraway, D. (2016). *Staying with the trouble: Making kin in the Chthulucene*. Durham, NC: Duke University Press.

International Olympic Committee. (2013). *Olympic legacy*. Lausanne, Switzerland: Didwedo.

Ioris, A. A. R. (2015). *Water, state and the city*. New York, NY: Palgrave Macmillan.

Jacobs, A. (2016, July 26). Keep your mouth closed: Aquatic Olympians face a toxic stew in Rio. *New York Times*. Retrieved August 27, 2016, from http://www.nytimes.com/2016/07/27/world/americas/brazil-rio-water-olympics.html

Khazan, O. (2016, March 31). What happens when there is sewage in the water? *Atlantic*. Retrieved August 25, 2016, from https://www.theatlantic.com/health/archive/2016/03/what-happens-when-theres-sewage-in-the-water/476013/

Keith, B. (2016, June 15). Rio 2016 reveals new slogan—"a new world." *SwimSwan*. Retrieved August 25, 2016, from https://swimswam.com/rio-2016-reveals-new-slogan-new-world/

Konkel, L. (2016, August 15). What's in Rio's bay and beaches? *National Geographic*. Retrieved August 30, 2016, from http://news.nationalgeographic.com/2016/08/what-s-in-rio-s-bay-/

Levy, M. (2016, August 12). Sailor Evi van Acker sick after racing on polluted Guanabara Bay at Rio Olympics. *Sydney Morning Herald*. Retrieved August 26, 2016, from http://www.smh.com.au/sport/olympics/rio-2016/sailor-evi-van-acker-sick-after-racing-on-polluted-guanabara-bay-at-rio-olympics-20160811-gqqpi7.html

Linton, J. (2014). Modern water and its discontents: A history of hydrosocial renewal. *WIREs Water*, 1, 111–120. doi:10.1002/wat2.1009.

Meech, G. (2015, July 30). AP investigation: Olympic water in Rio "toxic" levels. *KOAA5*. Retrieved September 4, 2016, from http://www.koaa.com/story/29673536/ap-investigation-olympic-water-in-rio-toxic-levels

Mitchell, A. (2009). *Sea sick: The global ocean in crisis*. Sydney, Australia: Emblem.

Neimanis, A. (2012). Hydrofeminisms: Or, on becoming a body of water. In H. Gunkel, C. Nigianni, & F. Soderback (eds.), *Undutiful daughters: New directions in feminist thought and practices* (pp. 85–99). New York, NY: Palgrave Macmillan.

Neimanis, A. (2014). Alongside the right to water, a posthumanist feminist imaginary. *Journal of Human Rights and the Environment*, 5(1), 5–24.

Neimanis, A., Asberg, C., & Hayes, S. (2015). Posthumanist imaginaries. In K. Backstrom & E. Lovbrand (eds.), *Research handbook on climate governance* (pp. 480–490). Northhampton, England: Edward Elgar.

Oppermann, S. (2013). Feminist ecocriticism: The new ecofeminist settlement. *Feminismo/s*, 22, 65–88.

Schantz, O. (2015). Coubertin's humanism facing post-humanism—the future of the Olympic Games. *Sport in Society*, 19(6), 840–856.

Rio 2016 Organizing Committee for the Olympic and Paralympic Games. (2013, March). *Sustainability management plan: Rio 2016 Olympic and Paralympic Games*. Retrieved August 25, 2016, from http://rio2016.com/sites/default/files/Plano_Gestao_Sustentabilidade_EN.pdf

Watts, J., & Vidal, J. (2016, August 3). What's in the water? Polluted fears taint Rio's picturesque bay ahead of Olympics. *Guardian*. Retrieved August 25, 2016, from https://www.theguardian.com/sport/2016/aug/03/pollution-fears-taint-rio-bay-olympic-games

World Health Organization. (2016, June). Q&A on recreational water quality in Rio de Janeiro.

Young, J. (2016, March 2). Rio's "bay of all delights": The polluted waters of the 2016 Olympics. *Vice Sports*. Retrieved August 25, 2016, from https://sports.vice.com/en_us/article/vvwmy3/rios-bay-of-all-delights-the-polluted-waters-of-the-2016-olympics

14 · FLATTENING THE CITY

Assemblage Urbanism and the Moving Body

OLIVER J. C. RICK AND JACOB J. BUSTAD

The experience of living, working, playing, and being in cities is arguably the definitive state of the contemporary human condition, as more than half of the world's people now live in urban areas (54% in 2014 according to the United Nations). Yet the importance of focusing on the urban is not only a function of a global concentration of populations. Following Brenner, Madden, and Wachsmuth (2011), we would also suggest that this "incipient process of 'complete' or 'planetary' urbanization" has particular impacts for understanding not only the shifting experience of contemporary urban contexts but also the rapidly urbanizing dimensions of all global life (p. 226). By this we mean not that the world's populations will all live in increasingly extant city spaces but that the dominant impact cities have on economic, political, social, cultural, technological, and importantly "physical" networks relies on the structuring of the extensive and extending assemblages in their service. As Amin and Thrift (2002) highlight, the dualistic thinking of cities as separated from suburban, ex-urban, or rural sites is a model that regularly fails to capture the state and effects of a "planetary" urbanization. Instead, cities can only be understood by grasping their existence as an outcome of various flows and sites across many spatialities and temporalities. Understood in this sense, the impact of the urban on structuring global social systems should direct study toward these sites, yet this chapter is not simply a call for scholars in the sociology of sport or physical cultural studies to be "more urban" in regards to the particular focus of their research. We are suggesting a deeper affinity between studying the urban and understanding the active body. The significance of the active body in understanding the urban and the shaping influence of the urban on the active body should not be understated. Fundamentally, the city, and its urbanizing effects, is the result of movement.

As Jensen (2006) suggests, the movement-centered nature of the urban is related to three factors:

> First of all, the contemporary city is one characterised by increased flows of people, symbols, and material goods. Second, the contemporary city is linked to multiple global-local networks of such flows. Third, as the intensity and number of connections are on the rise, this is a situation of "accelerating mobility" in the city. (p. 144)

As such, the city and its extending connections function as snapshots of material flows moving at various speeds. From the seemingly obdurate structures of the city as they slowly decay and are made anew, to the local interactions and movements of urban actors, as well as the flows of resources around the world directed by urban centers, attempts to enable and limit movement are at the center of an urbanized human existence. In this way, to study the active urban body, enmeshed in an assemblage of emerging connections, is to engage with how human physical practices play a role in the very ontogenesis of our increasingly urbanized contemporary social conjuncture. Equally, and taken in complement, a study of the urban active body is the necessary framework to understand the active body writ large. Indeed, if we are truly in a time of planetary urbanization, then there is no way to extricate the active body from its urban impacts and effects.

In this chapter, we embrace this necessity to engage the active body as part of the city in its essential movements, and we argue for a particular theoretical and methodological approach to do so. We will propose studying the city as an assemblage of human and nonhuman actors that together "enact" (Farías, 2010; Latour, 2005) the city as a form of social life and in which the active body—and its attendant cultures and lived experiences—takes on specific and intensified dimensions. "Flattening the city" therefore reflects a particular point of reorientation to studying the active city that shares affinities with affect theory, nonrepresentational theory, posthumanism, new materialism, and most centrally, assemblage thinking. In particular, we incorporate the concepts of assemblage urbanism (McFarlane, 2009, 2011a, 2011b) in developing an approach that, while not a definitive statement or treatise on how to study the urban, does offer a novel and (we hope) ultimately valuable approach to studying active bodies in contemporary cities.

Assemblage as a theoretical construct is most often traced back to the work of Deleuze and Guattari (1987) and the Deleuzean definition of assemblage as "a multiplicity constituted by heterogeneous terms and which establishes liaisons, relations between them" (Deleuze & Parnet, 1987, p. 52). However, assemblage has also been incorporated into a distinctive approach to engaging and analyzing the dynamics of urban environments. Thus as McFarlane (2011c) explains, while assemblage conceptually denotes the relations among people, places, and things in the remaking of the social world, it also entails "an approach, an orientation that operates as a way of thinking of the social, political, economic or cultural as a relational process of composition, and as a methodology attuned to practice, materiality and emergence" (p. 23). In this project, the incorporation of assemblage as the fundamental concept of assemblage urbanism means that this approach is specifically focused on the interactions between and across specific actors, associations, and practices and particular social and material environments. As Farías (2010) explains, there are several specific intellectual, academic, and political rationales for the approach offered by assemblage urbanism—this includes the natural affinity between "assemblage thinking" and cities, as the city exists as a "multiplicity" that is always unfolding in and through the relations among people, places, and things (p. 373). As further detailed below, this reorientation does not mandate the rejection of previous approaches to studying the city or human movement in urban environments, including political economy and critical urban theory. However, we argue that an alternate approach can provide a different starting point, with a different set of theoretical constructs and assumptions, as well as diverse expectations for possible outcomes.

In order to demonstrate the necessity of shifting toward this model of the urban and what it can mean for altering the study of active urban bodies, we have included three main sections in this chapter: "Studying the City," "Flattening the City," and "Moving Bodies in the Flattened City." These sections demonstrate the evolution of urban studies that underpins the model we are advancing, the basic tenets of this "flattened city" model, and some of the initial implications this may have for studying active bodies and physical cultures in cities. In outlining the broader advances in urban theory that have increasingly oriented toward ideas of assemblage urbanism, we seek to offer a mapping of the scholarly modes we have built on. In addition, we have provided an identification of key areas where studies of the active body have intersected with these theoretical advancements. From this grounding, we are able then to stake out the key dimensions to the model we are proposing and finally recognize key areas where the active body can be reengaged within this framework. Taken together, they should demonstrate the impetus for our attempt to reconsider the city in physical cultural studies and how these can be mobilized within a coherent approach to opening up the fundamental function of the active city. These are considerations that require further development, and as such we also hope for this chapter to be a starting point to an expanded conversation—an initial step in a broad reconsideration of the city as an important site for physical activity practices, communities, and cultures.

STUDYING THE CITY: CRITICAL URBAN THEORY AND ASSEMBLAGE URBANISM

Cities have often been recognized as important sites of cultural research, as they feature intensified and condensed forms of social life and demonstrate the particular interactions of economic, political, technological, material, and ecological factors. More recently, urban scholars have emphasized the ubiquity of the urban experience, of "urbanism as a defining feature of the human condition . . . in and through which capital, politics, everyday social relations and environmental politics are simultaneously organized and fought out" (McFarlane, 2011a, p. 205). This impetus toward a focus on the city has been supported by research centered on the significance of urban environments for particular forms of society (Holston & Appadurai, 1999; Hubbard, 2007) as well as the deep urbanization of global populations (Nadarajah & Yamamoto, 2007).

Yet this increase in scholarly discussion regarding the city has also meant that the theoretical and methodological frameworks involved have both multiplied and become more contested as researchers struggle to illuminate the complexity and actively evolving nature of urban life across a broad range of models and approaches. In particular, these recent debates have involved the value and role of *critical urban theory* as an approach to the contemporary city as well as the emergence of a related but also different approach via *assemblage urbanism*. Here we provide a background of critical urban theory in order to then demonstrate both the shared affinities and critical points of departure between this mode of urban inquiry and that of assemblage urbanism on which we build our physical-activity-oriented model.

Critical Urban Theory

In many ways, the origins of critical urban theory can be associated with the often-intertwined strands of the Frankfurt school of critical theory and 20th-century cultural Marxism. In regards to the Frankfurt school, many urban scholars worked to develop theories that served as an alternative to positivist social sciences and emphasized the immanent critique of society and culture toward identifying conflicts and contradictions in the urban (Brenner, 2009, p. 200). Across a similar period of time, scholars such as Henri Lefebvre (1991; 2003), Manuel Castells (1977; 1978, 1983, 1989), Peter Marcuse (2002; 2011), and others inflected and incorporated particular strands of Marxist thought into their own theories regarding cities and urban spaces. In this mode, scholars often described the city as a fundamental product of capitalism, meaning that urban environments were simultaneously characterized by oppression and inequality, as well as the potential for radical social change. As Merrifield (2002) explains, this dialectic view of the city as resistance and reification meant that Marxist urbanists "denounce[d] the unfair plenty of the capitalist city at the same time as they [upheld] the virtues and latent potentialities of urban life" (p. 5). In this mode, scholars often described the city as a fundamental product of capitalism, meaning that urban environments were simultaneously characterized by oppression and inequality, as well as the potential for radical social change.

Critical urban theory therefore emerged as an immanent critique of a modernist and scientific approach to the city that asserted that the city was knowable through the application of objective statistical tools and scientific methods. Before World War II, modernism was often translated into city planning in response to the effects of industrialization, as the first planners sought to curtail the various "ills" of the industrial city and simultaneously promote growth in a rational and predictable manner. In the wake of global conflict, the precepts of modernist planning and design were further embedded within the "planning imaginary" as cities became the focal point of modernist architecture (Bridge & Watson, 2003). Following Glazer (2007), during this period,

> modernism put forth one big and all-embracing idea: the city as the functional envelope of urban needs, which can be designed and implemented in one grand plan. Just as modernism calls for "machines of living," or for manufacturing, or selling, as against the architecturally elaborated structures of age before modernism, so it calls for the city to be the newly made proper envelope for all these machines. (p. 15)

Thus critical urban theory sought to challenge what Merrifield (2002) describes as a preoccupation within urban planning for "the 'hard city,' for the city that can be 'mapped' or 'modeled,' chronologically represented"—a model of urban planning that was simultaneously "blind to underlying processes and social relations" (pp. 4–5). Critical urbanists offered an alternative, embracing the tools embedded in Marxist materialism and undermining any possible claims to absolute objectivity in both studying and planning urban development. In building from these frameworks, many urban scholars have continued to prioritize a structural and social-class-based theoretical approach while eschewing the particular commitments of classical Marxism. These theories therefore have developed through a focus on structural processes and corresponding levels of analysis, but without the ultimate reduction of these processes to a function of a base-superstructure class struggle—instead, critical urban theory has been defined as a "critique of ideology (including social-scientific ideologies) *and* the critique of power, inequality, injustice and exploitation, at once within and among cities" (Brenner, 2009, p. 198). For Brenner (2009), critical urban theory is characterized by four specific elements that collectively shape this approach to studying the city: first, that there is a value in and demand for critical theory as an abstract analysis of urban environments and economic and political realities; second, that critical theory is inherently reflexive and produced within and toward particular social contexts; third, that critical theory incorporates a fundamental critique of instrumental reason and of theory as the institutionalization of best practices within a particular industry or field; and fourth, that critical theory serves to accentuate the contested nature of the actual and the possible by emphasizing the ways in which the city is produced by capitalism while also seeking out possibilities for emancipatory politics and social change (Brenner, 2009). In this mode, critical urban theory often functions as a form of "radical urban political economy" that Brenner et al. (2011) suggest is oriented to analyze issues fundamentally at the center of an advanced or late urban capitalism.

Critical urban theory therefore continues to inform contemporary urban studies, as well as a wide range of adjacent areas of studying urbanization, and remains as an invaluable approach that illuminates key dimensions of urban life. Indeed, a preponderance of studies that have looked at urban active bodies have drawn from either Marxist, neo-Marxist, or critical theory frameworks. The work of Scherer, Koch, & Holt (2016) in their study of an inner-city sport-for-development program is a good example, as it draws directly from Marx as well as neo-Marxist scholars such as David Harvey. Indeed, this is just one of many other examples (Bennett, Alexander, & Persky, 2013; Carter, 2011; Scherer & Davidson, 2010; etc.) that all embrace elements of a critical urban theory framework.

However, for some scholars, the adherence to a structural framework of the city has been limiting. As Farías (2010) states, "The obduracy of contemporary conceptions of the city in the field is striking. . . . [The] city has been understood in highly stable and bounded ways" (p. 9). As a result, researchers have sought to develop new theoretical frameworks that can move "beyond the strong structuralistic programme" (p. 1) and more readily embrace a complex, processual, and emergent notion of the city.

Assemblage Urbanism

Following McFarlane (2011c), the incorporation of theories of assemblage within urban studies and urban geography has often drawn from Deleuze and Guattari (1981) in their conceptualizing assemblage as "agencement," or "the alignment of different elements" (p. 24). Therefore, within these fields, assemblage urbanism places emphasis on "indeterminacy, emergence, becoming, processuality, turbulence, and the sociomateriality of phenomena" (McFarlane, 2011c, p. 24). Additionally, as an intellectual approach, assemblage urbanism has drawn from the introduction and incorporation of actor-network theory (ANT) as articulated in the work of Latour (2005), a framework that also engages with Deleuze and Guattari's conceptions. Following Farías and Bender (2010), the opening forays of ANT into urban studies have demonstrated the potential for a recognition of the conventions and "blind spots" of other approaches to urban research, including the influential but increasingly dated paradigm of Marxist political economy and the ongoing advancements of a critical urban theory (p. 1). This means that by focusing on the variously discursive, material, and "nonrepresentational" aspects involved in the remaking of urban environments, assemblage urbanism attempts to acknowledge the reality of larger processes and structures while also more intently focusing on the small, everyday, and informal ways in which the city is made to matter. As such, McFarlane (2011c) explains, the approach to interacting with and studying urban environments constituted by assemblage urbanism is predicated on "assemblage" as a concept used to

> emphasize the labour through which knowledge, resources, materials and histories become aligned and contested: it connotes the processual, generative and practice-based nature of [the] urban . . . as well as its unequal, contested and potentially transformative character. (p. 1)

Building on this, three principles of assemblage urbanism can be drawn out. The first of these guiding principles is an ontological conceptualization of the city as a "multiple object," following from Mol's (2002) research on how objects are always "enacted" within particular conditions and in relation to specific human and nonhuman elements. That is, the city is understood not epistemologically as a unitary social construction but instead ontologically, "acknowledging that different realities are being enacted here and there, now and then" (Farías, 2010, p. 13). This approach again emphasizes the continual and unceasing reshaping of urban social worlds, or what Amin and Thrift (2002) refer to as the process of "concrescence," or the ways that different entities and elements encounter and associate with each other in displaying the inherent heterogeneity of cities (p. 27). This means that instead of referring to "the city" as a bounded physical territory or as an economic unit and economic actor, and rather than relying on dual, bifurcated, and "Dickensian" (Burns, 2008) conceptions of the city, assemblage urbanism demonstrates and accentuates the multiple and processual realities of urban life. As Farías (2010) explains,

> The city is literally different things, has multiple different forms, gathers multiple different publics, fulfills multiple different functions, triggers multiple different practices, and so on. . . . [It] is made of multiple orders of value and groups of people often running parallel to one another. (p. 19)

In order to apprehend and describe this multiplicity, assemblage urbanism therefore focuses on the "urban assemblages" that are constituted in and through practices, actors, associations, and materialities. Importantly, this does not mean that an urban assemblage is the sum or total of a group of interrelated elements; rather, urban assemblages are the processes of this relation, the ways in which things come together. Thus urban assemblages refer to the processes "through which the city becomes a real-estate market, a filmic scene, a place of memory ... to one particular enactment of the city" (Farías, 2010, p. 15). Further, urban assemblages again stress the "work" involved in the reconstitution of multiple realities through associations and practices in that the focus is not on assemblage as a noun but on the active "assembling" of human, material, technological, and biological elements (Latour, 2005). As McFarlane (2011c) states, this approach to thinking about and engaging with urban milieus thus recognizes that the processes of neoliberalization and late capitalism, for example, are a part of the remaking of cities but that these processes do not constitute the totality of urban practices, forms, and realities.

As part of the framework of thinking about urban assemblages in relation to critical urban theory, the second principle or underlying assumption within assemblage urbanism most relevant to this analysis is the conceptualization of power. When theories of Marxist political economy and critical urban theory have focused on the power of the state and of political and economic elites, these theories have also often entailed a top-down conception of power in which governing bodies and rulers are endowed with more power than others and operate with and through knowledges hidden from the masses. Assemblage urbanism seeks to recognize the interactions between actors and institutions but disagrees with an understanding of power as a resource held only by those in privileged structural positions (Farías, 2011). Instead, urban assemblages are accompanied by a notion of power as an "immanent force," as something that "works on subjects as well as through them—at one and the same time" (Allen, 2008, p. 65). In short, and as Allen (2008) explains, this shift replaces the focus of critically investigating who has power and how much power they have and instead focuses on "how power exercises us"—that power is open ended and allows for and results in the unexpected while also simultaneously serving as a "normalizing" force that provides stability and regularity in our everyday lives and routines (p. 66).

The third principle of assemblage urbanism most relevant to this chapter is inextricably linked to the conception of power as an immanent force as well as to the attempt to move beyond the prevalent politics of critical urban theory. Again, while criticisms of assemblage urbanism have often referred to the possible "naive objectivism" of focusing on actors, associations, and materials instead of large-scale processes, Farías (2011) explains that any charges of objectivism could actually be applied to approaches that do not engage with the multiplicity of the city but rather "assume having privileged access to the real urban . . . and [assume] that by unveiling these hidden structures, the strength of the powerful will be combated" (p. 366). Instead, a focus on urban assemblages prioritizes the situated relations among people, places, and things and the different operations and effects of power constituted in and through this "assembling." This supports a crucial political dimension to the approach in that it recognizes that while economic and social asymmetries and inequalities are a regular and persistent feature of urban environments, cities are also the site of new spaces and forms of politics. In this way, thinking about and interacting with urban life through the assemblage framework also widens the field of political engagement and works to delimit the conception of politics provided through political economy (Amin & Thrift, 2002, p. 57).

When drawn together as an approach, assemblage urbanism engages the urban with "a sort of anti-structural concept that permits the researcher to speak of emergence, heterogeneity, the decentered and ephemeral in nonetheless ordered social life" (Marcus & Saka, 2006, p. 101). Building onto an existing body of critical urban theory, it provides "a different theorization and lexicon of urbanism that seeks not to displace existing urban theory, but to add to it" in studying the practices and associations of everyday urban life (McFarlane, 2011c, p. 184). Certainly urban studies scholars continue to produce research that has not adopted the framework advanced by McFarlane, but others have embraced it explicitly or at least its core tenets. Some have sought to directly address the model of assemblage urbanism (Cook & Ward, 2012; McCann, 2011; McCann, Roy, & Ward, 2013), while others have found affinity with its tenets in studying the "informal" city (Parnell & Robinson, 2012; Vasudevan, 2015), the "nonhuman" city (Franklin, 2017), urban transportation (Harris, 2013), green urbanism and sustainability (Blok, 2013; Ranganathan, 2015), and technology in the city (Lepawsky, Akese, Billah, Conolly, & McNabb, 2015; Shepard, 2013). Additionally, a selection of scholars has engaged with assemblage urbanism to study urban bodies, such as Lancione's (2013) study of homelessness in cities and, of specific interest for this chapter, those who have studied urban active bodies. We can point to Rosen-Kremer and Aravot's (2011) study of "Cycling in the Virtureal" and more recently Barnfield's (2016) research into recreational running in Bulgaria that additionally engages with nonrepresentational theory.

FLATTENING THE CITY

In contrast with critical urban theory, assemblage urbanism works to reorient theory and method around the city "as a thing in the making" in which there is a "capacity for its different people, spaces, activities, and things to interact in ways that exceed any attempt to regulate them" (Simone, 2009, p. 3). It does not deny or shed the idea of the city as having structures but pays more attention to how structuring can take place and the open possibilities that exist where these attempts are incomplete. Therefore, by redeveloping "the way we see, understand and thus live the world" (Dewsbury, 2011, p. 148), assemblage urbanism blurs the core tenets of critical urban theory with an approach to the city that embraces its complexities and looks to engage with its emergent properties.

As Brenner et al. (2011) explain, assemblage urbanism therefore often forgoes an examination of the "context of contexts" in and through the larger social processes that are the purview of critical urban theory. Instead, assemblage urbanism recognizes the explanatory value of these theories while also engaging in fieldwork that is primarily focused on the practices, actors, and materialities involved in the enactment of particular "urban assemblages" (Acuto, 2011). This means that while assemblage urbanism allows for a focus on the operations and relations of power, it also provides an alternate perspective on how power operates through social relations. That is, rather than attempt to only interpret the empirical findings of urban research through the theorized frameworks of large-scale social processes, this approach instead seeks to describe the practices and associations that make up the small-scale formations of urban experience as they variously underpin and undermine overarching attempts to structure the city. Rather than ignoring or silencing the presence and evidence of social inequality in urban environments, assemblage urbanism engages the actual practices, processes, and "sociomaterial orderings" of the "asymmetrical" social, economic, and political formations of contemporary cities (Farías, 2011, p. 341).

Building on this underlying framework provided by assemblage urbanism, to "flatten the city" is to embrace a conceptualization of the urban that is defined not by a collection of various actors and spaces but rather through the interaction of objects and actors. Instead of enforcing a hierarchical model that posits the city as a totality that can be prefigured in terms of structure and action, the flattened city insists that any and all interactions can play a significant role in shaping the city and our experiences in it and of it. Thus while the historical development of the city—the processes that have previously altered urban conditions—has real impacts on the present and future of the city, we would emphasize that the possibility for change (whether material, cultural, and/or political) is always assumed. Following Dewsbury (2011), in this mode, studying the city becomes "less about what it is then, and more about what it can do, what it can affect and bring about" (p. 150). Thus to "flatten the city" demands adhering to two key epistemological and ontological directives.

First, flatten agency—reconceptualize agency as something that is not allocated to only those in positions of relative power and/or as something that is only possessed by human actors. As a starting point, this should be a recognition that humans are not the singular, exceptional, or end point of a social process. This approach means always taking seriously the role of nonhuman actants, as "from the outset, neither humans nor non-humans have prefigured dominance in terms of agency" (Anderson, 2009, p. 123). In this view, agency is not understood as an intangible quality that is assigned or acquired. Agency is not measured by an invisible meter above each actor or objects that display how much or how little agency is held; it is a description of the ability to independently act. Flattening the city thus necessitates an understanding of "distributive agency," in which all objects hold the capacity for agency, and that this agency is demonstrated in surprising and unexpected fashions. As Bennett (2005) states,

> [A] distributive theory of agency does not deny that human persons are capable of reflective judgments and thus are crucial actants in many political transformations. But it attempts a more radical displacement of the human subject from the center of thinking about agency. (p. 454)

The ability of the human to act becomes cast in constant relation to other bodies/environments/policies and is understood only as a particular form of agency held by all things, not as an exclusive ability of humans. It is an attempt "to destabilise the monopoly of the human agent within its material environments" that allows the action of nonhuman elements to be drawn into the city in more meaningful ways, not just as passive backgrounds for human action (Roberts, 2012, p. 2513). The researcher must take seriously that the actors intimately involved in the making and remaking of the city may hold a range of agentic possibilities and are rarely translators of human intention. This means that relegating the nonhuman to a mere function of human action is to strip it of its potential in the formation of the social (Latour, 2005). We do not suggest anthropomorphizing objects—making characters out of rock formations or electrical grids—but instead posit that we should understand that due to the unpredictability of interaction, intention is rarely fully realized.

Second, flatten scale—remove prior ideas regarding the scale of human action and intent, as well as our conceptions of the spatial organizing of human and nonhuman interaction. This means that we remove the commonsense view of cities as located along a continuum from the "global" to the "local"—it is to disband the perspective of cities as viewed through a "well-ordered zoom." As Richard Smith (2010) articulates, while scale is a useful concept when analyzing the formation of cities and urban experience, problems arise when scales are conceived and utilized within analysis as *a priori* concepts. Indeed, this issue of avoiding any predetermined context, structure, or framework is (via Latour) a defining feature of both ANT and assemblage urbanism. Smith (2010) explains that "when the term 'scale' is used by social scientists it is done so to indicate an established state of affairs, a foundation, a given, an unquestionable frame-work from which one can subsequently begin to account for what is happening in a given situation" (p. 75). In a broad sense, then, flattening the city entails first a necessary awareness that urban experiences are often enacted through unexpected interactions among peoples, places, and things as much—if not more—as they are structured through spatial organization and institutional policy and procedure. As such, the researcher cannot take an *a priori* "map" of the terrain into the field, as while historical factors play a part in the current con-texts of urban development, these factors cannot be given privilege to explain how and why experiences unfold in the contemporary context.

Applied in conjunction, these steps demand that rather than assume the input of actors and objects, we must study each equally, describing the full assembled "ecology of things" (Bennett, 2010) shaping a particular setting from near or far. The active city is now flattened in that prefigured relationships, forms of power, and sites of privilege have been stripped away. Instead, this approach demands openness to how we conceive of the city and seeks to eschew assumptions in the practice of studying the active city. Agency and scale should not be ignored, but where they are made to matter, attention should be paid to understand how they have been structured. Also where preconceived ideas of how the urban functions are challenged, these unexpected dimensions to the city should be taken seri-ously. The flattened city approach then allows for an increasingly detailed and focused understanding of how structure can be formed in practice but demands first and foremost engaging with the city as follows:

1. *Emergent* urbanities are necessarily a product of the relational interactions between a range of actors and places, which together create the city as always more than the sum of its parts. Therefore,
2. *contingent* structures (such as institutions and policies) are indeed an impor-tant aspect of the city as a process of living, but these structures are never fixed and this process is never static.
   As a result
3. unexpected interactions among peoples, places, and things mean that the mak-ing of the city is constituted through both human and nonhuman actants and materialities, and these interactions occur in surprising and novel ways and provoke the further "remaking" of the urban.

Taken together, these tenets will require a willingness to engage with com-plexity and empirical "messiness" in our study of urban experience—there is not a single or simple story to be told of any urban human movement, even when the conceptualization and dissemination of a strong singular narrative may be professionally or politically expedient. While we must put forth the "stakes" of our conclusions, we must be fully committed to the idea that any analysis of the urban is incomplete. It means understanding that "places are not passive stages on which actions occur, rather they are the medium that impinge on, structure and facilitate these processes" (Anderson, 2009, p. 123), and this has implications for every aspect of this study, from theory, to method, to politics.

MOVING BODIES IN THE FLATTENED CITY

As we have asserted, consideration of the "flattened city" approach outlined in this chapter provides the opportunity to redevelop how we study, think about, and engage with moving bodies in urban environments by shifting some fundamental conceptions of the city and the necessary steps to researching the city. However, in our view, this approach is also made more imperative in that it brings together two as-yet-disparate fields of research within the study of human movement and physical culture—the study of the city and Deleuzean theories of assemblage that share an affinity with other nonrepresentational and new materialist theoretical frameworks—that will also impact how we conceive of the body. That is, on the one hand, the approach offered in this chapter is a response to the valuable work of scholars who have predominately operated from a background within or related to critical urban theory when studying sport and physical activity in the city (Bairner, 2011; Bale, 1996, 2001; Bélanger, 2000; Borden, 2001; Friedman, Andrews, & Silk, 2004; Silk, 2004; Silk & Andrews, 2006; Wilcox & Andrews, 2003; etc.). Yet at the same time the, model offers a means through which to reconsider the active body that can contribute to the valuable work on embodiment that has been done in PCS and the sociology of sport to date (Allen-Collinson, 2009, 2011; Pink, 2011; Thorpe, 2016; etc.) and the few who have tried to center this discussion in an urban setting (Ghannam, 2011; Jones, 2012; Zhang & Silk, 2006).

Specifically, we are going to suggest two key areas of our understanding of the active urban body that this model impacts. First, it is a model that essentially decenters the human body, stripping down the idea of human exceptionalism to explore a more distributed model of significance for nonhuman actors. And secondly, it is a model that necessarily demands embracing the concept of the body as affective within a deep relational framework to be able to fully understand the multiplicity of its connections to the city.

Decentering the Body

Initially, as with the decentering of the human subject when flattening agency, within the flattened city, the body now also becomes decentered, or at least stripped of its position of privilege. Instead, the body can be seen as part of a broad array of materiality that is central to how the urban functions. Connecting with the impetus to pursue the idea of an ecology of things in which agency is distributed, the body is best understood as part of a distributed materiality of the city. The body is constantly reconfiguring in unexpected and ever-developing relations with a wide range of actors and felt material environments. The body, much like the city, is not a platform for action but the outcome of a relational process alongside the other material that makes up the city. Indeed, this shift in how we conceive of the body is key to the politics and function of the approach we have advocated in this chapter, because as Bennett (2005) suggests, "Though human reflexivity is indispensable for transforming political life, on many occasions and in a variety of ways the efficacy of political change is not a function of humans alone. It is better understood, I think, as the conjoined effect of a variety of kinds of bodies" (p. 454). Human bodies should therefore not be seen as being in a separate category from the other bodies and materialities of the city but can be more fully understood as fundamentally embroiled in this tangible process of deeply felt interaction that is the urban in its constant becoming.

Studying the body as a singular and defined entity is a concept that becomes imploded as the active body cannot be detached from its instantaneous connections: geographically near and far, temporally immediate and extended, representational and affective. The body must be conceptualized as deeply enmeshed in a tangle of connections, interacting with a variety of actors in a range of registers. In this model, then, the city can be seen as an emergent expression of "the distinctive efficacy of a working whole made up, variously, of somatic, technological, cultural, and atmospheric elements" rather than as an ordered expression of human intentionality and structuring/resisting of power in dialectic conflict (Bennett, 2005, p. 447). For Bennett, as for us, embodied aspects of the urban are an important part of the city multiple but should not be given outsized importance or a sense of undue distinction from the rest of the city.

Bodily Affects

Through decentered consideration of these embodied dimensions of the city, the possibilities of the "flattened city" also open up the potential to further grasp the affective dimensions of the body in deeply felt, relational, and material ways. As Latham and McCormack (2017) suggest, "Exercise, sport, and fitness complicate questions of urban materiality because they are profoundly affective," an affective dimension best described as the interactive capacity and necessarily interdependent nature of bodies. This would not be the first time affect has been utilized as a theory for exploring the extracognitive and embodied dimensions of physical activity writ large. Certainly it is important to recognize the work of Markula (2008) and others (Evers, 2006; Roy, 2013; Thorpe & Rinehart, 2010) in pioneering these considerations. However, we suggest that the approach to affect that connects to and informs assemblage urbanism makes a substantive shift from many of these approaches by shifting where to begin with a concept of affect. Instead of starting with an individual focused expression and interpretation of the nonrepresentational—how the body affects and is affected—the approach to affect that we draw into the flattened city looks more at the relational forces that exist between actors. As Andrews, Chen, and Myers (2014) define it,

> In simple terms, affect is a mobile energy; an intensity which is the result of the relative movements and interactions between things, from atoms and molecules to fully formed human bodies and non-human objects, in space. Whilst initially affect is a purely physical non-cognitive event, it gives rise to less-than-fully conscious experiences. These are felt sensations or "feeling states" that, preceding full cognition (thoughts and emotions for example), manifest on a somatic register as vague but intense "atmospheres" or "vibes" which impact on an individual's capacity for engagement and involvement (and thus for themselves to affect others). (p. 214)

Working from this definition, the city assemblage becomes the ideal space through which to explore the "utter ubiquity of affect as a vital element of cities" (Thrift, 2004, p. 57) that shapes the physical experience, providing the impetus and constraint for our actions. When we move through the city, we encounter a range of actors; our movements are constrained and enabled by certain materialities, policies, people, and a range of other actors assembled together. We interact across the flattened city in a number of registers, shaping the city in its becoming as we are shaped by it. As such, attending to an idea of nonconscious affective relationality while studying conscious forms of interaction allows us to more fully describe how a range of actants can assemble to emergently form the city.

The body in motion as part of the flattened city is now reoriented theoretically. It is cast into being one of many assembled actors in the city; interacting through a range of modes and forming complex outcomes, moving bodies become dislocated from their anthropocentrism and relocated among a vast, vital "materiality that is self-transformative and already saturated with agentic capacities" (Coole, 2010, p. 92). The human body in motion then gets re-placed into the urban, "figured as neither mechanistic nor teleological but rather as alive with movement and with a certain power of expression" (Bennett, 2005, p. 447). Yet rather than diminishing the significance of studying the body as it moves in the urban, this makes its study more intense. As a result, the in-process, felt, relational, and emergent nature of human action in shaping the urban assemblage requires a more intensely ongoing and detailed study. This study and its shift in approach will also demand fundamental changes to the methods of studying the active urban body and urban physical cultures.

The "flattened city" approach detailed here therefore requires an interrogation or eschewing of previously held assumptions about the active body in the city. As bodies move in and through the urban, they engage in forms of interaction that highlight the (inter)actions of policy, place, people, and others across the city. They do not determine these connections and actions but become a particularly insightful means through which to understand the city and the actual functions of the assemblage as much as they also open up ways of considering the active body.

CONCLUSION

As this chapter has demonstrated, the constant redevelopment of cities as a particular locus of lived experience means that the ways in which scholars approach and comprehend urban environments are also necessarily subject to continuous debate and reconfiguration. For the authors, this demand for novel ways of studying, writing about, and engaging with specific contexts of and in the city reflects a larger trend in urban studies. As Brenner (2009) explains, "The meanings and modalities of critique can never be held constant; they must, on the contrary, be continually reinvented in relation to the unevenly evolving political-economic geographies of this process and the diverse conflicts it engenders" (p. 204). By incorporating the tenets of assemblage urbanism, an approach to flattening the city not only provides one such form of reinvention; it also reflexively questions the nature and purpose of urban theory. Following Farías (2011),

> Assemblage urbanism seeks to build on a radically relational understanding of power–knowledge, which messes up the object and standpoint of critique. . . . [Assemblage urbanism] promote[s] a more open and explorative form of engagement with the world; in a word, inquiry, not critique. (pp. 365–366)

Thus, in closing, we would like to emphasize that the model we are promoting is about more than a renewed and revised approach for scholars interested in studying active urban bodies. It is also about insisting on the openness of urban development and recognizing the always-embedded possibility for change and transformation, even—and perhaps especially—when confronted with the seeming obduracy of structural institutions and processes. For the authors, this has required a rethinking of our conceptualization of the urban as *experiential* and a reorientation toward *experimental* theories and methods, as well as our place within these issues as both citizens and researchers. This approach therefore features an inherent opposition to scholarly rigidity that reflects not only the interdisciplinary and multimethodological approach of the authors but also the "creative tension" between and within different theories and methods that is at the core of studying the active body (Andrews, 2008). Flattening the city reminds us that the city and its human and nonhuman inhabitants can and should be understood as unfinished and in situ, and this chapter suggests that the study of urban moving bodies can benefit from an incorporation of this framework.

REFERENCES

Acuto, M. (2011). Putting ANTs into the mille-feuille. *City, 15*(5), 552–562.

Allen, J. (2008). *Lost geographies of power.* Malden, MA: Wiley-Blackwell.

Allen-Collinson, J. (2009). Sporting embodiment: Sports studies and the (continuing) promise of phenomenology. *Qualitative Research in Sport and Exercise, 1*(3), 279–296.

Allen-Collinson, J. (2011). Intention and epochē in tension: Autophenomenography, bracketing and a novel approach to researching sporting embodiment. *Qualitative Research in Sport, Exercise and Health, 3*, 48–62.

Amin, A., & Thrift, N. (2002). *Cities: Reimagining the urban.* Cambridge, MA: Polity Press.

Anderson, J. (2009). Transient convergence and relational sensibility: Beyond the modern constitution of nature. *Emotion, Space and Society, 2*(2), 120–127.

Andrews, D. L. (2008). Kinesiology's inconvenient truth and the physical cultural studies imperative. *Quest, 60*(1), 45–62.

Andrews, G. J., Chen, S. and Myers, S. (2014). The "taking place" of health and wellbeing: Towards non-representational theory. *Social Science and Medicine, 108*, 210–222.

Bairner, A. (2011). Urban walking and the pedagogies of the street. *Sport, Education and Society, 16*(3), 371–384.

Bale, J. (1996). *Landscapes of modern sport.* Leicester, England: Leicester University Press.

Bale, J. (2001). *Sport, space and the city.* London, England: Routledge.

Barnfield, A. (2016). Public health, physical exercise and non-representational theory—a mixed method study of recreational running in Sofia, Bulgaria. *Critical Public Health, 26*(3), 281–293.

Bélanger, A. (2000). Sport venues and the spectacularization of urban spaces in North America: The case of the Molson Centre in Montreal. *International Review for the Sociology of Sport, 35*(3), 378–397.

Bennett, J. (2005). The agency of assemblages and the North American blackout. *Public Culture, 17*(3), 445–465.

Bennett, J. (2010). *Vibrant matter: A political ecology of things.* Durham, NC: Duke University Press.

Bennett, L., Bennett, M., Alexander, S., & Persky, J. (2013). The political and civic implications of Chicago's unsuccessful bid to host the 2016 Olympic Games. *Journal of Sport & Social Issues, 37*(4), 364–383.

Blok, A. (2013). Urban green assemblages: An ANT view on sustainable city building projects. *Science and Technology Studies, 26*(1), 5–24.

Borden, I. (2001). *Skateboarding, space and the city: Architecture and the body*. London, England: Berg.

Brenner, N. (2009). What is critical urban theory? *City, 13*(2–3), 198–207.

Brenner, N., Madden, D. J., & Wachsmuth, D. (2011). Assemblage urbanism and the challenges of critical urban theory. *City, 15*(2), 225–240.

Bridge, G., & Watson, S. (2003). Chapter 42—city interventions. In G. Bridge & S. Watson (eds.), *A companion to the city*. Oxford, England: Blackwell.

Burns, E. (Writer). (2008). The Dickensian aspect [Television series episode]. In *The Wire*. New York, NY: Home Box Office.

Carter, T. (2011). Interrogating athletic urbanism: On examining the politics of the city underpinning the production of the spectacle. *International Review for the Sociology of Sport, 46*(2), 131–139.

Castells, M. (1977). *The urban question: A Marxist approach* (A. Sheridan, trans.). London, England: Edward Arnold.

Castells, M. (1978). *City, class and power*. London, England: Macmillan.

Castells, M. (1983). *The city and the grassroots: A cross-cultural theory of urban social movements*. Berkeley: University of California Press.

Castells, M. (1989). *The informational city: Information technology, economic restructuring, and the urban regional process*. Oxford, England: Blackwell.

Cook, I. R., & Ward, K. (2012). Relational comparisons: The assembling of Cleveland's waterfront plan. *Urban Geography, 33*(6), 774–795.

Coole, D. (2010). The inertia of matter and the generativity of flesh. In D. Coole & S. Frost (eds.), *New materialisms: Ontology, agency, and politics* (pp. 1–43). Durham, NC: Duke University Press.

Deleuze, G., & Guattari, F. (1981). Rhizome. *Ideology and Consciousness, 8*, 49–71.

Deleuze, G., & Guattari, F. (1987). *A thousand plateaus: Capitalism and schizophrenia*. Minneapolis: University of Minnesota Press.

Deleuze, G., & Parnet, C. (1987). *Dialogues II*. New York, NY: Columbia University Press.

Dewsbury, J. D. (2011). The Deleuze-Guattarian assemblage: Plastic habits: The Deleuze-Guattarian assemblage. *Area, 43*(2), 148–153.

Evers, C. (2006). How to surf. *Journal of Sport & Social Issues, 30*(3), 229–243.

Farías, I. (2010). Introduction: Decentering the object of urban studies. In I. Farías & T. Bender (eds.), *Urban assemblages: How actor-network theory changes urban studies*. London, England: Routledge.

Farías, I. (2011). The politics of urban assemblages. *City, 15*(3–4), 365–374.

Farías, I., & Bender, T. (2010). *Urban assemblages: How actor-network theory changes urban studies*. London, England: Routledge.

Franklin, A. (2017). The more-than-human city. *Sociological Review, 65*(2), 202–217.

Friedman, M., Andrews, D., & Silk, M. (2004). Sport and the façade of redevelopment in the post-industrial city. *Sociology of Sport Journal, 21*, 119–139.

Ghannam, F. (2011). Mobility, liminality, and embodiment in urban Egypt. *American Ethnologist, 38*(4), 790–800.

Glazer, N. (2007). *From a cause to a style: Modernist architecture's encounter with the American city*. Princeton, NJ: Princeton University Press.

Harris, A. (2013). Concrete geographies: Assembling global Mumbai through transport infrastructure. *City: Analysis of Urban Trends, Culture, Theory, Policy, Action, 17*(3), 343–360.

Holston, J., & Appadurai, A. (1999). Introduction: Cities and citizenship. In J. Holston (ed.), *Cities and citizenship*. Durham, NC: Duke University Press.

Hubbard, P. (2007). Positioning cities in the world: Towards a politics of flow. In P. Taylor, B. Derudder, P. Saey, & F. Witlox (eds.), *Cities in globalization: Practices, policies and theories* (pp. 182–195). New York, NY: Routledge.

Jensen, O. (2006). "Facework," flow and the city: Simmel, Goffman, and mobility in the contemporary city. *Mobilities, 1*(2), 143–165.

Jones, P. (2012). Sensory indiscipline and affect: A study of commuter cycling. *Social & Cultural Geography, 13*(6), 645–658.

Lancione, M. (2013). Homeless people and the city of abstract machines: Assemblage thinking and the performative approach to homelessness. *Area, 45*(3), 358–364.

Latham, A., and McCormack, D. P. (2017). Affective cities. In M. Silk, D. L. Andrews, & H. Thorpe (eds.), *Routledge handbook of physical cultural studies* (Routledge international handbooks). New York, NY: Routledge.

Latour, B. (2005). *Reassembling the social: An introduction to actor-network theory*. Oxford, England: Oxford University Press.

Lefebvre, H. (1991). *The production of space*. Oxford, England: Blackwell.

Lefebvre, H. (2003). *The urban revolution*. Minneapolis: University of Minnesota Press.

Lepawsky, J., Akese, G., Billah, M., Conolly, C., & McNabb, C. (2015). Composing urban orders from rubbish electronics: Cityness and the site multiple. *International Journal of Urban and Regional Research, 39*(2), 185–199.

Marcus, G. E., & Saka, E. (2006). Assemblage. *Theory, Culture & Society, 23*(2–3), 101–106.

Marcuse, P. (2002). *Of states and cities: The Partitioning of urban space*. Oxford, England: Oxford University Press.

Marcuse, P. (2011). *Cities for people not for profit: Critical urban theory*. New York, NY: Routledge.

Markula, P. (2008). Affect[ing] bodies: Performative pedagogy of Pilates. *International Review of Qualitative Research, 1*(3), 381–408.

McCann, E. (2011). Veritable inventions: Cities, policies and assemblage (report). *Area, 43*(2), 143.

McCann, E., Roy, A., & Ward, K. (2013). Assembling/worlding cities. *Urban Geography, 34*(5), 581–589.

McFarlane, C. (2009). Translocal assemblages: Space, power and social movements. *Geoforum, 40*(4), 561–567.

McFarlane, C. (2011a). Assemblage and critical urbanism. *City, 15*(2), 204–224.

McFarlane, C. (2011b). Encountering, describing and transforming urbanism: Concluding reflections on assemblage and urban criticality. *City, 15*(6), 731–739.

McFarlane, C. (2011c). *Learning the city: Knowledge and translocal assemblage* (1st ed., RGS-IBG book series, 56). Malden, MA: Wiley-Blackwell.

Merrifield, A. (2002). *Metromarxism: A Marxist tale of the city*. New York, NY: Routledge.

Mol, A. (2002). *The body multiple: Ontology in medical practice*. Durham, NC: Duke University Press.

Nadarajah, M., & Yamamoto, A. T. (2007). Introduction. In M. Nadarajah & A. T. Yamamoto (eds.), *Urban crisis: Culture and the sustainability of cities* (pp. 3–12). Tokyo, Japan: United Nations University Press.

Parnell, S., & Robinson, J. (2012). (Re)theorising cities from the global south: Looking beyond neoliberalism. *Urban Geography, 33*(4), 593–617.

Pink, S. (2011). From embodiment to emplacement: Re-thinking competing bodies, senses and spatialities. *Sport Education and Society, 16*(3), 343–355.

Ranganathan, M. (2015). Storm drains as assemblages: The political ecology of flood risk in post-colonial Bangalore. *Antipode, 47*(5), 1300–1320.

Roberts, T. (2012). From "new materialism" to "machinic assemblage": Agency and affect in IKEA. *Environment and Planning A, 44*(10), 2512–2529.

Rosen-Kremer, O., & Aravot, I. (2011). Cycling in the virtureal. *Spaces and Flows: An International Journal of Urban and Extraurban Studies, 1*(1), 55–74.

Roy, G. (2013). Women in wetsuits: Revolting bodies in lesbian surf culture. *Journal of Lesbian Studies, 17*(3–4), 329–343.

Scherer, J., & Davidson, J. (2010). Promoting the "arriviste" city: Producing neoliberal urban identity and communities of consumption during the Edmonton oilers' 2006 playoff campaign. *International Review for the Sociology of Sport, 46*(2), 157–180.

Scherer, J., Koch, J., & Holt, N. (2016). The uses of an inner-city sport-for-development program: Dispatches from the (real) creative class. *Sociology of Sport Journal, 33*(3).

Shepard, M. (2013). Minor urbanism: Everyday entanglements of technology and urban life. *Continuum, 27*(4), 483–494.

Silk, M. (2004). A tale of two cities: Spaces of consumption and the façade of cultural development. *Journal of Sport & Social Issues, 28*(4), 349–378.

Silk, M., & Andrews, D. L. (2006). The fittest city in America. *Journal of Sport & Social Issues, 30*, 315–327.

Simone, A. (2009). *City life from Jakarta to Dakar: Movements at the crossroads*. New York, NY: Routledge.

Smith, R. G. (2010). Urban studies without "scale": Localizing the global through Singapore. In I. Farías & T. Bender (eds.), *Urban assemblages: How actor-network theory changes urban studies*. London, England: Routledge.

Thorpe, H. (2016). Athletic women's experiences of amenorrhea: Biomedical technologies, somatic ethics and embodied subjectivities. *Sociology of Sport Journal, 33*(1), 1–13.

Thorpe, H., & Rinehart, R. (2010). Alternative sport and affect: Non-representational theory examined. Special issue: Consumption and representation of lifestyle sport. *Sport in Society, 13*(7–8), 1268–1291.

Thrift, N. (2004). Intensities of feeling: Towards a spatial politics of affect. *Geografiska Annaler, Series B: Human Geography, 86*(1), 57–78.

Vasudevan, A. (2015). The makeshift city: Towards a global geography of squatting. *Progress in Human Geography, 39*(3), 338–359.

Wilcox, R., & Andrews, D. (2003). Sport in the city: Cultural, economic and political portraits. In R. Wilcox, D. Andrews, R. Pitter, & R. Irwin (eds.), *Sporting dystopias: The making and meanings of urban sport cultures* (pp. 1–16). New York: State University of New York Press.

Zhang, T. and Silk, M. (2006). Recentering Beijing: Sport, space, subjectivities. *Sociology of Sport Journal, 23*(4), 438–459.

15 · WHAT CAN NEW MATERIALISMS DO FOR THE CRITICAL STUDY OF SPORT AND PHYSICAL CULTURE?

(Who Does This Book Think It Is?)

RICHARD PRINGLE

The subtitle for this chapter has been appropriated from Deleuze and Guattari's (1987) *A Thousand Plateaus* and their chapter "10,000 B.C.: The Geology of Morals (Who Does the Earth Think It Is?)" (p. 39). Their amusing chapter begins, "The same Professor Challenger who made the earth scream with his pain machine, as described by Arthur Conan Doyle, gave a lecture after mixing several textbooks on geology and biology in a fashion befitting his simian disposition" (p. 40). The chapter reviews Professor Challenger's (he is aptly named) lecture to reveal geology as having the capacity to act, think, and affect. In a similar sense, I accept the vitality of this current coedited book and assume it will affect readers' thoughts, feelings, and actions: So who does this book think it is?

My prime task in this chapter is to reflect on the question, "What can new materialisms do for the critical study of sport and physical culture?" This type of questioning is something of a mantra within new materialisms—that is, to wonder what something "does" rather than what it "means." This is not to suggest that meanings are necessarily disconnected from what something does, but it broadens focus away from human understandings toward concerns with how bodies of various kinds act and affect each other. Such questioning, when focused on what a new body of knowledge might do, is open to considerable speculation. Indeed, given that new materialisms "are in process and they are not one thing" (St. Pierre, Jackson, & Mazzei, 2016, p. 99), I recognize the problems of attempting such a task. In this chapter, nevertheless, to help structure my speculations, I reconsider the justifications that have underpinned the return to materialist thinking to consider what changes might transpire. In doing so, I pay particular respect to the body, methodological issues, and possibilities for novel research approaches. Most importantly, I consider how this theoretical (re)turn could impact our critical scholarly endeavors, or as Simone Fullagar (2017) asks, How will new materialism "critically challenge rather than unknowingly reiterate, normative assumptions that negate or ignore different embodied practices of living and thinking" (p. 248)? I begin, however, by discussing issues associated with labeling something as "new."

WHAT DOES THE "NEW" IN "NEW MATERIALISMS" DO?

There is often discussion and debate with respect to the naming of emerging fields of study, as has been witnessed in the passionate discussions concerning physical cultural studies (see Mary Louise Adams et al., 2016). These debates, as poststructuralists readily know, reflect that nomenclature is political, and it is therefore important to examine how designations *affect* other bodies (i.e., human bodies and bodies of knowledge). In this section, I examine the notion of the "new" in the materialist turn.

The "new," as detailed in the introduction to this volume, refers in part to the promotion of a variety of ontological perspectives that have become known as flat or monist views of reality: ontologies that reject assumed dichotomies between the material and discursive or between physical and social realities. These non-dualist views concomitantly reject the notion of an internal-idealist ontological perspective and its associated assumption that realities are constructed inside individuals' heads. The material turn therefore offers new ways of viewing reality that "[open] up many fundamental questions about . . . epistemology, ethics and politics" (Alaimo & Hekman, 2008, p. 7). The monist perspectives fundamentally encourage a shift beyond concerns with language, discursive realities, and anthropocentric views to give greater consideration to nonhuman and inanimate objects. This focus on objects/matter acknowledges the capacities of matter to act—or be affected and affect other materialities—which directs research efforts toward the interactive or assembling processes that produce realities, humans, and sociocultural histories.

Elizabeth St. Pierre drew from Jacques Derrida to note that when you think with a new "concept you bring with it the entire structure in which it is thinkable" (cited in Guttorm, Hohti, & Paakkari, 2015, p. 15). The promotion of new materialism, as evidenced in this edited collection, therefore promotes innovative ways of seeing, thinking, and doing research, all of which promote fresh means of understanding existing social issues and the potential for social/material changes. New materialisms, in this light, can be viewed as a gift to critical researchers, and—in extending the topic of conversation from the introductory chapter—this text can be understood as academic gold. The "new" in new materialisms augers well for a potentially exciting shift in research direction.

Labeling something as "new" nevertheless has other connotations, such as the notion that it has a distinct beginning while also hinting that it is better than or an improvement on the "old." This leads to two associated issues worthy of consideration. The first concerns the assumption of a distinct beginning, as if the new materialist project developed "without the benefit of a simple recourse to either an earlier sociological materialism or . . . post-structuralism" (Fox & Alldred, 2016, p. 14). Yet such assertions appear dismissive of the value of the forces that prompted "new" materialist thinking. Deleuze and Guattari's (1977) influential text *Anti-Oedipus: Capitalism and Schizophrenia*, for example, synthesized the ideas of three influential materialists—Freud, Marx, and Nietzsche—to derive their own unique materialist views on capitalism and the political economy of desire. Deleuze and Guattari's work, in turn, has connections with the writings of Grosz, DeLanda, and Braidotti, for example. So, if required, a rhizomatic sketch of the interactions between various materialist thinkers could be drafted.

Sara Ahmed (2008) more bluntly remarks that an underpinning premise of new materialism has imagined feminism as "routinely anti-biological, or habitually 'social constructionist,'" yet such an argument, she claims, "offers a false and reductive history of feminist engagement with biology, science and materialism" (p. 24).[1] Ahmed refers to the text *Our Bodies, Ourselves* (first published in 1973) as one of several examples of earlier feminist proactive and political engagement with biology. In recognizing that matter has mattered "in different ways, for different feminisms, over time," she suggests that "we should avoid establishing a new terrain by clearing the ground of what has come before us" (p. 36). Ahmed concludes that if researchers are going to make sweeping claims about previous studies, then they should examine this work more closely and act respectfully before being "so willing to deposit our hope in the category of 'the new'" (p. 36).[2]

A similar rejoinder could also be raised in relation to poststructuralism in its various guises, given that it has been somewhat reductively framed by the "anthemic catchphrase 'there is nothing outside of text'" (Lyngdoh, 2017, p. 1). This catchphrase suggests that poststructuralists deny the existence of the "real-world"—a claim that is also drawn on to justify the need for "new" materialisms. Although there is no denying that poststructural approaches privilege text and semiotic systems, a closer reading of the original sources—rather than how they have been typically drawn on (and often by secondary sources)—reveals links of various strengths to materialism. Even Derrida, to whom the anthemic catchphrase is typically misattributed, has been regarded as a linguistic material-ist (see Colebrook, 2011; Lyngdoh, 2017). Moreover, concerns with representa-tionalism and materiality were abundantly evident in the work of Gilles Deleuze and Michel Foucault.

Foucault (1980), for example, was aware that his unique approach to histori-cizing sexualities could be misconstrued "as evading the biologically established existence of sexual function" (p. 151). As such, he pointedly clarified that the purpose of his study was to illustrate how the workings of power were connected "to bodies, functions, physiological processes, sensations and pleasures" (p. 151) to govern life, hence his neologism "biopolitics" (expressed without a hyphen). Foucault (1980) was adamant that the functioning body should not be effaced and argued for a biocultural understanding:

> What is needed is to make it visible through an analysis in which the biological and the historical are not consecutive to one another ... but are bound together in an increasingly complex fashion in accordance with the development of the mod-ern technologies of power that take life as their objective. Hence, I do not envi-sion a "history of mentalities" that would take account of bodies only through the manner in which they have been perceived and given meaning and value; but a "history of bodies" and the manner in which what is most material and most vital in them has been invested. (pp. 151–152)

Despite Foucault's materiality, his prime research problem, as Barad (2003) notes, was not in understanding how the biological and historical (or social) were bound together. Nevertheless, his vision in developing a historical under-standing of bodies with a specific emphasis on "what is most material and most vital" has become one of the key problems that has underpinned new materialist thinking.

In this light, rather than viewing new materialism as a clean break from post-structuralism, some view it as an offshoot (rhizomatically speaking). Bruno Latour, for example, understands actor-network theory as "an empirical ver-sion of poststructuralism" (Law, 2009, p. 145), whereas Stacy Alaimo and Susan Hekman (2008) note that the writers in their edited collection, *Material Femi-nisms*, are not rejecting poststructuralism but adding value to it by taking matter seriously:

> A central element of that attempt, however, is to build on rather than abandon the lessons learned in the linguistic turn. The new settlement we are seeking is not a return to modernism. Rather, it accomplishes what the postmoderns failed to do: a deconstruction of the material/discursive dichotomy that retains both elements without privileging either. The theorists assembled here have been

working to revise the paradigms of poststructuralism, postmodernism, and cultural studies in ways . . . to radically rethink materiality, the very "stuff" of bodies and natures. (p. 6)

Such a position reflects that poststructural tools that have served critical researchers well in the past do not have to be abandoned within new materialist thinking but necessarily require some reworking.

The second and related issue concerns the hint that the "new" is improved and better. I accept that *new* does not necessarily mean "improved" or "superior," yet given that it hints of these notions (e.g., advertisements often talk of "new and improved" products), it is worthy of comment. My concern with labeling a theoretical concept as "new" is that it appears to lead to misunderstandings of how such concepts develop. Deleuze and Guattari (1987) argue that all theoretical concepts are developed as responses to specific problems and that these concepts can only be fully understood in relation to their specific problem-position. They accept, therefore, that researchers working in different discursive/material contexts develop differing theoretical tools for differing problems. In this light, just as we would not bother comparing the value of a spanner with a hammer, it appears somewhat nonsensical to suggest that the tools of new materialism are an improvement from the "old" ones. Bruno Latour (2004) bluntly states, "There is no greater crime than to address with the equipment of an older period the challenges of the present one" (p. 231). There is value, however, in understanding the respective problems that differing theorists have grappled with in order to understand why differing theoretical concepts developed and how they have been useful. Indeed, the problems that new materialist thinkers aim to address have been comprehensively introduced within this coedited text.

Although I *heartily* acknowledge the promise of the ideas promoted within new materialisms while also accepting that the "new" *signals* a desire "to try to think differently" (St. Pierre et al., 2016, p. 100), given my concerns surrounding the idea of the new, for the remainder of this chapter, I will simply refer to new materialisms as a (re)turn to materialisms or, as Coole and Frost (2010) propose, "renewed materialisms" (p. 4). Alternatively, I sympathize with Rick Dolphijn and Iris Van der Tuin's (2012) suggestion that Baruch Spinoza, given his groundbreaking ideas from the 1600s, "might very well be considered the first (the foremost) new materialist" (p. 151).

HOW WILL THE SOCIOLOGY OF SPORT AND PHYSICAL CULTURE BE AFFECTED BY THE (RE)TURN TO MATERIALISM?

In this second section, I speculate on how a reinvigoration of materialist thinking will potentially affect the sociology of sport and physical culture fields, those assembled within these fields, and the social/material issues that critical researchers aim to examine and shape. In the first instance, readers facing the challenge of *meeting* the novel conceptualizations of Bruno Latour, Rosi Braidotti, Manuel DeLanda, Karen Barad, Gilles Deleuze, Félix Guattari, and the like will likely experience a range of responses, from "confusion, as well as anxiety" (Buchanan, 1997, p. 73), to delight and intrigue. For many, however, I envisage that there will also be an experience of affective resonance with the problems that materialist scholars engage with. As Sara Ahmed (2008) notes,

> I have had numerous conversations with friends, colleagues, participants at conferences, which have involved the use of quite casual forms of expression, which evoke a position that is not held by the speaker. One example of this is the expression, "I don't think everything is just social," which has been repeated to me several times. (p. 25)

Although Ahmed is critical of what this type of expression signals (i.e., a misreading of earlier feminist writings), it nevertheless reflects that numerous people are experiencing similar thoughts. I have also heard similar expressions in conferences and from work colleagues. At first, such comments were discussed in somewhat private and hushed tones, as if the commentator was aware that such thoughts could be met with derision and accusations of essentialism. I remember, as an example, sitting in the audience of a leisure studies conference in 2005 when a "biological" question was met with disparagement by the audience. The question followed a feminist presentation concerning the gendered use of leisure spaces in schools, which males were found to heavily dominate. The deliberately provocative question was "How much do you think hormones might explain your findings?" The presenter did not need to reply, as she was drowned by laughter: Under the dominance of social constructionist thinking, the biological question could only be conceived as a joke. This anecdote reveals, in part, how one's theoretical lens can obscure, as well as focus, various ways of thinking.

The renewal of materialist thinking will likely encourage sociologists of sport and physical culture to take biological questions more seriously. This poses several challenges. First, there is a need to reorient negative views toward biology.

Biology has undoubtedly been used to legitimate various inequalities, support discriminatory practices, and even justify the use of violence (Cox & Pringle, 2015; Vertinsky, 1990). In this respect, Holly Thorpe (2016) deemed the critique of biological essentialism as "politically necessary" (p. 2), yet she and other material feminists now suggest that it is timely to forge new understandings and relationships with biology. At the same time, contemporary scientific understandings of "biology" have changed significantly with talk of an "epigenetic revolution" (Waggoner & Uller, 2015, p. 177) and the concomitant shift away from determinism toward an understanding of biology as dynamically interactive within physical and social environments. Yet such observations reflect further challenges for sociocultural researchers, such as the difficulties of how to understand and critically interpret scientific knowledge and how such knowledge can be drawn on and assembled to produce "biocultural" understandings. As Simone Fullagar (2017) contends, the prime challenge of materialism is to develop research approaches that disturb dualisms such as biological/sociocultural and nature/nurture to produce knowledge about the biocultural body that challenges normative assumptions.

The importance of the examination of the body and the challenging of normalized thinking is something that has been promoted since the 1980s in sociocultural examinations of sport and movement (e.g., see Loy, 1991). Yet these studies, as Pirkko Markula details in chapter 1, primarily examined the human body as a discursive construction or as a representation of various identities. Thus rather than challenging dualistic notions of the mind and body or the social and biological, the examination of the body appeared somewhat ironically devoid of its materiality. The challenge of how to blur bodily matters with sociocultural concerns to produce knowledge of the biocultural remains.

Thorpe (2016) notes that to date there has been little agreement about how to overcome these very real challenges. Yet some researchers, including those in this text, offer valuable insights. Rather than ignoring biological knowledge, some have made inroads by examining how individuals make sense of scientific knowledge in relation to their experiences of sport and exercise (e.g., Jette, Vertinsky, & Ng, 2014; Thorpe, 2016). Such research, although anthropocentrically focused on meaning making, encourages new understandings of the relationships between biology and culture. In a similar vein, other researchers have gained critical understandings of how biological knowledge of the moving human body is produced in the "laboratory" (e.g., Gibson, 2015; Williams & Gibson, 2017). This important ethnographic research encourages forays into interdisciplinarity by aligning scientific views of movement with critical sociocultural concerns. A return to materialism will also encourage greater concern not just with biology but with other "material" sciences, such as geology, genetics, and geography, as the chapters in this text by Douglas Booth (chapter 11) and Kyle Bunds and Michael Giardina (chapter 12) reflect.

Yet the challenge of materialism is associated not simply with boundary crossings with science, as difficult as that may be, but with working with "new" ontologies that blur differing dualisms, such as material/social and quantitative/qualitative. Patti Lather (2016a), for example, draws from Karen Barad's concepts of relational entanglements to suggest that the "two-way traffic between the sciences and the humanities becomes thinkable as 'always already entangled,' interrupting the binary of science as either social constructivist or realist" (p. 126). I accept, accordingly, that it is the promotion of innovative understandings of agency, ontologies, and epistemologies that will pave inventive ways of doing (post)qualitative research to produce new knowledges.

At risk of oversimplifying the diverse materialists' ontologies, Nick Fox and Pam Alldred (2016) suggest that one of the underpinning ontological ideas of materialism is the focus on *relations* between various objects. Rather than assuming that an object exists as some form of prior essence, as naive realists might think, the focus is on how objects emerge through interactions with each other. Deleuze and Guattari's (1987) concept of assemblage is reflective of this idea. The assemblage refers to the connections, organizations, and play of forces among "bodies, . . . actions and passions, [or] an intermingling of bodies reacting to one another . . . [and] of acts and statements" (p. 88). The focus, correspondingly, is directed not to a body or a preexisting notion such as an assumed identity but to the "assemblages that coalesce to produce both events and the apparent reality of the relations they comprise" (Alldred & Fox, 2015, p. 908). Actor-network theory (see Simon Darnell, chapter 10) similarly treats "everything in the social and natural worlds as a continuously generated effect of the webs of relations within which they are located." As such, this approach "assumes that nothing has reality or form outside the enactment of those relations" (Law, 2009, p. 141), whereas Barad (2007) draws from quantum mechanics to similarly suggest that the "particle" does not simply exist in any fixed state but only becomes identifiable via the interactive attempts of scientists to measure the particle. Barad posits that the key elements in doing science (e.g., the object or particle under study, the measuring tools, and the researcher) "emerge from the experimental entanglement and do not pre-exist them" (Hollin, Forsyth, Giraud, & Potts, 2017, p. 930). It is in this light that Barad's concept of onto-epistemology makes sense, as ways of knowing are entangled with matter via the interactions of researchers so that the division between the "empirical (about knowledge) and ontology (about being)" breaks down (St. Pierre et al., 2016, p. 99).

These innovative onto-epistemological frameworks are already producing changes in how research is being "done" in the critical fields of sport and physical culture—changes that broadly relate to methodological approaches and the ethical study of divergent topics. I speculate that these changes are likely to multiply. I comment, in turn, on each in the following.

Methodological Changes

An onto-epistemological framework recognizes that the researcher emerges in the act of doing the research. The researcher, as such, is an entangled aspect of the method, and the research act can be recognized as material practice. Michael Giardina and Joshua Newman (2011), without select reference to the ontological turn, similarly promote the importance of recognizing that "the researcher's embodied self . . . produces the very cultural physicalities he or she experiences" (p. 530). In this manner, they argue that "we need to privilege bodily copresence within the theory, method, and practices of physical cultural studies" (p. 523).

The onto-epistemological turn, however, goes beyond privileging the *copresence* of the body within research by accepting that the body (of the researcher) emerges and is entangled with other bodies (including bodies of knowledge) in the research act. Simone Fullagar, in chapter 7 of this text, adopts such a perspective by accepting that the "research process is itself an assemblage of forces, relations and practices" and is therefore "concerned with how embodied matter contributes to ways of knowing." Her novel representational style of the "results" highlights the material entanglements in the production of depression and recovery.

The boundaries between the researcher, the research objects/subjects, affective flows, power, and matter (supposedly) dissolve within this onto-epistemological shift. Yet how the research act can achieve such a fully "dissolved" state is still in the process of playful development. Indeed, within this early stage of experimenting with method, I am reminded that "what the 'new' ushers in . . . is a re-imagining of what method might *do*, rather [than] what it *is* or *how to do it*" (St. Pierre et al., 2016, p. 105).

Nevertheless, differing writers have differing ideas on this issue of methodological experimentation. Nick Fox and Pam Alldred (2016), for example, have advocated for the advantages of using a "defined" method and have drawn from Deleuze and Guattari's notion of assemblage to produce details of how to design, collect/analyze data, and write up the research via the materialist concept of a research assemblage. Such an approach focuses on the affective flows within the research assemblage, whereas Elizabeth St. Pierre and colleagues (2016) adopt a more Deleuzean philosophical approach—that "thought does not need a method" (p. 105)—and question the need for a prearranged method: "We wonder whether conventional, pre-given methodologies can be useful in this new work, given that it calls for experimentation and ethical experimentation at that" (p. 106). I admire their desire to experiment and sidestep methodological recipes but speculate, given that many of us work within neoliberal research machines (i.e., universities), that the requirement to publish will necessitate the need for "legitimized" materialist methods. In the interim, this research space is ripe for experimentation (e.g., see Cathy van Ingen's 2016 paper "Getting Lost as a Way of Knowing") and promotion of innovative methods (e.g., see, within this text, Christopher M. McLeod and Matthew G. Hawzen's entanglements with object-oriented ontology [chapter 3] and Gavin Weedon's diffractive historiographical approach [chapter 4]).

One methodological change that might transpire given the onto-epistemological turn is a rethinking of quantitative research. Under the recent internal-idealist ontological regime, quantitative research was linked with positivism and was typically dismissed within critical approaches of sport and physical culture as reflective of hegemonic reductionist scientism. The excitement of the ontological turn, according to Patti Lather (2016b), is the "opening up of measurement, assessment, and quantitative methods to (post)critical inquiry" (p. 502). This "opening up" includes querying the potential of big data and digital analytics, as Shannon Jette, Katelyn Esmonde, David Andrews, and Carolyn Pluim explore in chapter 5 of this text. Lather adds, however, that a shift toward quantitative approaches will require "efforts to deepen student training in both qualitative and quantitative methodology in more philosophical directions regarding what it means to think in terms of new knowledge formations that refuse the humanities–science split" (p. 504). No doubt such efforts to "deepen student training" may also require the (re)training of the teachers, not least their values toward quantitative approaches. Nevertheless, given that legislators and policy makers tend to respond to the materiality of numbers, an opening up to quantitative methods may be one way that critical sport researchers can make more of a difference (Pringle & Falcous, 2016).

Foci of Study/Ethics

As already evidenced in this book, and therefore much easier to speculate about, research topics on diverse forms of matter will gain greater attention. Matters examined in this book, for example, include protein and waste (Samantha King), water (Kyle Bunds and Michael Giardina), the beach (Douglas Booth), pollution (Mary McDonald and Jennifer Sterling), scientific technologies (Shannon Jette, Katelyn Esmonde, and David Andrews), and the biocultural body (Pirrko Markula and Marianne Clark). This broader focus stems from the rejection of the notion that "humans act/behave and the rest of the world (the environment) is acted/behaved upon" (McGregor, 2014, p. 220). This more modest appraisal, as Douglas Booth reflects in chapter 11 of this text, offers "a sense of optimism that an ontology that directs attention to the essential and fragile interaction of human and nonhuman matters might cultivate a new ethics of care for the environment."

The need for a new ethics has indeed been suggested by Elizabeth St. Pierre, Alecia Jackson, and Lisa Mazzei (2016) as one of the key driving forces that has underpinned the recent ontological turn, given the increased acceptance of the "slow, creeping human impact on and destruction of our planet" (p. 101). Concomitantly, this onto-epistemological turn may be more effectively known as the *ethico-onto-epistemological* turn, "which makes it clear that how we conceive the relation of knowledge and being is a profoundly ethical issue, as is the relation between the human and the nonhuman" (St. Pierre et al., 2016, p. 99). This forefronting of ethics should be regarded as the embedding of political concerns within materialist approaches. Hence the critique "that new materialism ignores the unique specificity of human agency and the transformatory capabilities of our species" can be regarded as a misreading of "the emancipatory potential of posthumanist thinking" (Cudworth & Hobden, 2015, p. 134).

Will a Return to Materialisms Make a Critical Difference?

The prime critical issue is whether the ethical/political matters addressed in materialist research can achieve the transformative impact desired. This is a particularly relevant question given the recent concerns raised about the effectiveness of critical research. Edwards and Fenwick (2015), for example, note that critical researchers have "more or less successfully avoided changing the existing reproductions of power and inequalities despite . . . [their] critiques" (p. 1385). In a similar manner, Latour (2004) laments that in such troubling times, critical research appears incapable of being relevant:

> Wars. So many wars. Wars outside and inside. Cultural wars, science wars, and wars against terrorism. Wars against poverty and wars against the poor. Wars against ignorance and wars out of ignorance. My question is simple: Should we be at war, too, we, the scholars, the intellectuals? Is it really our duty to add fresh ruins to fields of runs? Is it really the task of the humanities to add deconstruction to destruction? . . . What has become of the critical spirit? Has it run out of steam? (p. 225)

Latour's concern with the appearance of critique "running out of steam" rests, in part, on what he views as the somewhat repetitive attempt of critical researchers to reveal the "social construction of scientific facts" (p. 227). Yet deconstruction, as he says, is not enough. His concerns, more specifically, rest on how the assumption of a (false) division between scientific facts and constructionism has driven critique away from matters of concern. Latour nevertheless also offers hope for a new path forward for critical research by arguing that if the critical mind, if it is to renew itself and be relevant again, is to be found in the cultivation of a *stubbornly realist attitude* . . . a realism dealing with what I will call *matters of concern*, not *matters of fact*. (p. 231)

The realist attitude Latour calls for is not the realism of the scientific approach but the need for empirical research focused via a relational ontology (as promoted in the material/semiotic tools that this book presents as new materialism). The hope of this ontological turn is not for a generalized solution to broad social problems (given that materialist research focuses on the specific) but a hope that a focus on "things and how they are assembled" can generate "more ideas than we have received" (Latour, 2004, p. 248), that looking at things afresh will produce new understandings and ways of doing, thinking, and being—in short, new realities.

There are, of course, no guarantees of transformation through adopting materialism, and there are, as expected, existing critiques concerned with notions of openness, uncertainty, experimentation, the ephemeral nature of studying seemingly fleeting assemblages, and antihumanism—which "is obnoxious to those who take people to be morally special" (Law, 2009, p. 147). Yet after having drawn from Latour (2004), it seems futile to engage in a countercritique as if "there is a single intellectual and political space to be won" (Law, 2009, p. 150). So, to conclude, I am hopeful that the material webs that have been woven in this book can link, or perhaps interfere, with other critical networks to produce a difference where it matters most.

NOTES

1. Joshua Newman, Holly Thorpe, and David Andrews, in their introductory chapter, similar reflect a debt of gratitude to feminist scholars for their formative efforts in locating the body as "radically embodied, fleshed, and materially consequential."

2. Of interest, a search via Google Scholar reveals an article published by James Pratt (1922) titled "The New Materialism." The article examines a resurgence of interest in materialism (20 years after many apparently believed that this theoretical approach was dead) and examines connections between the brain (matter) and consciousness with discussions linking to critical and naive realism, idealism, energy, and kinetics.

REFERENCES

Adams, M. L., Helstein, M. T., Kim, K. Y., McDonald, M. G., Davidson, J., Jamieson, K. M., . . . Rail, G. (2016). Feminist cultural studies: Uncertainties and possibilities. *Sociology of Sport Journal*, 33(1), 75–91.

Ahmed, S. (2008). Open forum imaginary prohibitions: Some preliminary remarks on the founding gestures of the new materialism. *European Journal of Women's Studies*, 15(1), 23–39.

Alaimo, S., & Hekman, S. J. (eds.). (2008). *Material feminisms*. Bloomington: Indiana University Press.

Alldred, P., & Fox, N. J. (2015). The sexuality-assemblages of young men: A new materialist analysis. *Sexualities, 18*(8), 905–920.

Barad, K. (2003). Posthumanist performativity: Toward an understanding of how matter comes to matter. *Signs: Journal of Women in Culture and Society, 28*(3), 801–831.

Barad, K. (2007). *Meeting the universe halfway: Quantum physics and the entanglement of matter and meaning*. Durham, NC: Duke University Press.

Buchanan, I. (1997). The problem of the body in Deleuze and Guattari, or, what can a body do? *Body and Society, 3*(3), 73–91.

Colebrook, C. (2011). Matter without bodies. *Derrida Today, 4*(1), 1–20.

Giardina, M. D., & Newman, J. I. (2011). Physical cultural studies and embodied research acts. *Cultural Studies ↔ Critical Methodologies, 11*(6), 523–534.

Coole, D., & Frost, S. (2010). Introducing the new materialisms. In D. Coole & S. Frost (eds.), *New materialisms: Ontology, agency, and politics* (pp. 1–43). Durham, NC: Duke University Press.

Cox, B. D., & Pringle, R. (2015). "Muscles for motherhood": A genealogical analysis of medicalized ways of knowing female footballers in New Zealand, 1921 and 1973–1975. *International Journal of the History of Sport, 32*(18), 2135–2149.

Cudworth, E., & Hobden, S. (2015). Liberation for straw dogs? Old materialism, new materialism, and the challenge of an emancipatory posthumanism. *Globalizations, 12*(1), 134–148.

Deleuze, G., & Guattari, F. (1977). *Anti-Oedipus: Capitalism and schizophrenia* (R. Hurley, M. Seem, & H. R. Lane, trans.). Minneapolis: University of Minnesota Press.

Deleuze, G., & Guattari, F. (1987). *A thousand plateaus: Capitalism and schizophrenia*. Minneapolis: University of Minnesota Press.

Dolphijn, R., & Van der Tuin, I. (2012). *New materialism: Interviews & cartographies*. Ann Arbor, MI: Open Humanities Press.

Edwards, R., & Fenwick, T. (2015). Critique and politics: A sociomaterialist intervention. *Educational Philosophy and Theory, 47*(13–14), 1385–1404.

Foucault, M. (1980). *The history of sexuality, volume I: An introduction* (R. Hurley, trans.). New York, NY: Vintage.

Fox, N. J., & Alldred, P. (2016). *Sociology and the new materialism: Theory, research, action*. London, England: Sage.

Fullagar, S. (2017). Post-qualitative inquiry and the new materialist turn: Implications for sport, health and physical culture research. *Qualitative Research in Sport, Exercise and Health, 9*(2), 247–257.

Giardina, M. D., & Newman, J. I. (2011). What is this "physical" in physical cultural studies? *Sociology of Sport Journal, 28*(1), 36–63.

Gibson, K. M. (2015). *A witness to fitness: Contextualising processes in exercise physiology* (Unpublished doctoral dissertation). University of Toronto, Canada.

Guttorm, H., Hohti, R., & Paakkari, A. (2015). "Do the next thing": An interview with Elizabeth Adams St. Pierre on post-qualitative methodology. *Reconceptualizing Educational Research Methodology, 6*(1).

Hollin, G., Forsyth, I., Giraud, E., & Potts, T. (2017). (Dis)entangling Barad: Materialisms and ethics. *Social Studies of Science, 47*(6), 918–941.

Jette, S., Vertinsky, P., & Ng, C. (2014). Balance and biomedicine: How Chinese Canadian women negotiate pregnancy-related "risk" and lifestyle directives. *Health, Risk & Society, 16*(6), 494–511.

Lather, P. (2016a). Top ten+ list: (Re)thinking ontology in (post)qualitative research. *Cultural Studies ↔ Critical Methodologies, 16*(2), 125–131.

Lather, P. (2016b). Post-face: Cultural studies of numeracy. *Cultural Studies ↔ Critical Methodologies, 16*(5), 502–505.

Latour, B. (2004). Why has critique run out of steam? From matters of fact to matters of concern. *Critical Inquiry, 30*(2), 225–248.

Law, J. (2009). Actor network theory and material semiotics. In B. Turner (ed.), *The new Blackwell companion to social theory* (pp. 141–158). Hoboken, NJ: Blackwell.

Loy, J. W. (1991). Introduction—missing in action: The case of the absent body. *Quest, 43*(2), 119–122.

Lyngdoh, S. S. (2017). Textualized body, embodied text: Derrida's linguistic materialism. *Journal of Indian Council of Philosophical Research, 35*(1), 1–14.

McGregor, C. (2014). From social movement learning to sociomaterial movement learning? Addressing the possibilities and limits of new materialism. *Studies in the Education of Adults, 46*(2), 211–227.

Pratt, J. (1922). The new materialism. *Journal of Philosophy, 19*(13), 337–351.

Pringle, R., & Falcous, M. (2016). Transformative research and epistemological hierarchies: Ruminating on how the sociology of the sport field could make more of a difference. *International Review for the Sociology of Sport.* Advance online publication. doi:10.1177/1012690216654297.

St. Pierre, E. A., Jackson, A. Y., & Mazzei, L. A. (2016). New empiricisms and new materialisms: Conditions for new inquiry. *Cultural Studies ↔ Critical Methodologies, 16*(2), 99–110.

Thorpe, H. (2016). Athletic women's experiences of amenorrhea: Biomedical technologies, somatic ethics and embodied subjectivities. *Sociology of Sport Journal, 33*(1), 1–13.

van Ingen, C. (2016). Getting lost as a way of knowing: The art of boxing within Shape Your Life. *Qualitative Research in Sport, Exercise and Health, 8*(5), 472–486.

Vertinsky, P. A. (1990). *The eternally wounded woman: Women, doctors, and exercise in the late nineteenth century.* Manchester, England: Manchester University Press.

Waggoner, M. R., & Uller, T. (2015). Epigenetic determinism in science and society. *New Genetics and Society, 34*(2), 177–195.

Williams, O., & Gibson, K. (2017). Exercise as a poisoned elixir: Inactivity, inequality and intervention. *Qualitative Research in Sport, Exercise and Health,* 1–17.

AFTERWORD

GRANT FARRED

The work of the new is to challenge. It is to confront us. It is to do more than make us take up an unprecedented series of difficulties or to posit a heretofore unthought range of possibilities. Both of these engagements are, needless to say, in and of themselves important. However, what matters most is that the work of the new explicates why what is designated as such is of consequence. What is "new" about the "new?" How is it that the new undoes all, or at least some, of what has gone before? How does the new threaten what obtains epistemologically—and obtains with stubbornness, inertia, resilience, and a sense of intellectual right (we could as easily name it "entitlement," so hard won is its status and so determinedly is that status protected)—in the seemingly insurmountable now? Why is what the new does impatient with, if not intolerant of, the old order, why will no order endure this (soon-to-be) previous mode of doing things? Is it merely subject to the logic that "everything has its season"? That "all things must come to an end"?

Possibly. However, what the new—that which is genuinely innovative and paradigm-shifting—does not seek to do (nor should it) is upend things such as the state of a field or the standing of a discipline simply for its own sake. If the new is to take the place of the old, to stand in its stead (replace it), then it is incumbent upon the new to do the work of, to invoke an awkward phrase (a phrase that itself betrays a certain—necessary—anxiety, but an anxiety that is not without its epistemological advantages), "self-legitimation." That is, the new must make the case for itself. It must argue for itself; it must explain—as fully as possible—why its methods, its insights, its modes of inquiry, and the archive it is generating are (alone, singularly, dare one say?) capable of completely changing the order of things. What is more, it must prove that what the new is doing is being done for the better. That is, the new order of things reveals what was constitutively lacking in its predecessor, and it, the new, as such not only addresses those lacks and deficiencies; it offers, at the very least, a more incisive way of thinking what its predecessor does not or cannot. It might not provide the "solution" as such, but it raises the possibility that as a mode of inquiry, such an outcome might result.

A difficulty arises here in no small measure because of the ways in which the cultural studies project is at once indebted to and haunted by, they might very well amount to the same thing; but, maybe not (so the distinction functions under the sign of erasure), the work of a founding figure such as Raymond Williams. Because of the effects (legacies, inheritances, and so on) of how Williams figures transitions between epochs (modes of thinking), it has become nigh on impossible to envisage a historical or disciplinary moment in which the disarticulation of the old by the new was an absolute break. Williams (1978), in his (later) critique of ideology (articulated in *Marxism and Literature*), provides the categories "dominant, residual and emergent" in order to think about "the genealogy of historical moments." (Such a naming of the genealogical, of course, draws on the way that Michel Foucault [1999] delineates it in his lectures, *Society Must Be Defended*. For Foucault, genealogy arises out of a combination of a resisting body knowledge, a knowledge most likely suppressed by the dominant order, and "local activism"—that is, as a politics aimed at undoing the order that it is opposing and the thinking that informs and shapes this politics.) In his work on ideology, Williams argues that the ways of the new are grounded in, arise out of, and bear traces of the old; as such, Williams explicates—maps, we might prefer—the pathway and the trajectory by which that which is now "dominant" came to ascendance.

We can thus safely assume that the temporality of the "dominant" must, *a priori*, be constitutively understood as, to some extent or another, autoimmune. That is, the "emergent" is always *a priori* detectable in the extant dominant. In his pharmacological metaphor (a discourse deeply rooted in the human body and especially, according to some critics, indebted to the emergence of the AIDS crisis in the 1980s and '90s), Jacques Derrida, in his 1983 essay "Plato's Pharmacy," outlines the "anatomy" of autoimmunity. Every "dominant" contains within itself the very conditions for its own undoing. Every dominant is composed of, as it were, nothing less than the very "substance" of its own death as long as we recognize, of course, the ways in which the emergent to come—the emergent *l'avenir*—contains within itself irrefutable (whether or not it is repressed, denied, and ignored) manifestations of the dominant. Every dominant succeeds itself, in part, as part of what is *l'avenir*. Every residual bears testament to what was. As such, every dominant is constituted, perforce, by much more than itself.

Marx, another formidable presence in the cultural studies project, is in critical moments preoccupied with the politics of autoimmunity. Marx is intrigued, to phrase the matter poorly, with the promise of (self-)undoing that inheres in every political. It is a promise Marx (with Engels) makes most famously, of course, very early in *The Communist Manifesto*. In a distinctly Shakespearean turn of phrase, Marx and Engels (2010) conjecture, with some glee, that "what the bourgeoisie, therefore, produces, above all, is its own grave-diggers. Its fall and the victory of the proletariat are equally inevitable" (p. 17). In a language redolent with the tragedy that marks the final scenes in *Hamlet*, Marx and Engels assure us that there is nothing for the bourgeoisie to do but undo itself and for the "proletariat" to emerge triumphant—all in good time, of course, because the "grave-diggers" operate on their own timetable. Not ours, regrettably.

The German Ideology bears traces of an autoimmune strain. In addition, *The German Ideology* might rank as the most acerbic (the Young Hegelians are put to the sword, delightfully) and funny in Marx's oeuvre. (One would be hard pressed not to chuckle at the moniker "Saint Max" that Marx and Engels append to their erstwhile Young Hegelian colleague Max Stirner; perhaps nothing stirs the inclination to derision as much as the refutation of the mimetic—that is, when two former Young Hegelians take their own to task.) In *The German Ideology* too, there is the promise of self-undoing; the threat of an undoing that emerges from within seems to lurk everywhere, especially in parts 1 and 3.

What Williams offers with his tripartite ideological structure is not so much a different iteration of autoimmunity à la Marx and Derrida but a certain "linearity" that, it must be said, has about it something of the teleological: that sense that history—indeed, time itself—must be understood chronologically, that one moment "leads," as such, to the next. However valuable the discreteness of Williams's categories are, what they signal is the impossibility of a complete or "radical" rupture. That is, there can be no absolute attenuation of one "moment" or mode and its displacement or usurpation by a historical force that is entirely distinct, or "new."

Williams is much more interested in what persists, in how one historical modality or articulation mutates (sometimes surprisingly, sometimes predictably) into something different. And yet despite its difference, what persists—those "indestructible residues," those ineluctable traces—remains strangely, ominously, hauntingly, prospectively familiar. As such, the dominant ideology can legitimately proclaim itself to be in command; however, while it is eminently recognizable as such, it all the while reveals visible or audible iterations of that out of which it emerged, of that which it has superseded, of that which it is supposed to have put to rest. In what the dominant recognizes as constitutive, the dominant is given a prescient glimpse of what it will or might be. It sees its future in itself.

This collection does a fine job of evincing a fidelity to Raymond Williams. It is determined to argue—inventively, occasionally garnished by a kind of intellectual joie de vivre and breadth of inquiry—for what exactly is "new" in this beast known as "new materialisms." However, it never does so at the expense of what it understands itself to be thinking against. That is, this project undertakes the difficult task of explicating its status—of establishing the grounds for its coming into being—while simultaneously (as though all materialisms inflect, infect, overwrite each other) attending to those forces that "persist." Let us name them, these forces, as resilient in addition to being residual.

It is never, then, what we might name a "fashionable" matter, this new materialism. It is never simply a matter of "out with the old and in with the new." It is, rather, a genealogy of the new. One can only imagine what might emerge out of this "new." We can only speculate as to what its residues will be, as to how the traces of what is articulated in this collection will manifest themselves in the materialisms *l'avenir*. There is every reason, on the basis of this collection, to be excited about exactly such a prospect. As such, this collection constitutes something on the order of an epistemological gift to future materialisms, or, perhaps more urgently, to extant materialisms in need of critical review. As such, this collection belongs to the future even—or because, it may be more felicitous to argue—as it is indelibly marked by the present. And in the spirit of Raymond Williams, it is insistently reflective about—but not nostalgic for or dismissive of—residual materialisms.

REFERENCES

Foucault, M. (1999). *"Society must be defended": Lectures at the Collège de France, 1975–1976* (D. Macey, trans.). New York, NY: Picador.
Marx, K., & Engels, F. (2010). *The communist manifesto*. Hollywood, FL: Simon & Brown.
Williams, R. (1978). *Marxism and literature*. Oxford, England: Oxford University Press.

NOTES ON CONTRIBUTORS

MARY LOUISE ADAMS is a professor in the School of Kinesiology and Health Studies and the Department of Sociology at Queen's University, where she teaches courses on sport and culture, the history and sociology of fitness, contemporary issues in sexuality, and the politics of knowledge. She is the author of *Artistic Impressions: Figure Skating, Masculinity and the Limits of Sport* (2011) and *The Trouble with Normal: Postwar Youth and the Making of Heterosexuality* (1997). She has written on issues related to the history of sexuality, queer and feminist social movements, the politics of HIV, and cultural studies of sport and movement.

DAVID L. ANDREWS is a professor of physical cultural studies in the Department of Kinesiology at the University of Maryland, College Park. His research contextualizes sport and physical culture in relation to the intersecting cultural, political, economic, and technological forces shaping contemporary society. His previous books include *Sport-Commerce-Culture: Essays on Sport in Late Capitalist America* (2006); *Sport and Neoliberalism: Politics, Consumption, and Culture* (edited with Michael Silk, 2012); and *The Routledge Handbook of Physical Cultural Studies* (edited with Michael Silk and Holly Thorpe, 2017). He is presently completing a book titled *Making Sport Great Again: The Uber-Sport Assemblage, Neoliberalism, and the Trump Articulation.*

KIRI BAXTER is a PhD candidate in sociology at Goldsmiths' College, University of London. Baxter has a master's in sport, culture, and society from the University of Brighton and is the winner of an academic excellence scholarship. She currently works in education in the United Kingdom and has worked in the sports industry for over 10 years, gaining national recognition for her work in sports development. Baxter has expertise in project management and monitoring and evaluation. Her career and academic background have primarily focused on working with a range of minority groups in physical activity intervention strategies. Her research now specifically concentrates on feminist epistemology and ontology, feminist notions of the body, the female embodied experience of sport, and feminist methodologies.

DOUGLAS BOOTH is a professor of sport studies in the School of Physical Education, Sport and Exercise Sciences at the University of Otago (New Zealand). He served as dean of the school for 10 years and is the author of *The Race*

Game (1998), *Australian Beach Cultures* (2001), and *The Field* (2005). Booth currently serves on the editorial boards of *Rethinking History* and the *Journal of Sport History* and is an executive member of the Australian Society for Sport History.

KYLE S. BUNDS is an assistant professor of sport and sustainable community development in the College of Natural Resources at North Carolina State University. His research broadly focuses on the interaction of sport and the environment. Bunds's research has been published in journals such as *Transportation Part D: Transport and Environment, Sociology of Sport Journal, Sport Management Review, Journal of Sport Management, Critical Studies in Media Communication, European Sport Management Quarterly, Sport in Society,* and *Communication, Culture, & Critique.*

JACOB J. BUSTAD is an assistant professor in the Department of Kinesiology in the College of Health Professions at Towson University. Prior to coming to Towson, he completed his PhD in kinesiology at the University of Maryland, College Park. He received his BA in mass communications and English from Buena Vista University in Storm Lake, Iowa, and his MA in American studies from the University of Kansas.

MARIANNE CLARK is a postdoctoral research fellow at Waikato University. Her research focuses on girls' and women's experiences of health and their moving bodies in diverse social, material, and discursive spaces. She is the coeditor of *The Evolving Feminine Ballet Body* (2018) and draws on feminist, poststructuralist, and new materialist perspectives as well as her own background in dance to explore innovative ways of understanding the moving body. Recently, her research interests have expanded to include the complex relationships between moving bodies and new health technologies and the sociomaterial implications of these relations for women's daily lives. Her research crosses the boundaries of health psychology, sport sociology, and critical digital health studies, and she has published more than 25 journal articles across a range of disciplines.

SIMON C. DARNELL is an assistant professor in the Faculty of Kinesiology and Physical Education at the University of Toronto. His research focuses on the social and political relationships between sport and international development and peace-building efforts, the development implications of sports megaevents, and the place of social activism in the cultures of elite sport. He is coeditor of *Routledge Handbook of Sport for Development and Peace* and coauthor of *The History and Politics of Sport for Development: Activists, Ideologues and Reformers,* both forthcoming in 2019. His research has also been published in a range of academic journals, including *Sociology of Sport Journal, International Journal of Sport Policy and Politics, Social Inclusion,* and *Compare: A Journal of Comparative and International Education.*

KATELYN ESMONDE is a doctoral candidate in physical cultural studies in the Department of Kinesiology at the University of Maryland, College Park. Her research focuses on gender, feminism, science and technology studies, theories of physical culture, and qualitative methods. She is currently researching the quantified self movement across physical culture.

GRANT FARRED is the author of a trilogy of works on sport and the event: *The Burden of Over-representation: Race, Sport and Philosophy* (2018), *In Motion, at Rest: The Event of the Athletic Body* (2014), and *Entre nous: Between the World (Cup) and Me* (forthcoming, 2019). His other works include *Martin Heidegger Saved My Life* (2015) and *What's My Name? Black Vernacular Intellectuals* (2004).

SAMANTHA FROST is a professor at the University of Illinois and currently the IPRH-Mellon Faculty Fellow in the Biohumanities, directing the Biohumanities Initiative that is part of an Emerging Areas in the Humanities Grant from the Andrew W. Mellon Foundation and hosted by the Illinois Program for Research in the Humanities. Her research focuses on the ways that our understanding of matter, materiality, or embodiment shapes our concepts of politics. Frost coedited, with Diana Coole, a collection of essays by contemporary theorists called *New Materialisms: Ontology, Agency, and Politics* (2010). The essays in this volume explore in a variety of ways how a focus on materiality shifts our sense of the terrain and stakes of politics. More recently, she authored *Biocultural Creatures: Toward A New Theory of the Human* (2016). While *Biocultural Creatures* elucidates the conceptual significance of the plasticity of the biological body, Frost's latest project elaborates the political possibilities made visible by acknowledgment of that plasticity. More specifically, Frost is examining research on epigenetic processes to articulate the ways that material, social, and symbolic environments shape our development and our capacities to engage and shape the worlds in which we live.

SIMONE FULLAGAR is professor and chair of the Physical Culture, Sport and Health research group at the University of Bath and adjunct professor at Griffith University, Australia. Fullagar is an interdisciplinary sociologist who has published widely using feminist poststructuralist and new materialist perspectives to critically explore active living policy, women's depression and recovery, and sport and alternative physical cultures (from cycle tourism, parkrun, and roller derby to dance).

MICHAEL D. GIARDINA is professor of media, politics, and physical culture and associate chair of the Department of Sport Management at Florida State University. He is the author or editor of more than 20 books, including *Sport, Spectacle, and NASCAR Nation: Consumption and the Cultural Politics of Neoliberalism* (with Joshua Newman, 2011), *Qualitative Inquiry at a Crossroads* (with

Norman K. Denzin, 2019), and *Physical Culture, Ethnography, and the Body: Theory, Method, and Praxis* (with Michele K. Donnelly, 2017). He is editor of the *Sociology of Sport Journal*, special issues editor of *Cultural Studies ↔ Critical Methodologies*, coeditor of three book series with Routledge, and the assistant director of the International Congress of Qualitative Inquiry.

MATTHEW G. HAWZEN is an assistant professor of sport sociology, management, and human performance in the sports administration program at Fairleigh Dickinson University in New Jersey. After earning his master's degree in kinesiology and health studies from Queen's University, Hawzen received his PhD in sport management from Florida State University in 2018 working with Dr. Joshua I. Newman as his supervisor. Hawzen's research and teaching draw on critical theories, qualitative methods, and cultural studies methodologies to examine cultural politics, biopolitics, and political economies of sport, physical activity, and the body in the United States. In addition to having presented his work at international academic conferences, Hawzen has published book chapters in *Sport and the Neoliberal University* (Rutgers University Press, 2018), *Routledge Handbook of Global Sports*, and *The Business and Culture of Sports* and journal articles in the *Sociology of Sport Journal, Communication and Sport, Transfers, Sport Management Review*, and *Journal of Sport & Social Issues*.

SHANNON JETTE is an associate professor in the Department of Kinesiology (School of Public Health) at the University of Maryland. Her research focuses on social, cultural, and historical aspects of knowledge production in the disciplines of kinesiology, medicine, and public health. She has published in such journals as *Sociology of Health & Illness* and *Health, Risk & Society*.

SAMANTHA KING is professor of kinesiology and health studies and gender studies at Queen's University. Her work has appeared in *Social Text, Ethnic and Racial Studies, Sociology of Sport Journal, Health Communication*, and the *International Journal of Drug Policy*, among other venues. Her book *Pink Ribbons, Inc.: Breast Cancer and the Politics of Philanthropy* (2006) is the subject of a National Film Board documentary by the same name. She is coeditor of *Messy Eating: Conversations on Animals as Food* (2019).

MARY G. MCDONALD is the Homer C. Rice Chair in Sports and Society in the School of History and Sociology at the Georgia Institute of Technology. A past president of the North American Society for the Sociology of Sport, McDonald has published numerous articles and book chapters and coedited (with Susan Birrell) *Reading Sport: Critical Essays on Power and Representation* (2000), which earned a Choice award as a top academic title. Her research focuses on American culture and sport, including issues of inequality as related to gender, race, class, and sexuality. She has edited special issues of the *Sociology of Sport*

Journal devoted to "(Post)identity and Sport" and "Whiteness and Sport." As the Homer C. Rice Chair, she directs the Ivan Allen College of Liberal Arts initiative in Sports, Society, and Technology.

CHRISTOPHER M. MCLEOD is an assistant professor of sport management in the Department of Kinesiology and Sport Management at Texas Tech University. McLeod received his PhD and MS in sport management from Florida State University and received his BS in physical education from the University of Otago. He studies sport markets, labor, and environmental sustainability using theories from sociology, economics, management, and science and technology studies. His philosophy is informed by actor-network theory and object-oriented ontology. McLeod has published 10 articles in journals such as *Sociology of Sport Journal*, *Sport Management Review*, *Journal of Sport & Social Issues*, and *Body and Society*. He is currently collaborating with Morethanbaseball, an organization dedicated to advocating for minor league baseball players.

PIRKKO MARKULA is a professor of sociocultural studies of physical activity at the University of Alberta, Canada. Her research interests include social analyses of dance, exercise, and sport, in which she has employed several theoretical lenses ranging from critical, cultural studies research to Foucault and Deleuze. She is the co-author of *Qualitative Research for Physical Culture* (with Michael Silk, 2011) and *Foucault, Sport and Exercise: Power, Knowledge and Transforming the Self* (with Richard Pringle, 2006); editor of *Feminist Sport Studies: Sharing Joy, Sharing Pain* (2005) and *Olympic Women and the Media: International Perspectives* (2009); and coeditor of *The Evolving Feminine Ballet Body* (with Marianne Clark, 2018), *Endurance Running: A Socio-Cultural Examination* (with William Bridel and Jim Denison, 2016), *Women and Exercise: Body, Health and Consumerism* (with Eileen Kennedy, 2011), *Critical Bodies: Representations, Identities and Practices of Weight and Body Management* (with Sarah Riley, Maree Burns, Hannah Frith, and Sally Wiggins, 2007), and *Moving Writing: Crafting Movement in Sport Research* (with Jim Denison, 2003).

JOSHUA I. NEWMAN (PhD, University of Maryland) is director of the Center for Sport, Health, and Equitable Development and professor of sport, media, and cultural studies at Florida State University. He has published two books and more than 90 articles and chapters on issues related to social inequalities, cultural politics, and political economics and ecologies of sport and physical activity. His book *Sport, Spectacle, and NASCAR Nation* (with M. Giardina, 2011) was awarded the North American Society for the Sociology of Sport (NASSS) Outstanding Book for 2012 and was named as a Choice Outstanding Academic Title in 2013. His work has been published in top international journals, such as the *Sociology of Sport Journal*, *Journal of Sport Management*, *Body and Society, Qualitative Inquiry,* and the *Journal of Sport & Social Issues.* He is a NASSS Research Fellow and serves on the editorial boards of *Communication & Sport,* the *International Review for the Sociology of Sport, Sociology of Sport Journal, Qualitative Research in Sport, Exercise and Health,* and *Journal of Global Sport Management.*

CAROLYN PLUIM (formerly Vander Schee) is associate professor at Northern Illinois University. Her research interests are focused around the intersections of sociology of education, curriculum studies, and educational policy, specifically as these relate to school health policies, practices, and pedagogies. In her work, she explores the ways in which contemporary school health policies are negotiated and experienced by students and school personnel. A central theme running throughout her research is the relationship between discourse and social dynamics as it bears on sociological understandings of health, illness, and the body and influences the responsibilities and obligations of public schools. She is the coauthor, with Michael Gard, of *Schools and Public Health: Past, Present, Future* (2014).

RICHARD PRINGLE is professor of the sociology of sport and physical education at Monash University, Melbourne. He is on the editorial boards of the *International Review for the Sociology of Sport* and *Curriculum Studies in Health and Physical Education.* He is a critical qualitative researcher who examines diverse sociocultural, pedagogical, and theoretical issues associated with sport, exercise, health, physical education, bodies, and gender relations.

OLIVER J. C. RICK is an accomplished researcher and assistant professor at Springfield College. He graduated from the University of Maryland with a PhD in physical cultural studies and has gone on to develop a research agenda focused on urban physical activity. This has included studying the role of recreation and active transportation in shaping urban formation as well as a project focused on Chinese urban identity and physical activity choices.

JENNIFER J. STERLING is a lecturer in the Department of American Studies at the University of Iowa. She completed her PhD in kinesiology (physical cultural studies) at the University of Maryland and joined the sport studies faculty at Iowa following a postdoctoral fellowship with the Sports, Society, and Technology (SST) program at the Georgia Institute of Technology. Her research interests revolve around the disciplinary intersections of sports studies, science and technology studies, and visual culture. In particular, her research explores how visualizations of physically active bodies (e.g., medical imaging, data visualization) shape understandings of the body.

HOLLY THORPE is professor in the Faculty of Health, Sport and Human Performance at the University of Waikato, New Zealand. Her research interests include the moving body, action sports, youth culture, gender, women's health, and sport for development. She continues to find much inspiration in the challenges of working across disciplines, engaging with social theory, and exploring feminist methodologies. Thorpe has published more than 60 articles and chapters on these topics and is the author of *Snowboarding Bodies in Theory and Practice* (2011) and *Transnational Mobilities in Action Sport Cultures* (2014) and has coedited various journals and collections, including *Women in Action Sport Cultures* (2016), the *Routledge Handbook of Physical Cultural Studies* (2017), and *New Sporting Femininities* (2018). She currently serves as associate editor for the *Journal of Sociology* and is coeditor of a new series titled New Femininities in Digital, Physical and Sporting Cultures with Palgrave Macmillan. Thorpe has been a recipient of Fulbright and Leverhulme Fellowships and received the 2018 New Zealand Royal Society Early Career Research Excellence Award for Social Sciences.

GAVIN WEEDON is with the Department of Sport Science, School of Science and Technology at Nottingham Trent University. His current research pursues the ecological dimensions of sport, health, and exercise cultures with a focus on supplementation, (post)genomic health and fitness technologies, and body-health-environment relations.

INDEX

Page numbers in *italics* refer to figures.